EINSTEIN
ON POLITICS

EINSTEIN
ON POLITICS

HIS PRIVATE THOUGHTS
AND PUBLIC STANDS ON
NATIONALISM, ZIONISM, WAR,
PEACE, AND THE BOMB

EDITED BY
DAVID E. ROWE AND
ROBERT SCHULMANN

PRINCETON UNIVERSITY PRESS
PRINCETON AND OXFORD

Copyright © 2007 by Princeton University Press
Published by Princeton University Press, 41 William Street, Princeton, New Jersey 08540
In the United Kingdom: Princeton University Press, 3 Market Place, Woodstock, Oxford-
shire OX20 1SY

Library of Congress Cataloging-in-Publication Data

Einstein on politics : his private thoughts and public stands on nationalism, Zionism, war,
peace, and the bomb / edited by David E. Rowe and Robert Schulmann.
    p.   cm.
  Inclubes bibliographical references and index.
  ISBN-13: 978-0-691-12094-2 (cloth : alk. paper)
  ISBN-10: 0-691-12094-3 (cloth : alk. paper)
    1. Einstein, Albert, 1879–1955—Political and social views.   2. Einstein, Albert,
1879–1955—Ethics.   3. Einstein, Albert, 1879–1955—Religion.   I. Rowe, David E.,
1950–   II. Schulmann, Robert J.
  QC16.E5E5157   2007
  530.092—dc22   2006100303

British Library Cataloging-in-Pulication Data is available

This book has been composed in Sabon with Eurostile family display

Printed on acid-free paper. ∞

press.princeton.edu

Printed in the United States of America

10 9 8 7 6 5 4 3 2 1

CONTENTS

LIST OF TEXTS

PREFACE

As the most celebrated scientific figure of the twentieth century, Albert Einstein needs no introduction. His achievements and their impact on modern physics have been the subject of countless studies, just as his unique career—in which he rose from the obscurity of a Swiss patent office to prominence on the front pages of leading newspapers around the world—has been described in books and articles pitched to almost every conceivable audience. Alongside this dominant image of Einstein as a lonely, singular scientific genius stands another: the familiar visage of the elderly humanist and pacifist. These two facets of Einstein's fame—as scientist and humanitarian—closely correspond to two fairly distinct periods in his career: his most creative period, from 1900 to 1919, and the longer thirty-five-year span of his public fame.

For the latter years, countless reports bear witness to his selfless character and the sacrifices he made not only to promote human welfare in general but to aid powerless individuals who sought his assistance. Viewed by many as a man whose altruism and sense of humanity give him the stature of a Gandhi or an Albert Schweitzer, he continues to fascinate in part because of his ambivalence in the pursuit of justice. As he turned outward, voicing public warnings against the rise of militarism, fascism, and nuclear catastrophe, he also stepped back inside himself to take a stoical, almost ascetic view of life filled with personal reflections on religious and ethical issues.

While the public and private lives of Einstein, the scientist and world citizen, have long been followed with avid interest, it is only within the last two decades or so that his legacy as a political figure has begun to receive closer attention through biographical studies such as those by Jamie Sayen (1985), Albrecht Fölsing (1993), Thomas Levenson (2003), and Hubert Goenner (2005). These books, together with the ongoing publication of the *Collected Papers of Albert Einstein*, have signaled a need to document the unfolding of Einstein's political thought in the context of the rapidly changing times in which he lived. Indeed, recent scholarship has finally begun to lift the veil of mystery surrounding Einstein's public and private lives, thereby revealing a far more complex and provocative personality than had ever been considered before.

There can be little point in trying to cordon off an authentic, private Einstein from the public figure who emerged on the world stage after 1919. One of our chief aims in this collection of texts is to convey the unity of what Einstein himself described as an apparent contradiction: "My passionate interest in social justice and social responsibility has always stood in curious contrast to a marked lack of desire for direct association with men and women" ("What I Believe," chapter 5). Though he never relented in his lifelong flight from the merely personal, we hope to show that private thoughts and public passions sprang from the same source.

Our still-popular image of Einstein as a wise-cracking, wild-haired, saintly old man was, at least to some degree, a concomitant of his extraordinary fame. An integral part of his persona found expression in a lifestyle that demonstratively shunned worldly fame, but not the influence and responsibility that went with it. At the same time, his public image was burnished by his tendency to dispense wisdom in the form of quaint aphorisms, their charm heightened by suspension in what came to be regarded as a homespun Einsteinian universe independent of space and time.

That Einstein's politics were rooted in an abiding sense of moral responsibility also provides a partial explanation for the powerful hold on the public his political pronouncements continue to exert to this day. As part of that mystique, his writings and utterances have usually been recorded with little or no attention to their historical setting. Thus, while a number of the texts chosen for the present volume have appeared in earlier anthologies of his writings, many have never before been properly identified or contextualized. In some cases, in fact, it is unlikely that the original context can ever be recovered due to the casual methods employed by earlier editors. Moreover, the texts in the most popular Einstein collections have invariably been shorn of the specific historical circumstances that prompted their creation. With the exception of *Einstein on Peace* (N & N 1960), the authoritative, 700-page volume compiled by Otto Nathan and Heinz Norden, and to a lesser degree Siegfried Grundmann's work (1998) on the Berlin period, all previous collections of Einstein's nonscientific writings have paid scant attention to the issues he confronted and the ways he went about tackling them. Thus our aim in the present source book is to present Einstein's public and private musings on the political problems of his times in a new light that will prove useful for a general readership as well as for historians. Most of his articles and speeches are presented in extenso, while a significant number of letters are excerpted.

In an effort to restore the context and texture of Einstein's political thought as well as to document the range of issues he addressed, we have adopted a largely thematic structure in this book. Its ten chapters take up the major nodes of his political engagement, some of which he pursued for relatively brief periods, others throughout the course of his life. Within the subsections of each chapter, the texts are presented chronologically. Furthermore, the chapters themselves have been arranged with an eye toward retaining the chronology of Einstein's political activities. The commentaries are intended to show as precisely as possible which factors influenced his writings and actions at a given historical moment.

As described below, internationalism and Zionism constitute the two dominant themes in the writings found in the first eight chapters of this source book, which cover the period 1914–1950. After this we turn back the clock in the final two chapters to consider how Einstein's views evolved on another series of issues that took on increasing importance during the last decade of his life: freedom, socialism, civil rights, and the threat of nuclear holocaust. Einstein was deeply troubled by the course of American foreign policy during the Cold War and even made damning comparisons between the military mentality of the Cold Warriors and the recently extinguished Prussian version he had known earlier. He was also an outspoken critic of those who advanced their careers by exploiting Americans' obsession with Communist infiltration on the home front. The last years of Einstein's life in the United States made him a hero for many on the left, at the same time as mainstream American public opinion came to admire him as an exotic yet familiar stranger whose face would grace an array of postage stamps, coffee mugs, bumper stickers, and T-shirts.

In the balance of this preface we offer a brief look at Einstein's political persona and long-term agenda. The Historical Introduction presents a more probing account of the public and private dimensions of Einstein's political life, providing background information on the principal themes that emerge in the book's ten chapters.

"Clear understanding of the prevailing objective conditions"

Einstein had little use for conventional political organizations. Preferring to remain outside the arena of professional politics, he gave his name to numerous causes, but never joined a political party nor felt himself accountable to any constituency. Fiercely independent, he was willing to take unpopular stands that sometimes left even his allies disgruntled or bewildered. These traits have contributed to the stubbornly persistent misconception that Einstein was simply a naive idealist who pontificated to no real purpose, an isolated genius cut off from the realities of contemporary political events. *Time* magazine, while elevating him to "Person of the Century," only reinforced this stereotype of Einstein as "the embodiment of pure intellect, the bumbling professor with the German accent, a comic cliché in a thousand films." Even when he "denounced McCarthyism and pleaded for an end to bigotry and racism," his American admirers regarded him as "well meaning if naive" (*Time*, 3 January 2000) while his detractors thought him outright dangerous.

When idealistic views remain unchecked by a realistic assessment of the everyday world, these may fairly be called naive. But this charge does not apply to Einstein. The same pragmatism that marked the early stages of his personal and professional lives came to shape his political views, which he constantly adapted and adjusted to changing circumstances while never abandoning principle. In rebutting a critic of his views on world government shortly after the end of the Second World War, he stressed that one's strength of conviction must be "based upon clear understanding of the prevailing objective conditions" ("Reply to Sumner Welles," chapter 8). His humane and democratic instincts were always tempered by a willingness to evaluate ideas in the light of their consequences in everyday life.

Nor was he simply a loner, cut off from human society. Quite the contrary: over the course of his public life, he often acted in concert with like-minded intellectuals, including such famous names as Sigmund Freud and Bertrand Russell. During and after the First World War the cohort of his political allies encompassed such prominent personalities as Paul Langevin, David Hilbert, and Romain Rolland. In his American period, two similarly distinguished individuals—Stephen Wise and Leo Szilard—joined him in political actions. Perhaps even more significant were the efforts of several others, once prominent but scarcely remembered today: Georg Nicolai, Heinrich Zangger, Emil Julius Gumbel,

Rosika Schwimmer, and Emery Reves. His interactions with such figures followed his lifelong pattern of drawing on the expertise and political passions of informal advisers who swam against the prevailing political currents. These relationships even bore a similarity to his association at the turn of the century with nonconformist colleagues in the informal Olympia Academy where, after hours as a clerk in the Swiss Patent Office, he had thrashed out topics in philosophy and the natural sciences culled from a variety of unorthodox sources.

These alliances sometimes led to public pronouncements, but in most cases the larger agenda that Einstein pursued can only be discerned through private correspondence or remarks made "off the record": hence our rationale for combining Einstein's private thoughts and public passions in a single volume, one that seeks to probe the deeper motivations and influences that informed his political activities.

In his writings and correspondence—from his initial testing of the political waters to his statement on "Human Rights" one year before his death (chapter 10)—Einstein stressed above all the importance of human dignity and the need for creative freedom. His sense of social justice was defined by a fierce empathy with the underdog that served as the moral catalyst of his political engagement. And yet, his sense of responsibility for the defenseless and the underprivileged was offset by a jealously guarded independence that made him indifferent to the temptations of political influence.

The Core Values: Internationalism and Cultural Zionism

Alongside this steadfast commitment to human freedom, two central themes define his lifelong search for means to advance his moral purpose, and to our mind, these constitute the heart of his political legacy. The first was his heartfelt belief that intellectuals had a moral obligation, clearly and truthfully, to strive for international solidarity and to address the fundamental causes of national hostility. Only thus could political leaders be forced to deal with the scourge of war, the single greatest challenge facing humanity in the twentieth century. The second and no less important theme for Einstein was his personal embrace of the cultural Zionist movement as a model for restoring dignity to the powerless. He deeply hoped that this movement would not only provide a spiritual homeland for the oppressed Jews of Eastern Europe but would also serve as a symbol of humanity's search for a world community based on mutual solidarity and as an "ideal form of human interdependence" (Stachel, 74).

Taken together, these two themes go a long way toward defining Einstein's political persona; indeed, they constitute the core elements that recur in various forms throughout the ten chapters of this book. As a pragmatic idealist, however, he recognized that politics was the art of achieving what is possible in a given situation. Indeed, this facet of his political persona—along with the powerful moral messages he sought to convey—make his legacy truly worthy of closer examination, especially at a time when political idealism has lost nearly all credibility. By studying the ways in which Einstein chose to advance his clearly articulated agenda we can see how far removed he really was from the mythic image of a bumbling, naive idealist.

As the documents in this volume amply illustrate, Einstein responded with considerable sensitivity to the tumultuous events of his day: from the First World War and its aftermath, including the resurgence of virulent anti-Semitism in early Weimar Germany that presaged the rise of Nazism, to the international efforts to defeat fascism, the horrors of the Holocaust, and the tensions and hopes that led to the founding of the state of Israel. An outspoken pacifist at the end of the Weimar era, he came to recognize the futility of passive resistance in the face of growing German militarism. Through his work with the League of Nations, Einstein learned firsthand about the obstacles and frustrations that stood in the way of even the most modest international endeavors. He remained a steadfast promoter of political causes that aimed to curtail the powers of national governments and, in particular, their ability to wage wars. Nonetheless, as the menace of Nazism threatened to engulf Europe, he helped mobilize the American research effort that produced the first atomic bombs. He was, however, dismayed when these terrifying weapons were used against Japanese civilians. It was in the wake of Hiroshima and Nagasaki that Einstein began an earnest campaign not only to control nuclear weapons but to build the framework for a world government with enough power and authority to contain national military conflicts and violence.

Einstein's retrospective assessment of his contributions in the political sphere was typically hard-headed. Admitting that he had not "made any systematic effort to ameliorate the lot of men," he nevertheless prided himself on speaking out "on public issues whenever they appeared to me so bad and unfortunate that silence would have made me feel guilty of complicity" ("Human Rights," chapter 10). In fact, he responded forcefully at three critical historical junctures with controlled bursts of intense political activity.

Responding to Crises

The first of these, from 1919 to 1923, was marked by the collapse of imperial Germany, struggles to establish a new concert of European nations, and growing anti-Semitism that poisoned relations within German society. Einstein's elation at the beginning of this period was palpable when he wrote his sister Maja and her husband Paul Winteler two days after the Kaiser's abdication: "That I could live to see this!! No catastrophe is too great not to be gladly risked for such magnificent compensation. Where we are, militarism and reactionary stupor have been thoroughly obliterated" (CPAE 8B, Doc. 652). In the years that followed, Einstein took his place on the world stage as the acknowledged leader of a scientific revolution. At the same time, he served as a symbol of hope for

Jews round the world, which he experienced firsthand on visits to New York, Tel Aviv, and Buenos Aires. In an era when German politicians were viewed with skepticism if not disdain by their French and British counterparts, Einstein was welcomed by their respective scientific communities. An articulate and effective emissary, he promoted reconciliation and renewed cultural contacts between the new Germany and its European neighbors.

As the Weimar Republic began to stabilize during the mid-1920s, Einstein drew hope that the seeds of democracy had finally begun to sprout. These fragile plants could not withstand the storms of the Great Depression, however, and in September 1930 Hitler's Nazi Party enjoyed its first success in a national election. Einstein quickly grasped that Germany stood on the cusp of an era of economic and political crisis. The ensuing period saw the decline of republican hopes and the gathering of popular support for extremists on the right and left, a contest from which the Nazis emerged victorious.

During these years, from 1930 to 1932, his second phase of intense activity, Einstein was extraordinarily active as a speaker and writer, particularly during his three lengthy stays in the United States. A long-standing admiration for Wilsonian democratic ideals continued to buoy his hopes, though he now expressed deep concern about American isolationism in the face of an impending collapse of the European political order. Still, his fundamental view of the American people had not changed from the assessment he gave a reporter in 1921: "America is already in advance of all other nations in the matter of internationalism. It has what might be called an international 'psyche.' The extent of America's leaning to internationalism was shown by the initial success of Wilson's ideas," which had met with much popular acclaim ("On Internationalism," chapter 1). It was with this image of the American character in mind that Einstein began reaching out in the early 1930s to audiences throughout the United States, emerging as a major political celebrity within pacifist circles. Yet even at this early juncture, one can see a noticeable shift away from "militant pacifism" as Einstein weighed the consequences of failure at the Geneva Disarmament Conference, which opened in 1932. By this time, he stood in close contact with the political publicist Emery Reves, later a leading figure in the world-government movement.

With Hitler's appointment as chancellor on 30 January 1933 and the sudden collapse of the opposition in the wake of the Reichstag fire one month later, the Weimar Republic had now been (legally) overthrown. In far-off Pasadena, California, Einstein responded by formulating his credo for political freedom—words he would repeat after applying for American citizenship—to explain why he would not return to Germany, a promise he kept for the remainder of his life.

Thus began a new period of emigration and reorientation, during which he tended to work more quietly behind the scenes with allies, such as Rabbi Stephen Wise, hoping to mobilize support for the fight against fascism as well as for the Zionist cause. Einstein repeatedly expressed his solidarity with President Roosevelt and publicly supported the policies of his administration, though he parted company with Roosevelt in warning of the danger to the Western democracies of a fascist victory in the Spanish Civil War and in private asides on the ingrained hostility of American elites toward the Soviet Union. Casting aside his earlier militant pacifism, he urged the United States to work together with the Allies to defeat the Axis Powers using whatever military technology happened to be available. Prompted by Leo Szilard, he even supported research that might lead to nuclear weapons, fearing that German scientists were already pursuing this very possibility. Throughout these years Einstein also sharpened his attacks on the British policy of curtailing Jewish immigration to Palestine. Even more outrageous to his mind was Great Britain's cynical effort to pursue there a policy of divide and conquer with Arabs and Jews.

Einstein's third period of intense political activity began almost immediately after the first atomic bomb fell on Hiroshima on 6 August 1945. He now returned to the public eye in the guise of "grandfather of the bomb," a title he preferred to exploit rather than dispute. During the next five years he became a leading spokesman for the world-government movement, virtually picking up where he had left off politically in 1933 when Hitler rose to power. Only weeks after the devastation of Hiroshima and Nagasaki, he again joined forces with Emery Reves, whose new book, *The Anatomy of Peace*, quickly emerged as the bible of the world-government movement. By promoting it on numerous occasions—including a widely read article prepared by the journalist Raymond Gram Swing for the November 1945 issue of the *Atlantic Monthly*—Einstein helped make Reves's book an overnight sensation.

At the same time, he kept world government on the agenda of the Emergency Committee of Atomic Scientists after being appointed as its chairman in May 1946. The ECAS served as an important platform for educational initiatives as well as efforts to place atomic energy under the United Nations' control. Historians have typically regarded these events as part of a chapter in American history that never really opened, largely because of escalating tensions between the United States and Stalin's Soviet Union. Despite the looming shadows of the Cold War, however, Einstein and his allies continued to press forward, suffering a major setback in 1947 when Russia rejected the terms of the Baruch proposal on the grounds that it attempted to preserve the American nuclear monopoly into the foreseeable future. Einstein's final period of concerted political activity essentially ended in September 1949 when President Truman announced that the Soviet Union had successfully tested a nuclear weapon, thereby breaking the American monopoly. With this breakthrough, the stage was set for an arms race that would polarize the political landscape throughout the world and dominate foreign relations for the next forty years. Undeterred, Einstein struggled on during his last six years to bridge the gulf that separated the superpowers drawing on his "feeble capacity" to promote "truth and justice at the risk of pleasing no one" ("Human Rights," chapter 10).

NOTE ON SOURCES AND METHOD

The texts in this volume were taken from both German and English-language sources. Most writings are presented in extenso, whereas many of the letters have been excerpted. In all cases, we have used ellipses to indicate where words or passages in the texts have been omitted. We have also introduced minimal emendations in punctuation where they serve to enhance the fluidity of the transcriptions.

Throughout his life Einstein wrote almost all of his speeches, letters, and comments in German. We have used translations that previously appeared in English-language periodicals, books or anthologies, taking pains to present them as they appeared in the original published source, though informing the reader where a republication may be found in the anthologies. All other translations have been provided by the editors.

Where texts that appear in collections remain unpublished, whether in German or English, we have tried to provide the sources in the Albert Einstein Archives at the Jewish National and University Library, Jerusalem. Thanks to the cooperative efforts of the Hebrew University, the Jewish National and University Library, and the Einstein Papers Project, a number of texts presented in this volume, as well as many documents cited in the commentary, are available from the Archives online in digitized form (www.alberteinstein.info/manuscripts/index.html).

ACKNOWLEDGMENTS

This book began five years ago with some rough-hewn ideas and the conviction that Einstein's political views deserved serious (re-)consideration. As we immersed ourselves in the sources, it soon became apparent that our initial framework was too narrow, and over time the project evolved into a volume nearly three times lengthier than the one we originally envisioned. Without the contributions and support of a number of individuals and institutions it never would have appeared in its present form. Among the many individuals who contributed to its realization, we would especially like to thank John Stachel, the founding editor of the Collected Papers of Albert Einstein, and Alice Calaprice, both of whom cast keen, critical eyes over the entire manuscript. We also benefited from the remarks made by Mary Jo Nye and Sam Schweber on an earlier version of the text as well as some helpful suggestions made by Allan Axelrad and Volker Remmert.

Sam Elworthy of Princeton University Press patiently guided the project through rough waters and safely into port, ably supported by Susan Berezin and Helen Schenck. We gratefully acknowledge the support of Giuseppe Castagnetti, Lindy Divarci, Dieter Hoffmann, and Jürgen Renn of the Max Planck Institute of the History of Science. We also received friendly assistance and advice from Roni Grosz and Barbara Wolff of the Albert Einstein Archives of the Hebrew University, Jerusalem. For the right to publish extracts from Einstein 1956 and N & N 1960, we are grateful to The Hebrew University of Jerusalem; for those from Einstein 1954, we thank Crown Publishers, a division of Random House, Inc. "The War is Won, but the Peace is Not," is published with the permission of the Philosophical Library, New York.

It remains to thank some key individuals whose personal support during the course of this project is most appreciated: Leo Corry, Virginia I. Holmes, Michel Janssen, Andreas Karachalios, William Kelley, Hans Kraepelien, Katherine Pandora, Leonard Rubin, Erhard Scholz, Alexis Schwarzenbach, and Scott Walter. Last, but obviously not least, our thanks go to Hilde and Andy, and Judit, Vera, and Kati for bearing with us throughout this sometimes bumpy ride.

EINSTEIN
ON POLITICS

HISTORICAL INTRODUCTION

Writing in the last months of the First World War, Einstein summed up his internationalist position with simple passion: "By heritage I am a Jew, by nationality Swiss, by conviction a human being and *only* a human being with no particular penchant for a state or national entity" (CPAE 8B, Doc. 560).

Readings in popular science in his early years had laid the groundwork. As a precocious teenager he devoured Aaron Bernstein's multivolume encyclopedic study of the natural sciences, which besides offering a wealth of insights into the world of physics, provided numerous examples of the critical link between scientific creativity and the political equilibrium among nations that was necessary for it to flourish. An early adulthood filled, as he described it, with "constant wandering" and the sensation of being "everywhere a stranger" (CPAE 9, Doc. 337) most certainly also contributed to an instinctive internationalism, as did his work in physics. Scientific research was, after all, premised on a belief in the transnational character of science, its collaborative aspect, and its dependence on verification by a band of brothers and sisters bound together by the search for truth and not the vagaries of national origin.

Adherence to the Zionist cause was the other equally crucial pole of attraction for Einstein after decades of searching for physical truths. He found solace in the dream of building a spiritual community that might restore pride to his Jewish brethren, the perennial outsiders. While at first glance the principle of Zionism appears to contradict a firm belief in internationalism, Einstein saw Jewish nationalism as a necessary step in the elevation of self-esteem. So understood, his efforts on behalf of this cause were part of a larger ethical-political vision that aimed to promote international understanding and peace, a position he staked out with almost messianic zeal in the mid-1920s ("Mission," chapter 3). Even in the wake of the Holocaust he held fast to the dream of a binational Palestinian state in which Jews and Arabs would live together amicably.

"To put an end to anarchy in international relations"

The First World War deeply offended Einstein's moral sensibilities. As he began to grapple with concerns that lay outside the scientific realm, he harnessed the quest for harmony and unity in physics equally to his growing political involvement. His initiation into politics was decisively influenced by the physiology professor and social critic Georg Nicolai, principal author of the "Manifesto to the Europeans," to which Einstein affixed his signature in October 1914. Nicolai served as the first of a number of associates who would help define Einstein's political course during his lifetime. In his *Biology of War*, Nicolai provided to a generation of pacifists the definitive rationale for internationalism: "Since culture and patriotism are both ideals, each of which must completely possess an individual, no one can serve both masters. Either patriot or *Kulturmensch*. ... He who aspires to be a cultural patriot must soon recognize that a national culture is connected by a thousand hidden threads to that of other countries and must collapse if international relations are severed" (Nicolai, 318–319). The appeal for Einstein was obvious. At the height of jingoistic fever in Germany, he wrote that individuals should be judged not by national origin but by their intentions and capabilities ("My Opinion on the War," chapter 1).

Wartime correspondence with colleagues in neutral Europe further attested to his concern to keep alive an international "republic of letters" in the face of what he considered a continent's descent into madness. Taking concrete steps at home, he joined the pacifist New Fatherland Association in 1915, presumably at the urging of Nicolai. In the same year he allowed his name to be put forward for the executive council of the Central Organization for a Durable Peace, yet another indication of how he was being drawn, however cautiously at first, into a web of political concerns. After the German military command reintroduced unrestricted submarine warfare in early 1917, thereby bringing the United States into the war, Einstein began to contemplate a strategic plan for ending the conflict and promoting a lasting peace. In a remarkable letter to Heinrich Zangger, written on 21 August 1917, he set forth six key principles for a future international organization that bore an uncanny resemblance to the blueprint for a League of Nations that Woodrow Wilson made public the following year.

Others in the "enemy camp" went further. After Great Britain introduced general conscription in early 1916, Bertrand Russell emerged as a leading voice among conscientious objectors, drawing huge crowds to his antiwar lectures. It seems fitting that both he and Einstein would join forces four decades later to issue the Russell-Einstein Manifesto calling for the control of weapons of mass destruction, an appeal that echoed their commitment during World War I.

Dismayed as he had been by the military cast of imperial German society, Einstein was hopeful in the aftermath of the war that the new republic would not yield to the temptations of its predecessor's "religion of power." Though old elites had not been swept away in Germany, Einstein now recognized that the minimal effectiveness to which the intellectual had been relegated under the empire could be overcome by appeals to an international press that created his popular fame and continued to further it. His views on moral and political issues, confined to his correspondence and the occasional publication during the war, would now be broadcast throughout the world and carry great weight. It is under these circumstances that Einstein found his voice, decisively ended his seclusion, and joined the political fray.

A new sense of confidence was evident in his political stance. Replying to Arnold Sommerfeld who found "everything unspeakably miserable and stupid" in the new Germany, Einstein expressed optimism that "Germans who love culture will soon again be able to be as proud of their fatherland as ever—and with more justification than before 1914" (CPAE 8B, Docs. 662 and 665). Unable to suppress a touch of spitefulness at the thought that exclusion of German academics from international conferences might teach them humility, he was fearful that the fragile fabric of intellectual cooperation would be irreparably rent. He participated in various initiatives to obtain funding for research in the defeated countries, chiefly in his capacity as a member of the Emergency Society for German Science and Scholarship.

His condemnation of the harsh terms of the Versailles Peace Treaty was coupled with the opinion that the Allies might be the lesser of two evils, indications of his evenhanded and independent assessment of the postwar situation. Particularly galling for him was the punitive Allied blockade of Germany after the war, as a result of which especially the children of Central Europe suffered devastating disease and malnutrition.

Einstein's canny realism is reflected in words he penned less than a year after the cessation of hostilities: "The greatest danger for future developments is, in my opinion, the potential withdrawal of the Americans; let's hope that Wilson can prevent it" (CPAE 9, Doc. 97). The same letter sounded what would become a perennial leitmotif: "I don't believe that humanity as such can change in essence, but I do believe that it is possible and even *necessary* to put an end to anarchy in international relations, even though the sacrifice of autonomy will be significant for individual states."

Confidence about the future waxed and waned with the volatile turn of events. Still hopeful in August 1919 that divisiveness within the ranks of the Western Powers might actually contribute to the formation of a League of Nations, he turned pessimistic in December: "My expectations for the League of Nations do not seem to want to materialize" (CPAE 9, Doc. 198). A month later the League was formed in accordance with the Versailles Treaty, though it remained terminally weakened by America's refusal to join and by its reliance solely on economic sanctions as a last resort. Appointed in 1922 to the League's International Committee on Intellectual Cooperation, Einstein attempted for a decade, in an on-again off-again relationship, to counter the intrusion of national interests into its ongoing work (chapter 4).

"More dignity and more independence in our own ranks!"

Though his personal encounters with the increasingly virulent anti-Semitism that reared its head almost immediately in postwar Germany were few, Einstein witnessed public humiliations of friends and especially of the large East European Jewish community in Berlin. The blind racial hatred directed toward these Jews shocked him: "Until seven years ago I lived in Switzerland, and as long as I lived there I was not aware of my Jewishness. . . . This changed as soon as I took up residence in Berlin. There I saw the plight of many young Jews. I saw how anti-Semitic surroundings prevented them from pursuing regular studies and how they struggled for a secure existence. This is especially true of East European Jews, who are constantly subject to harassment. . . . These and similar experiences have awakened my Jewish national feelings" ("How I Became a Zionist," chapter 3). Profound compassion for the underdog was complemented by another decisive factor. Anti-Semitism harbored an even uglier sentiment, one that particularly outraged Einstein. Gentile contempt for the Jew was one thing, but even more despicable was the behavior of his fellow German Jews, who in their desperate attempts to assimilate had sacrificed their dignity before Jew and Gentile alike by heaping scorn on their more vulnerable Russian and Polish brethren, the *Ostjuden*.

These socially and economically disadvantaged members of the Jewish community, numbering about 30,000 in Berlin, represented somewhat more than one-quarter of the total Jewish population of the city. Various measures contemplated by the Prussian government to control and even deport East European Jews after the war exacerbated the fear of many German Jews that they were the real targets of official displeasure. Einstein dismissed this fear with the acerbic observation that it was "a Jewish weakness . . . always and anxiously to try to keep the Gentiles in good humor" (letter to Felix Frankfurter, 28 May 1921). Given the depth of his feeling on the matter, it comes as little surprise then that Einstein's first public stance on a political matter was a protest of the official and unofficial discrimination practiced against *Ostjuden*, which appeared as an article in a liberal Berlin daily at the end of 1919 ("Immigration from the East," chapter 3).

Acutely Einstein came to realize "the excessive price at which the blessings of assimilation are bought by Jewish communities . . . a loss of solidarity, of moral independence and of self-respect" (Einstein 1931a, 25). The solution, he thought, was straightforward: "More dignity and more independence in our own ranks! Only when we have the courage to regard ourselves as a nation, only when we respect ourselves, can we win the respect of others, or put another way, the respect of others will then follow of itself" ("A Letter of Confession," chapter 3). By expressing solidarity with an East European Jewish community that had maintained authenticity through loyalty to its roots, their assimilated German brethren too might recapture a sense of communal worth (M. Brenner, 129–152). Einstein's conclusion: "Zionism [is] the only effort which leads us closer to [the] goal" of healing divisiveness and uniting all Jews (letter to Heinrich York-Steiner, 19 November 1929).

The fatal illusions of assimilation could best be avoided by undertaking tasks in common. Of these the most important was the restoration of Jewish national life in Palestine, the very heart of the Zionist enterprise. This undertaking had been launched most decisively at the end of the nineteenth century by Theodor Herzl, an assimilated Jew, who was Paris correspondent of a liberal Viennese newspaper. During the Dreyfus affair in France in 1894 he experienced at first hand Jewish vulnerability to what seemed to many the immutable hatred of the Gentile. In his work *The Jewish State*, he sketched a new vision of a land that Jews could call their own, in which they might gain acceptance by ceasing to be a national anomaly.

Two points of departure dominated discussion of the Zionist cause. One stressed the spiritual and cultural character of the movement, the other its political goals. Cultural Zionists directed their energies toward a Palestine that would stand in a "constant reciprocal spiritual relationship" with Jews in the Diaspora (Poppel, 97–98). This group sought relatively modest funding for small-scale enterprises committed to the social and economic development of Palestine as a cultural homeland of world Jewry. The political Zionists, on the other hand, believed that the political struggle had only just begun and stressed the importance of mass colonization and ambitious supporting budgets to advance the cause of a Jewish state there.

Einstein's adherence was clear from the beginning. His preference for Zionism's cultural aspect always outweighed purely political considerations. Shortly after his first and only visit to Palestine, in 1923, he wrote to a friend: "On the whole, the country [Palestine] is not very fertile. It will become a moral center, but will not be able to absorb a large proportion of the Jewish people" (letter to Maurice Solovine, 20 May 1923). Central to Einstein's conception of this moral center was the creation of an independent place of study and research, where the intellectual vitality of his brethren might find fulfillment free of the crushing burden of anti-Semitism. A Hebrew University presented the opportunity to correct a "deficient spiritual balance" among the large number of Jewish students who were unable to gain access to higher education. Einstein argued that its establishment was "not a question of taste but of necessity" (*Manchester Guardian*, 10 June 1921).

Prospects for the university initially looked poor. The European leadership of the World Zionist Organization in London was strapped for funds and embroiled in political bickering with its largest affiliate, the Zionist Organization of America. The Europeans favored the formation of a political entity in Palestine while the American Zionists supported a cultural settlement. In spite of Einstein's natural inclination to the "American position," when the London organization requested that he join its delegation on a fund-raising mission to the United States in 1921, he agreed "without even giving it five minutes' thought" (letter to Fritz Haber, chapter 3). Much as he had made a principled compromise in accepting Zionist nationalism in the face of his bedrock faith in internationalism, so Einstein, ever the pragmatist, accepted in the name of solidarity the political agenda of the World Zionist Organization though his sympathies lay elsewhere.

The trip to America, while modest in its fund-raising success, took on the character of a triumphal procession for Einstein, who was singled out for adulation by large crowds of recent East European Jewish immigrants. The "unbroken vitality of the masses of American Jewry" ("How I Became a Zionist," chapter 3) impressed him greatly, and he found similarly enthusiastic Jews two years later on his only visit to Palestine. Here too he found confirmation of his faith in Palestine as "the incarnation of a reawakening national feeling of community of all Jews" (On a Jewish Palestine, chapter 3). His sense of kinship with the Jewish people had little to do with conventional religious views or Zionist political aspirations but rather reflected his identification with the moral values embodied in Jewish cultural traditions—with an emphasis on ethical standards, social solidarity, as well as intellectual and artistic achievement.

"The talk here is of almost nothing but Einstein"

In science, just as in politics, success often depends crucially on timing. In Einstein's case, scientific *and* political events converged in a dramatic manner in November 1919 when the British scientific community announced the results from two eclipse expeditions. These findings confirmed his prediction for the bending of light rays in the proximity of the Sun, a major breakthrough for the general theory of relativity. News crossed the English Channel quickly, one conduit being Robert W. Lawson, a British physicist who spent the war years interned in Austria and afterward continued to cultivate friendly relations with scientists in the German-speaking world. Writing to Arnold Berliner, editor of *Die Naturwissenschaften*, Lawson gave this vivid account of the excitement in England in the wake of the historic announcement of 6 November:

> The talk here is of almost nothing but Einstein, and if he were to come here now I think he would be welcomed like a victorious general. The fact that a theory formulated by a German has been confirmed by observations on the part of Englishmen has brought the possibility of co-operation between these scientifically-minded nations much closer. Quite apart from the great scientific value of his brilliant theory, Einstein has done mankind an incalculable service. (N & N 1960, 27–29)

Lawson later helped promote Einstein's work in the English-speaking world by preparing translations of several of his scientific writings (including, in 1920, the English edition of Einstein's popular account of the special and general theories of relativity [CPAE 6, Doc. 42]). Yet while internationalists such as Lawson and Berliner were clearly elated by the prospect that Einstein's triumph might help to improve relations between their respective scientific communities, other more conservative scientists were far more subdued. Some took an especially dim view of the public's fascination with the relativity revolution. In Germany, Einstein's sudden fame tended to polarize opinion along familiar political lines; for those who could still not fathom having lost the war, the celebration abroad fostered suspicion and resentment at home. Before November 1918, Einstein's political views were largely unarticulated, but they became increasingly conspicuous afterward, especially in Berlin. As an internationalist in an era of rampant nationalist hostilities, he soon emerged as an unofficial spokesman for Germany's new political order, voicing his views on a wide range of issues and causes. Monarchists and conservatives detested these pronouncements, accusing Einstein of pandering to the masses. And while many found his politics distasteful, some even blamed him for launching a media spectacle to promote what they called his self-proclaimed relativity revolution.

Five months after the Kapp Putsch of March 1920 had shaken Berlin to its foundations, the city registered a mild aftershock in the form of a politically motivated anti-relativity campaign. This, too, passed quickly, but for a brief time the ruckus made Einstein waver as he tried to assess the strength of those forces who sought to blacken his name or even drive him out of Germany altogether. He was well aware that the politicization of relativity had much to do with a power struggle within the German physics profession itself. Behind the public controversy lay a long-standing tension between experimental physicists—many of whom felt that Immanuel Kant had said the last word on the epistemological status of space and time—and a generation of theoreticians who regarded Max Planck and Einstein as their leading lights. Einstein's highly mathematical style of theorizing, guided by daring thought experiments, had stirred ample controversy even before the power elite of German physics opened the doors of the Prussian Academy to him. During the war, when Einstein realized a crucial breakthrough with his generalized theory of relativity, resistance had already begun to harden along ideological lines.

German Anti-Relativists and Flirtations with the Media

Two experimentalists, Ernst Gehrcke and Philipp Lenard, would emerge as Einstein's most outspoken opponents. Both had nothing but contempt for the public's fascination with relativity, which they attributed to unsavory influences in the media that threatened to undermine the integrity of their discipline. As representatives of the scientific establishment, on the other hand, they were reluctant to cross the threshold into politics. Even Lenard, a vociferous enemy of the English scientific community, had no wish to utter in public what he and others were expressing privately, namely that Einstein's theory was antithetical to good, sound German science—that relativity, like its author, was *un-German.*

That fateful step was taken by one Paul Weyland. An engineer by training, Weyland emerged as a journalist and right-wing demagogue in postwar Berlin, a breeding ground for political malcontents. Armed with inside information obtained from Gehrcke, he gained financial backing for a series of lectures in the Berlin Philharmonic Hall that—so he claimed—would expose the unscrupulous behavior and evasive machinations of Einstein and his allies. Although he characterized these lectures as "scientific," Weyland's real aim was to exploit widespread anti-Semitism—as voiced by those who bemoaned Jewish influence in the Weimar Republic, especially on the press—by suggesting that Einstein's theory of relativity was part of a massive propaganda campaign aimed at duping the German people. Outlandish as that claim might seem, Gehrcke had long been saying that relativity was nothing but a form of "mass suggestion," a kind of scientific hoax. Thus, in launching his counter-campaign, Weyland was banking heavily on mobilizing anti-relativists such as Gehrcke who deeply resented the influential "Einstein clique" in Berlin.

Three days after the first meeting at the Philharmonic Hall, Einstein countered with a polemical article in the *Berliner Tageblatt* which argued that things would look different "were I a German nationalist, whether bearing a swastika or not, rather than a Jew of liberal international bent." He went on to attack both Gehrcke and Lenard, challenging them or any other critics to appear before a scientific audience at a forthcoming conference in Bad Nauheim. Writing to his friend Marcel Grossmann in mid-September, he gave an ironical assessment of the situation prior to this dramatic showdown: "This world is a remarkable house of fools. Presently every coachman and waiter is debating whether the theory of relativity is correct. Their conviction about that is determined by the political party to which they belong" (CPAE 10, Doc. 148). On the morning of 23 September, a huge crowd gathered in Bad Nauheim in anticipation of a spectacular debate between Einstein and Lenard. Max Planck presided on this occasion, a delicate undertaking given the preconference publicity. A special correspondent for the *Berliner Tageblatt* found the lectures "a hailstorm of differentials, coordinate invariants, elementary action quanta, transformations, vectorial systems, etc." (CPAE 7, 109). Afterward, only fifteen minutes remained for the general discussion, most of which was taken up by the Lenard-Einstein debate (CPAE 7, Doc. 46). As it turned out, both men were intent on saving face; consequently, neither had anything new to say. Einstein and a few of his friends tried to mollify Lenard, but the latter refused to shake hands with his rival or to accept Einstein's public apology for attacking him in the *Berliner Tageblatt* (Schönbeck 2003, 353). The press coverage of the confrontation was generally fair to both sides, though many reporters sensed that what they had witnessed was really a clash of worldviews rather than a scientific debate. It was the only time Einstein and Lenard ever met, but the repercussions from this single encounter would prove far more lasting than either could have then imagined. At the time, it seemed just another episode in the ongoing publicity spectacle.

Einstein was constantly in the limelight, particularly during his numerous travels abroad in the early 1920s when he was often asked to grant interviews to foreign correspondents. Occasionally these produced results that rankled German sensibilities, causing him to publish a retraction (see chapter 2 for an instance involving American sensitivities). As he gradually became wary of the pitfalls involved in dealings with the media, he strove to evade these by presenting his views as rationally and dispassionately as possible. Once he grew accustomed to the routine exaggerations of the popular press, he even became quite adept at handling the many reporters who dogged his every step. It was a skill that proved vitally important for his emergence as a significant voice in contemporary politics.

Although the theory of relativity had long been controversial within the scientific community, after 1920 its reception became closely tied to the delicate political situation in Europe. During his trip to Paris in the spring of 1922, Einstein observed first-hand how relativity divided French public opinion along familiar fault lines. As a distinguished historian at the Sorbonne told it: "I don't understand Einstein's equations. All I know is that the Dreyfus adherents claim that he is a genius, whereas the Dreyfus opponents say he is an ass. And the remarkable thing is that although the Dreyfus affair has long been forgotten, the same groups line up and face each other at the slightest provocation" (Frank 1949, 314).

Even before he had returned to Berlin, the ultranationalist physicist Johannes Stark published a scathing attack on Einstein in the *Deutsche Tageszeitung*:

Since the end of the war the French have suppressed the German people in the most brutal manner. They have torn away piece after piece from their body, have engaged in one act of extortion after another, they have placed colored troops to watch over the Rhineland, and they have made insufferable demands on the German people through the reparation commission. And just at this very time, Herr Einstein travels to Paris to deliver lectures. From a personal standpoint, this trip is completely understandable. He has, after all, proclaimed in an article for the *Berliner Tageblatt* [referring to the passage cited above] that he is a German Jew of pacifist and internationalist persuasion; he therefore lacks the feelings of a German faced with this French oppression. (4 April 1922; Grundmann, 209)

Stark went further by attacking Einstein's supporters and the Weimar government: "His friends, who are otherwise so quick to leap to his defense in the daily press, should have told him what is appropriate and what is not. In any case, the responsible governmental offices should have made it clear to him that this was absolutely not the proper time for a German citizen with official ties to engage in friendly scientific relations with the French" (ibid.).

Targets of Political Violence

Einstein's friends included Emil Julius Gumbel, a statistician whom he met during the war when they were fellow members of the New Fatherland Association. Historians of the Weimar Republic still cite him as the authoritative source on political violence during its early years. In 1922 Gumbel published *Four Years of Political Murder*, a continuation of an earlier brochure in which he had documented "that the German judicial system had left over three hundred political murders unpunished" (Gumbel 1922, 145). The author noted that he had expected this evidence would produce one of two possible responses: "either the judicial system would believe that I was speaking the truth" and would act accordingly, or "it would believe that I am lying, and would then punish me as a slanderer." But, in fact, neither of these reactions occurred: "Although the brochure in no way went without notice, there has not been a single effort on the part of the authorities to dispute the correctness of my contentions. On the contrary, the highest responsible authority, the Minister of Justice, expressly confirmed my contentions on more than one occasion. Nevertheless, not a single murderer has been punished" (ibid.).

Like Gumbel, Einstein was outraged by the brutal murder in early 1919 of far-left Spartacist leaders Karl Liebknecht and Rosa Luxemburg at the hands of right-wing paramilitary thugs. During the years that followed, the enemies of the republic targeted bourgeois political leaders, particularly those who had played a role in the signing of the Versailles Treaty. One of these was the foreign minister, Walther Rathenau, a brilliant intellectual whom Einstein genuinely admired. Yet unlike Einstein, Rathenau was a heartfelt patriot. Describing himself as a German of Jewish descent, he was vilified by those who considered the Weimar government a *Judenrepublik*. While some chanted

> Knallt ab den Walther Rathenau/Die gottverfluchte Judensau
> (Shoot down that Walther Rathenau/The accursed goddamned
> Jewish sow),

others acted, driving past his open automobile and firing the bullets that ended his life. Most Germans were shocked and left wondering who would be next. Among those targeted was Einstein, as he confided to his friend Maurice Solovine: "there has been much excitement here ever since the abominable murder of Rathenau. I myself am being constantly warned to be cautious; I've canceled my lectures and am officially 'absent,' although I never actually left. Anti-Semitism is very widespread" (Fölsing, 597).

In Heidelberg, Philipp Lenard refused to observe the day of national mourning for Rathenau. When word got around that he and his assistants were working as usual in the physics institute, demonstrators stormed the building and seized its director, an incident that left the Nobel laureate even more receptive to the message of the radical right. Still smarting from his encounter with Einstein at Bad Nauheim, Lenard began preparing a new version of his critique of relativity theory, accompanied by a word of admonition to his fellow German physicists with clear anti-Semitic overtones. He also alluded by inversion to the role played by the so-called Jewish press in promoting relativity. "A German press," Lenard asserted, "should have avoided the impression, which for informed persons can only be described as incredible, that German science was dependent on the judgment of the English. For the fact that those who observed the solar eclipse were Englishmen has nothing to do with judging the results . . . But frankly we must first have a well-prepared, truly German press" (Lenard 2003, 427).

By 1924 the relativity rumble had largely died down, but not without laying the groundwork for the Aryan physics movement that later emerged full force with Hitler's ascension to power (Beyerchen, 123–167).

Spinoza's God and Militant Pacifism

As he passed his fiftieth birthday in 1929, Einstein's life entered a distinctly new phase. For the first time he began to address a wider set of moral and political issues of a more philosophical nature, expressing views that had much in common with those of Baruch Spinoza and Arthur Schopenhauer, but also Freud. Yet as he turned inward, addressing a broad range of concerns far beyond his chosen field of expertise, the physicist Einstein entered the arena of world politics in a manner far more direct than ever before.

Among all his writings, including the autobiographical sketch (Einstein 1949), none provides a more intimate portrait of how Einstein saw the world than "What I Believe" (chapter 5). In this brief essay he enunciated three principles of central importance for his mature political views, formulated at a time when Europe was entering a new era of crisis. First, he firmly rejected autocratic forms of government—citing Mussolini's Italy and Stalin's Russia as explicit examples—while affirming his belief in democracy, despite the recent travails of the Weimar Republic. Second, he upheld the creative individual and not the political state as society's most precious asset. Third, he characterized the military mentality as the single worst manifestation of the "herd mind" in modern society. From this time onward, these three principles would be constantly recurring themes, as illustrated by the writings collected in chapters 9 and 10 of this book.

Einstein also wrote about the emotional source of religious and scientific thought, which he identified with mankind's sense of the mysterious ("Religion and Science," chapter 5). Both, he thought, had a common psychological bond, but he maintained that this was seldom appreciated due to the still primitive character of religious experience. His evolutionary approach to religious sensibilities was probably inspired, at least in part, by Freud's *The Future of an Illusion* (Freud 1927), a work he almost surely read. Another reader, Romain Rolland, afterward questioned Freud about the origins of the "oceanic feeling" that Rolland himself took to be the true source of religious experience (Freud gave his answer in *Civilization and Its Discontents* [Freud 1930]). To what extent Einstein followed these contemporary discussions remains unclear, but there can be no doubt that others read what he had to say about a higher form of religious experience which Einstein called "cosmic religion." Indeed, in the United States during the early 1930s his views on this subject touched off nearly as much heated controversy as anything he had to say about space, time, or politics.

Already in 1929, Boston's Cardinal O'Connell branded Einstein's theory of relativity a "befogged speculation, producing universal doubt about God and His Creation," while suggesting that it "cloaked the ghastly apparition of atheism" (Clark, 413). Another leading theologian, Dr. Fulton J. Sheen, called Einstein's views on religion stupid and nonsensical: "There is only one fault with his [Einstein's] cosmical religion; he put an extra letter in the word—the letter 's'." Alarmed by these attacks, New York's Rabbi Herbert S. Goldstein sent Einstein a telegram asking: "Do you believe in God?" He received this oft-quoted reply: "I believe in Spinoza's God, Who reveals Himself in the lawful harmony of the world, not in a God Who concerns Himself with the fate and the doings of mankind" (*New York Times*, 25 April 1929). This statement satisfied Rabbi Goldstein, who cited it as evidence that Einstein was not an atheist, but conservative religious figures continued to look askance at his cosmic religion (Jammer, 83).

Some Americans took issue with his political views as well. Einstein's presence in the United States was so unpopular with the Woman Patriot Corporation that this organization filed a sixteen-page memorandum with the State Department in November 1932 in an attempt to block him from entering the country. This effort led to an interview at the American Consulate in Berlin, a 45-minute encounter that ended abruptly when Einstein was pointedly asked: "Are you a Communist or an anarchist?" (Sayen, 7). According to the *New York Times*, he turned the tables on the interviewer by replying: "Is this an interview or an inquisition?" The incident occurred only a few days after he had sent a mocking "Reply to the Women of America" (chapter 5) poking fun at their scare tactics. Two decades later, as a naturalized American citizen, he would once again face the wrath of a new generation of right-wing zealots.

With the sudden collapse of the world economy and the rise of National Socialism, Einstein became deeply troubled by renewed European tensions and the threat of another devastating war. At this juncture he joined forces with other liberals and leftists who sought to head off fascism under the banner of militant pacifism. Inspired by the activism of Rosika Schwimmer, a Hungarian-born feminist and founding member of the Women's Peace Party, Einstein emerged as a leading spokesman for civil disobedience, urging young people in the United States and elsewhere to refuse military service. Although he had earlier made a strong moral case for pacifism in a letter to the mathematician Jacques Hadamard (see chapter 4), it is noteworthy that he did not throw himself headlong into the German pacifist movement at this time, perhaps owing to its divisive tendencies (Riesenberger, 233–252). Instead, he struck out on his own during the winter of 1930–31, speaking not to German audiences but to various groups in the United States.

As hopes for democracy quickly faded in Germany, Einstein pleaded with Americans not to ignore the mounting tensions in Europe ("America and the Disarmament Conference," chapter 5). At the same time, he appealed to the idealism of American youth, challenging them to fight militarism by refusing induction into the military. Einstein's approach to moral disarmament, or militant pacifism as he preferred to call it, had a decidedly pragmatic purpose, namely to launch a citizens' revolt against conscription that would eventually lead to a tidal wave of protests against compulsory military service. In his so-called "Two Percent Speech" (chapter 5), delivered in New York in late 1930, he claimed that if only this modest percentage of youth resisted induction into the military their efforts would suffice to bring down the whole draft system. These youths were not to think of themselves as martyrs for some higher cause, however; their idealism was to be harnessed to a viable political program. Speaking on that occasion to a group of pacifists, he warned them to avoid incestuous preaching:

> When those who are bound together by pacifist ideals hold a meeting they are always consorting with their own kind only. They are like sheep huddled together while the wolves wait outside. I think pacifist speakers have this difficulty: they usually reach their own crowd, who are pacifists already. The sheep's voice does not get beyond this circle and therefore is ineffective. . . . Real pacifists, those who are not up in the clouds but who think and count realities, must fearlessly try to do things of practical value to the cause and not merely speak about pacifism. Deeds are needed. Mere words do not get pacifists anywhere.

Einstein clearly saw the refusal to take up arms as a special form of civil disobedience fundamental to a political strategy that aimed to confront the problem of escalating military activity as France and Germany rearmed behind the scenes. This became particularly acute with the Japanese invasion of Manchuria in

September 1931. Even so, he had been slow to reach the conclusion that organized pacifists might exert significant political influence in a democratic society. Before this time, Einstein conceived of pacifism exclusively as a moral platform that could only be promoted through the influence of the world's spiritual leadership. Like Freud, he saw war as a barbarous human activity rooted in the aggressive instincts of the animal world. Thus, initially pacifism had a more universal character for Einstein, but the temper of the times called for action and he responded by promoting militant pacifism as a political tool. Critical for its implementation was the need for popular involvement. As he urged in a statement to the Kellogg League in mid-1931, "the ideal of pacifism" should be practiced "every week through social gatherings with cultural programs in which peace is especially emphasized. Only in this way can the ideal of world peace get under the skin of the common people" (Einstein 1933, 42). If enough idealistic people worldwide joined hands as war resisters, their governments' militaristic policies would be undermined. At the same time, he briefly held out hopes for a breakthrough at the 1932 Disarmament Conference held in Geneva, a major political undertaking that proved to be a fiasco for the pacifist movement (chapter 5).

"Disarm at one blow or not at all"

The issue of disarmament had long been recognized as central to European collective security arrangements as established by the Versailles Treaty. Although German politicians and diplomats had struggled since its signing to escape the treaty's harsh terms—including the reduction of Germany's army to a mere 100,000 men—they had not overlooked the fact that its measures also placed moral responsibilities on their wartime enemies. Those who, following Rathenau's lead, pursued a policy of meeting its terms continually reminded Allied representatives of their countries' obligation to disarm once Germany had been shorn of its military capability. Pacifists, too, could take heart at this prospect by pointing to the preface of Part V of the treaty, which dealt with arms limitations. This stipulated that German disarmament was to be undertaken "in order to render possible the initiation of a general limitation of the armaments of all nations" (Bennett, 53). By the mid 1920s, foreign minister Gustav Stresemann was able to convince his French and British counterparts that Germany had fulfilled the disarmament portion of its treaty obligations. This breakthrough led to the disbandment of the Allied Inspection Commission and opened the way for Germany's entry into the League of Nations. Thereafter complicated diplomatic maneuvering ensued during which internationally inclined policy makers tried to find a framework for meaningful negotiations on arms reductions.

After many false starts, the long-awaited disarmament conference finally opened in Geneva in the spring of 1932 with representatives from over fifty nations attending. As events would prove, the prevailing conditions of economic and political volatility made this a particularly inauspicious time to negotiate arms control. Germany's weak coalition, headed by Chancellor Heinrich Brüning, was already on the brink of collapse. The recent defeat of Aristide Briand's candidature for the presidency of the French Republic spelled the end of the spirit of Locarno that had improved Franco-German relations during the late 1920s. By now the dominant nationalist camp in Germany was loudly demanding the termination of one-sided disarmament conditions, just as French rightists were equally vocal about maintaining the security arrangements guaranteed by the Versailles Treaty. In the face of this impasse, British and American opinion remained divided, though there was great reluctance within the British Conservative Party to give any binding assurances to France, a move that might have helped push the stalled negotiations forward. Hawks, such as Winston Churchill, warned against any weakening of French military power, fearing potential loss of control in the Rhineland, while across the Atlantic the Hoover administration was in no mood to become mired in European politics just as it was facing economic problems that threatened to sweep away the very foundation of the capitalist system.

President Hoover did, however, propose a sweeping plan for disarmament in June 1932. Initially, this concept met with a positive response, though its promise quickly melted away after the conference went into recess. Philip Noel-Baker, a British pacifist who played a major role in Geneva, later blamed the backroom diplomacy of leading British Conservatives for killing Hoover's effort to break the deadlock (see Noel-Baker). While there may be much truth to this, such visionary plans probably had little chance of ultimate success at Geneva. Despite strong public support for disarmament, any meaningful agreement would have had to overcome not only the opinions of reluctant politicians but also the strong vested interests of their nations' military-industrial complexes. Negotiations at Geneva were further constrained by the lengthy reports filed by numerous technical experts on military equipment and arms. Many subsequent discussions dealt with measures aimed at bans of certain types of warfare, such as the use of chemical and biological weapons, or limiting the size of tanks and the number and caliber of field artillery.

Einstein was highly critical of this approach, arguing forcefully that such half-measures were doomed to fail. "To arm," he wrote, "is to give one's voice and make one's preparations, not for peace but for war. Therefore people will not disarm step by step; they will disarm at one blow or not at all" (Einstein 1954, 102–103). On this point he never had the slightest doubt, which accounts for his relentless efforts to fight what he later called the military mentality. Soon after the Geneva conference opened, he pronounced it a debacle, criticizing all parties at the negotiating table for their failure to rise above narrow national interests. Nevertheless, he realized that the problem lay not with the negotiators, but rather with the fact that they were the representatives of sovereign nation-states with nothing to fear but one another. Under such circumstances they could not possibly come up with a truly effective plan for international disarmament.

The Geneva conference brought Einstein into contact with all sorts of people from the world of politics and foreign affairs. One of these was the Hungarian-born publicist Imre Révész, better known as Emery Reves, the name he assumed after becoming a British citizen in 1940. How they got to know one another may never be known, but Révész had a habit of linking up with all kinds of interesting people. As a student in Zurich, he befriended fellow Hungarian John von Neumann, the great mathematician. Perhaps he made Einstein's acquaintance by sending him a copy of his Zurich dissertation, which dealt with the economic and political thought of Walther Rathenau (Einstein owned a copy, now housed at Hebrew University). Yet while the circumstances that first brought them together remain unclear, there can be no question about the nature of the political bond that joined them. For Révész was a liberal intellectual whose approach to the global issues of war and peace had much in common with Einstein's own.

In 1930, with militarism and totalitarianism on the rise, Révész founded the Cooperation Press Service, an independent propaganda machine of a most subtle kind. By publishing the views of leading political figures from the Western democracies in foreign newspapers, he hoped to give each party a voice abroad that could help pave the way for better understanding between their respective nations. A major objective of this enterprise was to overcome the jingoistic tendencies of the national media by infiltrating leading newspapers with the views of foreign policy makers and observers. It was an auspicious time to launch such an enterprise, and soon Révész was busy running his flourishing company out of two offices: one located on the Kurfürstendamm in Berlin, the other in Paris. It helped that he was fluent in ten languages, for eventually his enterprise spanned the globe, publishing the views of over one hundred leading political and cultural figures in some four hundred newspapers in seventy countries stretching over six continents. In the late 1930s his star contributor was

Winston Churchill, who later rewarded his publicist from the "wilderness years" with the rights to serialize his wartime memoirs for *Life* magazine.

As it happened, Révész's two principal rivals were also situated in Berlin: on the right stood Joseph Goebbels, whose career as a propagandist requires no further comment; on the left was Willi Münzenberg, the "Red Millionaire" who orchestrated propaganda for the Comintern (see McMeekin). Einstein knew Goebbels by reputation and had several direct dealings with Münzenberg's operation, most notably in connection with anti-Nazi propaganda in the wake of the Reichstag fire (chapter 9). But during his last three years in Berlin he attached himself to Révész's Cooperation Press Service, along with Thomas Mann and a handful of other intellectuals. A glimpse of the discussions that took place at that time can be gleaned from a letter Révész sent Einstein on 2 November 1932, shortly after he had visited him at Einstein's summer home in Caputh. In it he fired off a series of eleven key questions on disarmament and peace, some of which the physicist subsequently addressed in public or private writings. After the war they met again in Princeton to discuss strategy for launching one of the most successful publications of the immediate postwar era, Emery Reves's book, *The Anatomy of Peace*.

As a media star in his own right, Einstein was quite cognizant of the various competing propaganda machines in Berlin. Keeping this in mind, we can better appreciate the significance of his message to his fellow Berliners when he spoke at the opening ceremony of the German Radio Exhibition in August 1930. "Until the present day," he emphasized, "people only got to know each other through the distorting mirror of their own daily press. Radio reveals them to one another in most vivid form and, for the most part, from an ingratiating side. Thus it contributes to eradicating feelings of mutual alienation, which can so easily turn into mistrust and enmity" (chapter 5). Einstein's hopes that radio would help usher in a new democratic era were quickly crushed. Indeed, three years later the featured speaker at the German

Radio Exhibition was none other than the Nazi propaganda minister, Joseph Goebbels, who spoke in the presence of his party chief, Adolf Hitler.

Cast Out of the Temple

In December 1930 Einstein got another taste of modern communications media when he arrived in New York only to find himself confronted by a huge throng of reporters and photographers. In his diary he recorded that "the reporters asked particularly inane questions to which I replied with cheap jokes that were received with enthusiasm" (11 December 1930, Einstein Archives 29-134). A perhaps more discerning member of the press asked him what he thought of Germany's new political star. Intent on downplaying the Nazis' recent electoral success, Einstein responded: "Hitler is no more representative of the Germany of this decade than are the smaller anti-Semitic disturbances. Hitler is living—or shall I say sitting—on the empty stomach of Germany. As soon as economic conditions improve, Hitler will sink into oblivion. He dramatizes impossible extremes in an amateurish manner" (Einstein 1931b, 106–107).

Meanwhile, back in Germany the Nazis were again fanning the flames of anti-Semitism, making Berlin's most prominent Jewish intellectual a natural target. He was criticized in the *Völkischer Beobachter* for taking a Belgian ship to America (the SS-writer conveniently overlooked the fact that no German-flag boat made a call in Los Angeles, his final destination; Levenson, 387). That same year witnessed the publication of a bizarre collection of writings under the title *100 Authors against Einstein*, a compendium that made the earlier attacks by Gehrcke and Lenard look almost dignified by comparison. Einstein saw no reason to pay heed to rabid anti-relativists any more than he felt alarmed by spreading sympathy for the Nazis. As he saw it, both were political epiphenomena symptomatic of a growing disillusionment with economic problems that exacerbated social conflicts and fed a sense of despair. Berlin was a true cauldron for social unrest, but he continued to hold out hope that democratic forces would eventually prevail.

Einstein's personal relations with scientific colleagues at the Prussian Academy—Max Planck, Max von Laue, Fritz Haber, and Walther Nernst—had always been cordial, despite obvious differences in temperament and outlook. What he shared with them, despite all other differences, was above all a sense of belonging to an institution dedicated to promoting the ideals of a tiny intellectual elite. No one could have expressed this more lucidly than Einstein himself when he spoke on the occasion of Planck's 60th birthday. Invoking quasi-religious imagery that stressed the values shared by all members of the scientific community, he lauded Planck's special place in the Temple of Science, attained not by virtue of his achievements, but by the pursuit of the sublime that guided his work. To be sure, science also required the contributions of those who used it to test their intellectual powers or technical ingenuity. But such individuals lacked the deeper commitment of the pure researcher. "Were an angel of the Lord to come," he intoned, "and drive all the [former types] . . . out of the temple," only a few would remain inside, and "our Planck is one of them, and that is why we love him" (CPAE 7, Doc. 7). While acknowledging that such an angel would have difficulty deciding whom to drive out of the temple, Einstein had no doubts about the significance of those few who were allowed to remain within: for without them "the temple would never have come to be." Clearly this ascetic, idealized image of the scientists' realm—a notion that stood at the very heart of German academic life—resonated completely with Einstein's scientific identity. Indeed, in his tribute to Planck he cites Schopenhauer's opinion that one of the strongest motives leading to scientific and artistic creativity stems from a longing to escape the painful realities of the everyday world in a search for higher truths, a view he reiterated on numerous occasions. Thus it was fitting that Einstein was chosen in 1929 as the first recipient of the Planck Medal (see Plate 14).

It would be difficult to overstate the importance of this other-worldly, elitist outlook when assessing Einstein's views on worldly matters. For nearly two decades he brought a special allure to the Prussian Academy of Sciences, shining like an exotic ornament within this by now rather antiquated collective of high scholarly achievement. What he received in turn can best be discerned from the numerous warm letters he exchanged with Planck and others in that small circle of Berlin scholars to which he belonged until early 1933. The events of that year, in particular his stormy departure from the academy, were painful experiences that left lasting wounds.

Einstein's initial decision to sever all official ties with Germany was taken neither lightly nor hastily. Like many others, he took a "wait and see" attitude when Hitler was first appointed chancellor on 30 January 1933. He did not have long to wait, however. Just one month later the Reichstag fire provided the ideal pretext for the Nazi chancellor to demand emergency powers that would enable his government to deal with its traditional enemies, including those who had supposedly betrayed Germany during the years of the Weimar Republic. On his return to Europe in late March, Einstein immediately issued a forceful public statement condemning "the actions of brutal force and oppression taken against all free intellects and against the Jews" in Germany ("Statement on Conditions in Germany," chapter 6). While the right-wing press answered with predictable invective, even the more moderate papers adopted a tone of self-satisfaction in their coverage of the controversy over his resignation from the Prussian Academy and his decision to relinquish German citizenship. Once again, but now with the sanction of a state-sponsored ideology, science became embroiled in politics.

Philipp Lenard had high hopes that the Nazi revolution would lead to a revolutionary new approach to physics. Although he did not join the Nazi Party until 1937, Lenard had superb political credentials, thanks to his outspoken opposition to Einstein and "Jewish physics." In 1929 he published his *Grosse Naturforscher*

(Great Scientists), a patchwork of romanticized biographies that served to demonstrate that the history of physics owed nothing to the Jews. Indeed, this book became one of the canonical works of the Aryan physics movement, which adopted the aged Nobel laureate as its spiritual leader (Beyerchen, 123–124). When the Third Reich honored him in 1935 at a ceremony renaming his Heidelberg laboratory the Philipp Lenard Institute, his life story was recounted as another version of Hitler's *Mein Kampf*, a struggle to defeat the Jewish enemy from within, embodied by Einstein and his theory of relativity.

Two years earlier, Lenard aired his rancor against all those who had allowed themselves to be "duped" by Einsteinian pseudoscience:

> The foremost example of the damaging influence upon natural science from the Jewish side was presented by Mr. Einstein, with his "theories" mathematically blundered together out of good, preexisting knowledge and his own arbitrary garnishes, theories which are now gradually decaying, which is the fate of procreations alien to nature. In the process, one cannot spare researchers, even those with genuine achievements, from the charge that they indeed first let the "relativity-Jews" become established in Germany. They did not see—nor did they want to see—how very erroneous it was, even in a nonscholarly connection, to consider especially this Jew as a "good German" (Lenard 1933).

Lenard's tirade clearly targeted Planck and the Berlin establishment whom he and his allies blamed for the debacle in Bad Nauheim in 1920 that ended the first anti-relativity campaign. As in those days, Max von Laue once again rose to the challenge posed by Lenard and other anti-relativists who sought to vilify Einstein and smear his name. As president of the German Physical Society, in September 1933 Laue delivered a plenary lecture on Galileo's career, drawing an unmistakable parallel between the case of Einstein and Galileo's persecution by the Inquisition. (One can discern a sense of Einstein's own identification with

Galileo and his fate from the closing documents in chapter 2.) Lenard and Stark found this an outrageous affront, but their anger only grew soon afterward, after learning that the Berlin Academy had chosen Laue to fill Einstein's vacant chair. Both sent letters of protest to the Prussian Ministry of Culture in which they cited Laue's Galileo lecture as only the latest among his many misdeeds as the leading defender of Einstein and "Jewish physics."

"My awareness of the essential nature of Judaism"

Einstein was not one to cling to an idea in the face of hard facts that dictated failure. Already dismayed by the feeble negotiations at the Geneva Disarmament Conference, he simply discarded pacifism as a political strategy soon after Hitler was appointed chancellor. That cataclysmic event now put his long-standing concern for the fate of European Jewry in sharp relief. Having found his own moral compass by rejecting the assimilationist cravings of many German Jews, he broke once and for all with Germany and took up life in the United States.

Breaking faith with Zionism was a different matter. In spite of the fact that his hopes would be sorely tested by events on the ground in the Middle East, Einstein remained true to the Zionist dream of immigration to Palestine, albeit to its less intrusive cultural aspects. At the same time it is true that he was not particularly well versed in the day-to-day realities of the region, though on his only visit to Palestine in 1923 he consulted with one of his hosts on "some of the intricacies of the Arab question" (Kisch, 30).

In November 1917 in the Balfour Declaration, the British government stated that it viewed with favor the establishment in Palestine of a national home for the Jewish people, with the understanding that nothing would be done to prejudice the rights of existing Arab communities. Three years later the League of Nations created the British Mandate for Palestine. By March 1920

Arab riots, abetted to no small degree by a young militant nationalist, Haj Amin al-Husseini, escalated into a conflict of great violence between Arabs and Jews. In their report on the unrest,
which continued sporadically for more than a year, the British set
the pattern of equivocation that would dog their policy in Palestine for almost thirty years. Conceding justifications to both sides,
they managed to elicit the distrust of both parties. While the Jews
formed an illegal grassroots military organization, the Haganah,
and encouraged large-scale immigration with an eye to political
sovereignty, the Arabs, by force of arms, sought to deny them
their goal of a Jewish majority. In 1929, in a reprise of the riots at
the beginning of the decade, the Arabs, again under the leadership
of al-Husseini, now elevated to the position of Grand Mufti of
Jerusalem, unleashed the so-called Hebron riots.

This upheaval at the end of the 1920s exposed a faulty premise
in the Zionist calculation, particularly as it was perceived by
many European Jews. It was a premise that would come to haunt
Einstein. He and many others had taken at face value the slogan
that Palestine was a land without people for a people without
land. By the end of the decade, the acquisition of land through
purchase from absentee Arab landowners by private capital as
well as by the Jewish National Fund had created acute tension between Jew and Arab. The riots of August 1929 undermined any
pretense of coexistence between the two parties and set nationalist
passions ablaze.

Einstein's shock was palpable. In stretching his internationalist
principles to accommodate Zionist nationalism, he had not
counted on such deep-seated Arab hostility toward Jewish settlement nor its festering resentment of Jewish chauvinism and exclu-
sivity. The very basis for Zionist legitimacy lay in the preservation
at all costs of what Einstein called its "spiritualized" character.
During a reception given in his honor in the City Hall of New
York in 1930, he reiterated his view that the "unity of Jews the
world over is in no wise a political unity and should never become

such. It rests exclusively on a moral tradition. Out of this alone can the Jewish people maintain its creative powers, and on this alone it claims its basis for existence" ("The Jewish Mission in Palestine," chapter 3). The moral basis was reinforced by a subtle messianic tone: "Revitalized through the mysticism of Zionism," Jews might "through the study of their past, through a better understanding of the spirit of their race . . . learn anew the mission they can accomplish." ("Mission," chapter 3). At the same time this very same moral tradition demanded a just solution of the conflict between Arab and Jew, an end that seemed less and less attainable as the political situation in Palestine continued to deteriorate.

Foremost among those who had a far more cold-blooded perception of the Palestinian realities was David Ben-Gurion, secretary general of the General Federation of Labor in Palestine and later head of the Jewish Agency. Ben-Gurion recognized the depth and rationale of Arab objections to Zionism. He was aware of the tragic nature of a clash between two genuine claims to the same land. While prepared to go some distance in accommodating the Arabs, he was one of the first to plan for the creation of a shadow state and a shadow military force. Above all, he welcomed Jewish settlers as the true "army of Zionist fulfillment" (Segev, 255, citing Ben-Gurion's memoirs).

Einstein, on the other hand, believed that a solution might be found in a binational formula that prescribed a common state in historic Palestine shared between Jewish and Arab populations. Events on the ground dictated a different reality. The quota system imposed on immigration to the United States in 1924 diverted East European emigration to Palestine, a pressure counterbalanced by forceful appeals to the British Mandatory Authority on the part of the Arabs to halt the population inflow and land transfer. Mutual intransigence exposed the overly optimistic character of the binational formulation, supported on the Jewish side by a minority consisting for the most part of cultural Zionists. The need to provide security and the material means of advancement to a swelling Jewish population ensured the triumph of adherents of the political Zionist position.

For the Arabs the presumed economic advantages conferred by Jewish colonization were overwhelmed by political fears that they would cease being masters in their own house. In 1918 there were fewer than 60,000 Jews in Palestine; by the end of the 1930s, their number had grown to just under 500,000, roughly 30 percent of the total population. Einstein had underestimated both the goal and the determination of the Jewish settlers as well as the accelerating nationalist sentiments of Palestinian Arabs, writing off their differences as "more psychological than real" (letter to *Falastin*, 15 March 1930, chapter 3). In fact, the obstacles to accommodation were enormous and would increase with time.

Once again the British response served only to infuriate both parties. Fearful of compromising its access to the Suez Canal, and to the oil fields of Iraq and the Arabian Peninsula and to India beyond, Great Britain initiated a number of measures to curtail immigration of Jews and mollify the Arabs. None went far enough to satisfy the latter and did everything but reassure the former. Einstein too reserved his harshest criticism for what he considered the incompetent and mean-spirited administration of the Mandatory Authority. As early as 1929, soon after the Hebron riots, he advised against "leaning too much on the English" (letter to Chaim Weizmann, 25 November 1929). The increased pressure of Jewish immigration from Nazi Germany after 1933 and a stiffening of resistance by the Palestinian Arabs gave rise to the Arab general strike of 1936, and a year later to the formation of the British Peel Commission. In a sobering account of the dramatic situation, the commission concluded that the Mandatory Authority could not bridge the irreconcilable differences between Jew and Arab, and that binationalism was a dead issue. It further recommended that Palestine be partitioned, a position that was accepted with major qualifications by the Zionists and rejected outright by the Arabs. What followed was a tragic inversion of

Einstein's advice to avoid British mediation and deal directly with the other party. The immediacy of the relationship between Arab and Jew was defined increasingly by mutual terrorizing, what in the period between 1936 and 1939 came to be known as the Arab Revolt.

Undaunted, Einstein continued to lay out his principles of an understanding between the two peoples. In an address in New York in April 1938, entitled "Our Debt to Zionism," he first praised the role of Zionism in returning a sense of community and self-respect to the Jews, but then warned darkly against the "narrow nationalism" with which the movement was countering Arab violence. In doing so, he once again decisively threw in his lot with the cultural rather than the political Zionists: "I should much rather see reasonable agreement with the Arabs on the basis of living together in peace than the creation of a Jewish state. Apart from practical consideration, my awareness of the essential nature of Judaism resists the idea of a Jewish state with borders, an army, and a measure of temporal power no matter how modest" ("Our Debt to Zionism," chapter 6).

A Desperate Urgency to Find a Homeland

Hitler came to power in January 1933. The first years of his rule were marked by increasing personal violence against Jews and a progressive stripping of their livelihood, property, and civil rights, but it was only in early November 1938 that the Nazi regime initiated a coordinated nationwide policy of terror on its Jewish population, the so-called Night of Broken Glass. Showing a remarkably inopportune sense of timing, the British Foreign Office decided at this critical juncture to withdraw from the table its partition plan for Palestine. To add insult to injury, two weeks after the night of terror, 10,000 Jewish children from Germany and the Czech lands were denied entrance to Palestine even as immigration there increasingly became a matter of life and death. Einstein was stung by this rebuff, which he interpreted as an emphatic British retreat from the Balfour Declaration of two decades earlier. The children were admitted in a *Kindertransport* to England in December of the same year, but no systematic effort was undertaken by His Majesty's government to find alternative sanctuaries for Jews remaining in Europe. The final straw was embodied in the 1939 White Paper, which declared a limit of 15,000 a year for five years on Jewish immigration to Palestine and none after that, unless the Arabs relented. What Einstein perceived as a betrayal of Jewish and Arab interests in fact spelled a dead end for British colonial policy. Zionist rejection of the White Paper was matched by Arab demands that all Jewish immigration be halted at once. Terrorism on both sides now overshadowed the policy of both local parties. On the Jewish side, any pretense of accepting binational parity was abandoned as creation of a Jewish state absorbed all energies. For the Arabs, the hope of a permanent Arab majority was slipping away.

Einstein's reaction to these events was twofold: during the war years he urged American Jews to offer generous support to those who sought to escape the pincers of the Nazi war machine, as well as to Jewish settlers in Palestine and elsewhere (chapter 7). He also feverishly engaged in securing affidavits for European refugees seeking entry into the United States. In fact, the sheer number of the affidavits that he signed soon diluted their impact on the authorities. To his closest friend, Einstein wrote in the autumn of 1938 that he could not "give any more affidavits and would endanger those that are still pending if I were to give more. . . . The pressure on us from the poor people over there is such that one almost despairs, faced as one is with the depth of misery and the few possibilities of helping" (letter to Michele Besso, 10 October 1938). For the most part, he kept a low public profile during these years, though he staunchly supported Russian War Relief, recognizing the Soviet contribution to defeating the racist Nazi regime after Hitler's invasion of summer 1941. Appalled at new measures to exclude refugees initiated by the State

Department in his new homeland, he was able after lengthy entreaties and with the help of his friend and close political ally, Rabbi Stephen S. Wise, to prevail upon President Franklin D. Roosevelt to create the War Refugee Board in January 1944. This effort to grant more generous terms of entry to the United States for European Jews came far too late. Nazi plans for wholesale slaughter, formalized at the Wannsee Conference in January 1942 as the "Final Solution," had already been widely implemented two years later. Nevertheless, however inadequate the contribution, the Board served to relieve the suffering of about 200,000 Jews by the end of the war.

Rabbi Wise was Einstein's most direct link with the leaders of the American Jewish community. As the most prestigious figure in its moderate center, Wise steered a perilous course between those conservative members of the Jewish community who claimed that Jews should not pressure the American people and its government during wartime and others who insisted that public opinion must be mobilized to intervene on behalf of the dwindling remnant of European Jewry, as well as to further the goal of a Jewish state. Sensitive to the administration's primary objective of defeating Hitler militarily, both Wise and Einstein may have been too ready to accept FDR's vague assurances rather than demanding decisive action.

No such ambiguity affected Einstein's opinion of his former countrymen. A year after the end of the war, he informed a nationalistic colleague with whom he had once been on amicable terms: "The Germans slaughtered my Jewish brethren; I will have nothing further to do with them" (letter to Arnold Sommerfeld, 14 December 1946).

The Zionist leadership's decision in 1942 to embrace officially the goal of a Jewish state stood in the shadow of the ongoing Holocaust. David Ben-Gurion's resolve and that of his colleagues now took on special meaning—the rescue of their people from mechanized Nazi madness. The decision was further reinforced by the persistence of British exclusivist policy toward immigration, in large part a function of fears of growing Soviet influence in the Middle East as well as Arab alienation. In 1944 Einstein summarized the British position as appeasement, pure and simple, "yielding to interests partly British, partly Arabian" ("Palestine, Setting of Sacred History . . .," chapter 7).

Hearings were held before a congressional committee in early 1944 to decide on free entry of Jews into Palestine. When testimony was offered on the Arab perspective and published in the *Princeton Herald* by Princeton professor Philipp Hitti, Einstein and a colleague could not resist entering the fray. They blamed Arab leaders for fomenting hatred of Zionist settlers, claiming that "they act as all fascist forces have acted: they screen their fear of social reform behind nationalistic slogans and demagoguery." By so doing, the Arab people have failed to see what Jewish youth has accomplished: "They created new forms of cooperative settlements and raised the living standard of the Arabian and the Jewish population alike." Though it is certain that Einstein's coauthor, Erich Kahler, took the lead in formulating the responses (chapter 7), the surprisingly harsh and defensive tone of the articles seems in part due to Einstein's reaction to what he considered an implicit reproach. Hitti rejected the appeal for a spiritualized Jewish community dear to Einstein's heart and argued that the battle for Palestine was between competing ideologies, in the name of which each side was prepared to justify using force.

In an effort to embroil the Americans in their Palestinian nightmare, Great Britain formed the Anglo-American Committee of Inquiry on Jewish Problems in Palestine and Europe. Einstein was called to testify before the committee in mid-January 1946, and used this opportunity to heap a veritable torrent of abuse on British policy. He located the root of the problem in a policy of divide and conquer, fueled by British indifference to the incitement of the Palestinian peasants by Arab landowners and exacerbated by active British discouragement of attempts for Arab-Jewish cooperation. Nor were the political Zionists spared. Arguing that a Jewish majority in Palestine was not important, Einstein dismissed the goal of a Jewish state: "The state idea is not according to my heart. I cannot understand why it is needed. It is connected with many difficulties and a narrow-mindedness. I believe it is bad." That Arab and Jew live in harmony was a far greater guarantor of a just political arrangement ("Testimony at a Hearing . . .," chapter 7). Nevertheless, he remained a strong proponent of unlimited Jewish immigration into Palestine. Many of Einstein's recommendations were accepted by the committee, though its own recommendations to the British government fell on deaf ears.

Any remaining good intentions dissolved into chaos after the leveling of the King David Hotel by the Jewish terrorist organization, the Irgun Zvai Leumi, in which almost one hundred Jews, Arabs, and Britons were killed. Einstein summarized the result: men of reason would be pushed aside by "men of action" (letter to Hans Mühsam, 3 April 1946). A state of undeclared war now existed between the parties. Privately, some of his harshest barbs were saved for his fellow Jews: "With respect to Palestine we have advocated unreasonable and unjust demands under the influence of demagogues and other loudmouths. Our impotence is bad. If we had power it might be worse still" (letter to Hans Mühsam, 22 January 1947).

After more than a decade of debating their extrication from Palestine, the British decided in February 1947 to cut their losses and turn the mandate over to the United Nations. Nine months later, the U.N. General Assembly accepted a plan to partition Palestine. Until their departure in mid-May 1948, the British adopted a passive attitude in the ongoing civil war, though their sale of arms to the Arabs and support of the Arab Legion placed the Jews at a distinct disadvantage. With Arab armies and Jewish fighters still locked in combat in autumn 1948, Einstein admitted that "there is no going back, and one has to fight it out. At the same time, we must realize that the 'big ones' [the United States and the Soviet Union] are simply playing cat and mouse with us and can destroy us any time they seriously want to" (letter to Hans Mühsam, 24 September 1948).

Although never an advocate for a Jewish state in Palestine, Einstein accepted this outcome as the inevitable consequence of Britain's failure to create a workable political settlement in which Arabs and Jews could live together in peace.

On the death of President Chaim Weizmann of Israel in November 1952, negotiations were begun to woo Einstein as Weizmann's successor to this largely ceremonial office. Einstein's response to these overtures reveals the remarkable consistency with which he addressed the question of his relationship to Jewry since his first flirtation with Zionism at the end of 1919. Pleading advancing age and a lack of "natural aptitude and the experience to deal properly with people," Einstein expressed regret at having to decline "because my relationship to the Jewish people has become my strongest human bond, ever since I became fully aware of our precarious situation among the nations of the world." A practical consideration also intruded: "I also gave thought to the difficult situation that could arise if the government or the parliament made decisions which might create a conflict with my conscience; for the fact that one has no actual influence on the course of events does not relieve one of moral responsibility" (N & N 1960, 572–573).

Neither by ideology nor by his ideals was Einstein blinded to the complexities of the situation in Palestine. Nevertheless, his lack of familiarity with its realities led him to misjudge the situation, and in this he was not alone. Politically astute observers too, such as his erstwhile ally Chaim Weizmann, suffered their disappointments. Even before the destruction of European Jewry gave a desperate urgency to finding a homeland for the displaced and dispossessed, Weizmann and moderates such as Einstein failed to take the true measure of nationalist fervor on both sides of the divide in Palestine, though Einstein's point of departure, from which he never wavered, was the fundamental moral necessity of Arab and Jew to live with mutual respect (letter to Zvi Lurie, chapter 10). His conciliatory appeals, with their characteristic mixture of hardheaded pragmatism and refusal to make concessions of principle, were ignored. And still the struggle continues to reconcile security and recognition for Israel with the legitimate demands of Palestinian Arabs.

"Like shooting birds in the dark"

Along with the theory of relativity, Einstein's name is famously associated with the Manhattan Project that built the first nuclear weapons. As some saw it, the terrifying implications of $E = mc^2$, the famous formula he derived from special relativity, were revealed to the world with the destruction of Hiroshima and Nagasaki. Those familiar with the history of modern physics were less apt to draw this simple moral in view of the fact that nuclear fission was only discovered in 1938. Before then no one, including Einstein, had any inkling that there might be a way to exploit his purely theoretical finding to develop a technological means for converting mass into energy.

When physicists first began to probe the interior of the atom, they occasionally discussed the possibility of tapping the energy bound within it. Einstein, however, always expressed skepticism on this score. In 1935 he was invited to deliver a plenary lecture at the annual meeting of the American Association for the Advancement of Science held that year in Pittsburgh. Speaking before a large crowd that included some three dozen reporters, he rederived his famous formula. At the press conference afterward a reporter asked whether the huge amounts of energy stored in an atom might be released by bombarding it with heavy particles. To this, Einstein reportedly gave the wry reply that hoping for that would be "something like shooting birds in the dark in a country where there are only a few birds" (N & N 1960, 290).

Three years later, Lise Meitner helped correct that mistaken impression. Meitner first met Einstein in 1909 when as a young woman she heard him lecture on radiation theory in Salzburg (Rhodes 1988, 259–260). Nearly thirty years later Meitner had reason to recall that lecture again when trying to explain the massive release of energy that took place when uranium nuclei were bombarded with neutrons. Her former colleague, Otto Hahn, had discovered that a lighter element, barium, appeared as a reaction product, and Meitner surmised that this could only mean that the nuclei of the uranium atoms had split: what Hahn and his collaborator Fritz Strassmann had found was a laboratory technique that produced nuclear fission.

Word of this discovery quickly reached Copenhagen just as Niels Bohr was about to depart for a lengthy stay at Princeton's Institute for Advanced Study. Soon after his arrival, Bohr, working in collaboration with John A. Wheeler, realized that only the uranium-235 isotope could be split by slow neutrons. This represented a major setback since U-235 accounts for less than one percent of natural uranium; moreover, the technological problem of separating isotopes appeared arduous, perhaps not even feasible. Interviewed by the *New York Times* on his sixtieth birthday, 14 March 1939, Einstein remained skeptical. Commenting on the results obtained thus far, he stated that these "do not justify the assumption of practical utilization of the atomic energies released in the process." Three months later he learned otherwise from his friend Leo Szilard, who convinced him that a nuclear chain reaction was a distinct possibility.

From 1930 until the end of the Second World War, Einstein's political outlook was dominated by his staunch opposition to fascism, particularly the Nazi variant that coupled authoritarian militarism with anti-Semitism and racism. In "Ten Fateful Years" (the final document in chapter 6) he painted a gloomy picture of Europe on the brink of disaster, where "to the east of the Rhine free exercise of the intellect exists no longer," replaced by the terrorism of gangsters whose propaganda poisoned German youth with systematic lies. Little wonder that he became deeply alarmed when, shortly before the outbreak of the war, he learned from Szilard that German scientists might be engaged in research to produce a nuclear weapon.

At Szilard's urging, Einstein helped prepare the famous letter of 2 August 1945 addressed to President Roosevelt that has often been credited with launching the Manhattan Project. It was not until October, however, that this letter was handed over to the president by an economic adviser, Alexander Sachs, who had ties with the physicists. After their meeting in the White House, FDR authorized the formation of an Advisory Committee on Uranium chaired by the chief of the Bureau of Standards, Lyman J. Briggs. Over the next several months little was accomplished, however, and in June 1940 the Briggs Committee was disbanded after Roosevelt created the National Defense Research Committee. The NDRC was chaired by Vannevar Bush, who took a wait-and-see approach to nuclear research. But when Bush and others read the British government's MAUD Committee Report in October 1941 on the practicability of a "uranium bomb," the tide turned. On 6 December 1941, the eve of Pearl Harbor, the Manhattan Project was finally approved. Einstein, labeled a security risk in a confidential report issued by the U.S. Army, remained officially in the dark about nuclear research throughout the war.

Nuclear Nightmare as an Incentive for World Peace

Soon after Japan's surrender following the dramatic devastation of Hiroshima and Nagasaki by atomic bombs, Einstein was in the news again as politicians and scientists in the United States struggled to reach a consensus regarding the future control of nuclear energy. Although often connected with the Manhattan Project, he had no involvement whatsoever with nuclear research either during or after the war, nor was he actively engaged in the effort to move nuclear technology away from the military and place it under the control of a civilian organization, the Atomic Energy Commission. Einstein had never engaged in research on nuclear physics and some of his postwar pronouncements reveal that even after the release of the Smyth Report he had only a vague picture of the research that been undertaken in this field.

Nevertheless, the cultural and political shock waves that hit American shores in the wake of Hiroshima made Einstein an irresistible symbol for the generation of physicists whose work actually contributed to building this terrifying weapon. *Time* magazine employed this imagery on the cover of its 1 July 1946 issue, which showed Einstein's familiar face—the sad eyes and chaotic hair—superimposed over a mushroom cloud accompanied by the mysterious equation $E = mc^2$ (see Plate 22). *Newsweek* ran a feature story in March 1947 on "Einstein—the Man Who Started It All." As Spencer Weart wrote in *Nuclear Fear*, "When scientists and the press collaborated in making the old man a prime symbol of nuclear energy, they were using the symbolism to insist that scientists, however wise and well-meaning, had released forces far beyond the ordinary ken" (Weart, 112). What few noticed, however, was how deftly Einstein managed to use his flimsy credentials as "grandfather of the bomb" to garner media attention, though not for himself, to be sure, but rather for his long-term political agenda.

Serendipitous timing again played a major part in this next chapter of Einstein's political career. Only weeks after the war ended, he received a letter from Emery Reves along with a copy of his newly released book, *The Anatomy of Peace*. Recalling their earlier conversations at Einstein's summer home, Reves wrote: "I feel particularly indebted to you as I do not think that without your philosophical outlook this book could ever have been written" (24 August 1945). He also enclosed to Einstein a draft for a public appeal issued by former Supreme Court justice Owen J. Roberts that called for a massive effort to make the book available to the American public. The key passage in this appeal read as follows:

It happens that at this crucial moment of our history, a small book has been published, a very important book, which expresses with

the utmost clarity and simplicity what many of us have been thinking and saying. . . . We urge every American man and woman of
good will to read this book, to think about its conclusions, and to
discuss it with neighbors and friends, privately and publicly. A few
weeks ago these conclusions seemed important, but something
concerning the future. With the reality of atomic warfare, they are
of immediate, urgent necessity if civilization does not want to commit suicide. (Einstein Archives, 57-291)

Einstein was only too pleased to support this effort. Being familiar
with Reves's general political orientation from their earlier encounters in Berlin, he read *The Anatomy of Peace* with considerable anticipation. He was not to be disappointed: for here he
found a clear, cogent, and yet impassioned argument for a new
political world order. In his opening chapter, Reves described the
contours of this "Copernican world" beginning with the assertion
that: "Nothing can distort the true picture of conditions and
events in the world more than to regard one's own country as the
center of the universe" (Reves 1945, 1). Pointing to the fundamental flaw in the proposed charter of the United Nations, Reves
called for a true world government with enforceable executive
and judicial powers.

Galvanized by Reves's book, Einstein prepared to make his
first public pronouncements on the subject of nuclear warfare.
He contacted the radio talk show host, Raymond Gram Swing,
an advocate of world government, and with Swing's help the article "Einstein on the Atomic Bomb" was published in the November 1945 issue of the *Atlantic Monthly*. Swing and Einstein
were both cognizant of the unique opportunity this represented
to promote the world government-movement by exploiting Einstein's reputation as the scientific genius behind the bomb. Not
wishing to squander this newfound political capital, the physicist made only a rather weak disclaimer—that he did not consider himself the father of the release of atomic energy—a
remark that readers might easily dismiss as just another expres-

sion of Einstein's modesty. He struck a similar note with the following endorsement:

"I myself do not have the gift of explanation which would be needed to persuade large numbers of people of the urgency of the problems that now face the human race. Hence, I should like to commend someone who has this gift of explanation: Emery Reves, whose book *Anatomy of Peace* is intelligent, clear, brief, and, if I must use the absurd term, dynamic on the topic of war and need for world government."

As Einstein's first public statement on the atomic bomb, this article received a great deal of attention and considerable critical scrutiny. Many expressed strong disagreement with Einstein's views on the Soviet Union; others of more liberal persuasion dismissed the notion that Stalin's government would ever agree to cede its military power to an international authority. Using his company's network, Reves saw to it that the full text was published in eighteen countries outside the United States (letter from Reves, 27 November 1945). Demand for *The Anatomy of Peace* suddenly soared, climbing to the top of the best-seller lists in both the *New York Times* and *New York Herald Tribune* by March 1946 and remaining there for the next six months (Rozelle, 29). Eventually, worldwide sales reached 800,000 copies, and an estimated 50 million people around the world had the opportunity to read the three-issue condensation that appeared in *Reader's Digest*. Reves attributed this staggering success to the influence of Einstein's support (Gilbert, 15). For Einstein, Reves's book provided precisely the kind of manifesto needed to promote the cause of peace through world government. (See Plate 23 for Einstein's explicit endorsement of the book and its use in advertising the work.)

In March 1946 meanwhile, President Truman had appointed Bernard Baruch as the U.S. representative to the United Nations' Atomic Energy Commission. Just ten days later, the State Department issued the Acheson-Lilienthal Report, which proposed that atomic energy be placed under U.N. control, a proposal welcomed by nearly all physicists, including Einstein. When Baruch presented his version of the U.S. plan to the UNAEC in June, the Soviet delegate, Andrei Gromyko, began what turned out to be a lengthy series of fruitless debates. Nevertheless, at this juncture the advocates of world government were still riding a wave of optimism, as reflected in Einstein's remarks in a widely publicized interview from 23 June 1946, "The Real Problem Is in the Hearts of Men" (chapter 8).

At the same time, Einstein joined Szilard and other leading American physicists in pushing for a new Atomic Energy Commission under civilian control, a proposal that gained congressional approval despite strong resistance from the U.S. Army (see Smith). Throughout the ensuing months, however, U.S.-Soviet relations deteriorated rapidly, and in March 1947, after a long delay, Gromyko summarily rejected Baruch's proposals calling for international controls on nuclear research and development. Soon thereafter, addressing a joint session of Congress, FDR's successor announced the Truman Doctrine aimed at curbing Soviet aggression. The Cold War era had officially begun with fatal implications for the world-government movement.

Einstein was interviewed later that summer by Raymond Swing for a radio program that dealt with two bipartisan Congressional initiatives aimed at strengthening the U.N. charter. Ever upbeat, he expressed his confidence that these resolutions would "mark a turning point in international politics, if they find the vigorous support of large sections of the American people" (N & N 1960, 415). Three months later, in an "Open Letter to the General Assembly of the United Nations" (chapter 8), he called for reforms that would strengthen the organization's mandate and lead to a platform for world government. After the Soviets rebuffed this proposal, however, he recognized that the window of opportunity for negotiations was fast closing.

Many on the American side felt no great urgency to break the impasse so long as the United States maintained a monopoly on nuclear weapons, but they were caught by surprise when the

Soviets detonated an atomic bomb in 1949. A mood of panic set in, and on 31 January 1950 President Truman announced that the United States would undertake an all-out effort to develop the hydrogen bomb, a weapon vastly more destructive than its predecessor.

The arms race that Einstein had long feared had now become a reality. Two weeks later he issued a stern warning to the American people during a television broadcast hosted by Eleanor Roosevelt (chapter 8), echoing the central message he had sounded over the past five years:

> The first goal must be to do away with mutual fear and distrust. Solemn renunciation of the policy of violence, not only with respect to weapons of mass destruction, is without doubt necessary. Such renunciation, however, will be effective only if a supranational judicial and executive agency is established at the same time, with power to settle questions of immediate concern to the security of nations. Even a declaration by a number of nations that they would collaborate loyally in the realization of such a "restricted world government" would considerably reduce the imminent danger of war.

"Nothing is farther from my mind than anarchist ideas"

The Cold War ushered in a period of intense rivalry and suspicion between the United States and the Soviet Union. Domestically, formal and unofficial probes into an individual's political reliability became commonplace. The era of the loyalty oath and the security-clearance boards had arrived. Given his propensity to speak his mind and his fearless association with organizations under suspicion, Einstein presented a tempting target for those who viewed expressions of sympathy for the wartime Russian ally, however measured, as treasonable.

Claims that he was an extremist had been levied against Einstein early in his career. In response to the charge in 1919 that he was "a Communist and anarchist," he declared in an interview that "nothing is farther from my mind than anarchist ideas. I do advocate a planned economy, which cannot, however, be carried out in all workplaces. In this sense I am a socialist" (*Neues Wiener Journal*, 25 December 1919). Similar wild-eyed accusations of radicalism were floated after the Second World War. Infuriated, for example, by Einstein's call to break relations in late 1945 with Spanish leader Francisco Franco, an erstwhile ally of Hitler's Germany, Rep. John Rankin of Mississippi attacked Einstein on the floor of Congress as a "foreign-born agitator" who sought "to further the spread of Communism throughout the world" (Jerome, 94).

Often detractors skirted the issue of outright allegiance to Moscow by asserting that Einstein was but a hapless victim. Just months before the Russian regime exploded an atomic device, *Life* magazine presented a powerful visual display of Einstein's questionable loyalties by situating him prominently in a rogue's gallery of photos. The banner title read: "Dupes and Fellow Travelers Dress up Communist Fronts" (*Life*, 4 April 1949). Thus were assertions of political unreliability readily paired with claims of naïveté.

Opponents and undiscriminating media were not the only ones who made the case for Einstein's radical bias. Friends on the left also tried to appropriate him. With great fanfare, the editors of *Monthly Review*, a journal providing commentary from a socialist and Marxist perspective, published Einstein's celebrated essay "Why Socialism?" (chapter 9) in their inaugural issue of May 1949. In the course of reprinting this flagship piece annually, the editors, in a prefatory note to the 2000 issue, placed Einstein's advocacy of socialism at the very heart of his "humane and democratic" vision. Affirmation of this advocacy also came from Einstein's close friend and coeditor of *Einstein on Peace*, Otto Nathan. His unequivocal statement that "Einstein was a socialist"

(N & N 1960, vi) has done much to define the popular perception of Einstein as a tribune of the left.

While accusations from the right are easy to dismiss, claims made by those sympathetic to Einstein's political leanings should also be held up to scrutiny. On closer examination of the essay "Why Socialism?" (chapter 9) it becomes apparent that Einstein's powerful indictment of "the oligarchy of private capital" serves not so much to advance socialism as an economic system per se but to advocate a planned economy as a significant instrument for achieving desirable social-ethical ends. Central to Einstein's acceptance of a socialist system was his expectation that "under socialism, there was a greater possibility of attaining the maximum degree of individual freedom compatible with the public welfare than under any other system known to man" (N & N 1960, vi). Yet he was quick to point out that "socialism as such cannot be considered the solution to all social problems but merely as a framework within which such a solution is possible" ("A Reply to the Soviet Scientists," chapter 8). Unlike many of his contemporaries, he carefully weighed the benefits of a planned economy against the danger that an all-powerful and overweening bureaucracy might encroach upon the rights of the individual and overwhelm the classical liberal ideal of intellectual freedom. He concluded the essay by expressing the hope that the *Monthly Review* would, at a time when free speech and critical thought were endangered, provide the socialist agenda with a much needed forum: "Clarity about the aims and problems of socialism is of greatest significance in our age of transition. Since, under present circumstances, free and unhindered discussion of these problems has come under a powerful taboo, I consider the foundation of this magazine to be an important public service."

Unwavering in his attachment to the principles of universal social justice and solidarity with his fellow Jews, Einstein was, in other respects, steadfast only in his unorthodoxy and almost reckless independence. He sought a social system that rejected gross income inequality and the exploitation of the economically vulnerable, but placed equal if not more importance on the traditional liberal goal of self-realization of the individual. This essential value could best be safeguarded within a democratic framework, but ultimately its defense depended on the political and moral qualities of its citizenry. Fearful that these qualities were imperiled by the hysteria of the McCarthy era, he spoke out forcefully only to be labeled unfairly as a Red or naive dupe (Jerome, 144–146).

Lacking any exposure to the working-class movement in the waning decades of the German Empire, Einstein had been wary of the mass organizations of the left which sprang up or developed a new lease on life in the Weimar republic, while expressing his solidarity with some of their goals. Though he demonstrated philosophical sympathy for socialism with a human face, including many of its economic principles, he did not identify with the intellectual tradition of the European labor movement or the Marxist legacy. Instead, he placed his faith in appeals to reason by a liberal intelligentsia, which, in availing itself of the decorous and principled use of manifestos, might best guide the fortunes of Weimar Germany or the American republic. A longing for community and for a harmonious society reinforced his unease with "a political culture of problem-solving by negotiation, dispute, and majority vote" as well as with the Marxist concept of class struggle (Goenner 2003, 55).

The central issue always remained the free play afforded every individual to develop creative potential. At the height of the Great Depression in 1932, Einstein discussed the need to control the ill effects on society of unchecked egoism and competition while arguing that the state is only of real use to industry as a "limiting and regulative force" which keeps competition among workers within healthy limits ("Production and Work," chapter 9). In the following decade, when he supported more emphatically the idea of a socialist economy, he was careful to hedge his proposal with the caveat that "the concentrated power of the state be under the effective control of its citizenry." Constant political struggle and vigilance were crucial to the maintenance of individual freedom ("Is There Room . . ?," chapter 9).

Another element in Otto Nathan's characterization of Einstein as a socialist does not hold up. Observing that "he believed in socialism because, as a convinced egalitarian, he was opposed to the class division in capitalism," Nathan overlooked the fact that Einstein was rather more an elitist than an egalitarian. In "What I Believe" (chapter 5), for instance, Einstein stated quite unabashedly in 1930 that "what is truly valuable in our bustle of life is not the nation . . . but the creative and impressionable individuality, the personality—he who produces the noble and sublime while the common herd remains dull in thought and insensible in feeling." Hardly the thoughts of one attuned to the dynamics of class struggle. The mainly private writings collected in chapter 9 in the section "Intellectual Elitism and Political Idealism" indicate that, if anything, this strand of Einstein's political thought grew stronger with time.

From pointing out in 1920 that Bolshevik theories were "ridiculous" (letter to Hedwig and Max Born, chapter 9) to his observation five years later that the rulers of Russia would "have to change their methods [in order] to gain moral credibility with civilized nations" ("On a Document Collection . . .," chapter 9), Einstein maintained a curious if wary interest in the Soviet Union. Not surprisingly the high-water mark of his support of Russia was reached during the Second World War, most particularly after Stalin's forces brought the German war machine to a halt at Stalingrad: "We and our children owe a great debt of gratitude to the Russian people for having experienced such immense losses and suffering" ("Address to Jewish Council on Russian War Relief," chapter 9). After the war, however, his enthusiasm cooled considerably and he accused the regime of being "obsessed by the utopia of isolationism" (N & N 1960, 416).

Though never enthralled by the Soviet experiment, his steadfast refusal to join the chorus of anti-Soviet voices left him open to Cold Warriors' suspicions that he was just another fellow traveler. The philosopher Sidney Hook occasionally played provocateur in such encounters, which shed light on Einstein's essentially practical view of politics. Hook, one of America's most vocal anticommunists, devoted a whole chapter of his autobiography (Hook 1987) to "My Running Debate with Einstein," only to conclude that when it came to politics the brilliant physicist was no different from the muddled-headed leftists who were soft on Soviet communism. Hook made a particularly revealing foil for Einstein, as they shared many of the same values, both politically and intellectually, and yet could never see eye to eye. As a political philosopher, Hook wanted to refute Marxist claims for scientific socialism; Einstein never imagined that science and politics had anything to do with one another. Hook saw communist agents and their sympathizers in the U.S. as the greatest threat to Americans' freedom; Einstein thought that the greater danger to freedom in the United States was posed by those who grossly exaggerated the influence of Communists. Hook was a political philosopher intent on winning his argument; Einstein was a moral philosopher who was only interested in promoting his cause. Hook apparently never appreciated the difference, whereas Einstein surely did, cutting off debate once he saw that Hook's agenda would only lead to more acrimony and mistrust between them.

"The State exists for man, not man for the State"

Politics, for Einstein, was fundamentally a matter of how individual human beings choose to organize their lives. Thus, it should come as no surprise that even in his writings on political economy and socialism, the issue of political rights and human freedom took on an overriding importance that overshadowed all other considerations. This was the theme he returned to again and again, as, for example, when he wrote about the forthcoming Geneva Disarmament Conference ("The Road to Peace," chapter 5) that "the State exists for man, not man for the State . . . I believe that the most important mission of the State is to protect the individual and to make it possible for him to develop into a creative personality. The State should be our servant; we should not be slaves of the State." Einstein deeply valued human liberty and believed that the state had merely to provide for the basic material conditions that would enable talented individuals to pursue various forms of creative endeavor.

Closely connected with the state's responsibility to promote creative freedom was its obligation to protect the individual's rights, especially political rights. Thus a major precondition for a functioning democracy was the state's commitment to uphold these rights through a legal system that protected freedom of speech for all citizens, including those who wished to criticize the government. Among those who availed themselves of this newly granted freedom in Weimar Germany were two outspoken leftists, the statistician E. J. Gumbel and journalist Carl von Ossietzky. Einstein deeply admired both men, perhaps all the more so because he, unlike them, normally reserved his harsher political criticisms for private communications. Both were long targeted by Nazi propagandists, who succeeded in ending their careers even before Hitler came to power. Einstein's heartfelt words of praise for Ossietzky, Gumbel, and his friend Paul Langevin (chapter 10) reflect his own deep faith that the ultimate source of all that is valuable in human affairs resides in the noble-minded spirit of such rare individuals.

Mere words, of course, meant nothing to him if they were not supported by actions, which explains why he revered Gandhi as the greatest political personality of the era. Among political thinkers, he held Freud in the highest esteem, taking evident delight in their prearranged correspondence dealing with the psycho-biological roots of aggression and warfare (chapter 4). In an earlier letter, he praised Freud's writings for their "deep devotion to the great goal of the internal and external liberation of man from the evils of war. This was the profound hope of all those who have been revered as moral and spiritual leaders beyond the limits of their own time and country, from Jesus to Goethe and Kant" (chapter 9).

Einstein's antimilitarist sensibilities ran deep, and throughout the early 1930s he was one of the most visible advocates of pacifism and internationalism. Yet as events would reveal, these ideals were not absolute moral imperatives for him; circumstances could require going to war to defend a higher form of political life. Even before the Nazis came to power, Einstein clearly saw the futility of unilateral pacifism and the dire need for an international organization with strong executive and judicial powers. The fact that he only publicly abandoned pacifism after Hitler seized power hardly means that he was guilty of political naïveté. Consistency in one's long-term goals was critical, the modalities of achieving those goals dispensable, as Einstein emphasized on numerous occasions.

Some of his contemporaries, such as the Japanese pacifist Seiei Shinohara (chapter 10), seem to have understood this, whereas the biographer Ronald W. Clark did not. In a view that reinforces the standard canard that Einstein flip-flopped on the issue, Clark likened his pacifism to an article of clothing that after being "tucked away for a while . . . could be pulled from the drawer and worn once again, a garment for fine days" (Clark, 492). Just one example of abundant evidence to the contrary is this passage from a letter written during the war:

> During the twenties [early thirties, ed.] I advocated repeatedly refusal of military service. I am still of the opinion that to make war impossible is one of the most important goals of humanity. On the other hand I recognized that refusal of military service could not be endorsed any longer since in certain countries the resistance to compulsory military service became impossible. As long as these conditions prevail everyone has the duty to do his best to protect the rest of the world by doing his share of the fighting. The fight for the pacifist goal is more needed than ever in our time. The only means to reach this goal is, in my opinion, international organization for the enforcement of military security for the whole world. (Letter to John G. Moore, 30 March 1942)

Contrary to a commonly voiced criticism, Einstein was constantly weighing his political ideals and goals against the realities of the day. He knew the difference between principles and praxis, and his actions often reflected a recognition that political progress depended on a realistic assessment of concrete conditions and circumstances. By the same token, he had no use for high-minded political pronouncements that had no chance of achieving anything of practical value (letter to Abraham Muste, chapter 8). For the same reasons, he strenuously sought to avoid ideological bickering and protracted political debates, as illustrated by his dealings with Sidney Hook (chapters 9 and 10).

Like many émigrés, Einstein hoped the United States would act to save Europe from the barbarians who had seized power in Germany. He praised Roosevelt for staying the course, hoping that afterward the Allies would strip Germany of the Ruhr region, thereby reducing its economy to agricultural production. Numerous postwar pronouncements reveal the depth of his loathing for the German people, whom he held collectively accountable for the crimes of the war and, above all, for the Holocaust. Still, in assessing Einstein's broader political views it is important to note how he generally preferred to address a problem in abstract terms that embraced many different situations rather than dwell on the specific conditions that pertain in one country at a given time.

Clearly he weighed his words carefully in 1933 when he formulated his "Political Manifesto" (chapter 6) explaining why he preferred not to live in Nazi Germany: "As long as I have any choice, I will only stay in a country where political liberty, tolerance, and equality of all citizens before the law prevail. Political liberty implies the freedom to express one's political opinions orally and in writing; tolerance implies respect for any and every individual opinion." Since these conditions did not prevail in Hitler's Germany any more than in Stalin's Soviet Union, logic dictated that he would make his future domicile in a country with established democratic traditions.

Yet Einstein knew very well that this was no matter of ordinary logic but rather a gut-wrenching problem about how to arrange the next phase of his life. Though a convinced determinist, his language nevertheless conveys that on some level he believed he could and must choose. Moreover, from the political standpoint he obviously realized that he, unlike most others in a similar predicament, had the freedom to live wherever he wished. His message, then, clearly implied that an individual with the *power to choose* and the moral capacity to understand what is most valuable in human life will invariably choose freedom. And like so many for whom the bells of freedom tolled, they rang for him from the other side of the Atlantic. Home was now Princeton, New Jersey, within easy reach of New York City, and vice versa.

Until the dawn of the Cold War era, Einstein voiced continual admiration for the social and political culture of the United States. Not only were the country's laws and institutions based on sound egalitarian principles but even more importantly the habits and values of its people were democratic to the core. Following an examination on his application for citizenship in June 1940, he proclaimed in a radio broadcast that Americans were "not suited, either by temperament or tradition, to live under a totalitarian system . . . [and] many of them would find life not worth living under such circumstances" ("I Am an American," chapter 10). Though he commented on the disadvantaged lot of minorities and Negroes in America as early as 1931, it was only after the war that he began to speak out more insistently about the enduring legacy of slavery manifested in white America's feelings of superiority toward blacks. Einstein counseled open-minded people to analyze the roots of their racial prejudice, which he regarded as nothing more than a desire to maintain the traditional attitudes of white masters toward their African slaves.

"Political passions, once . . . fanned into flames, exact their victims"

Human freedom stood at the heart of Einstein's political legacy, but he surely never dreamed that he would spend the final years of his life defending the rights of U.S. citizens accused of high crimes against their own government. By the early 1950s, J. Edgar Hoover, Senator Joseph McCarthy, and the right wing of the Republican Party had seized center stage in American politics. After the FBI got wind of the espionage ring that enabled the physicist Klaus Fuchs to pass vital information from the Manhattan Project to Soviet agents, alarm bells went off in Washington. Hoover intensified the FBI's search for closet Communists and their sympathizers, while Congress began holding hearings that contributed to a national state of hysteria.

No one was to be spared, not even the world's most famous physicist. Few, in fact, knew that Einstein was one of Hoover's top suspects, having been linked by an informant to the confessed spy, Klaus Fuchs, a German-born Communist. Had Einstein himself been allowed to read the misinformation contained in his FBI file, his worst suspicions about the hysteria of his newly chosen compatriots would have been confirmed: "Dr. Einstein, since being ousted from Germany as a Communist, has been sponsoring the principal Communist causes in the U.S., has contributed to Communist magazines, and has been an honorary member of the U.S.S.R. Academy since 1927." Or, for another sampling: "Einstein is either the Chairman of, member, sponsor, patron or otherwise affiliated with 10 organizations which have been cited by the Attorney General . . . as being Communistic groups" (Jerome, xxi).

Government bureaucracies often get lost in their own paperwork, an inevitable problem when desk-bound employees are unable to peer over their piles of documents into the real world. The FBI under Hoover was surely no exception. In Einstein's case Hoover's staffers managed to compile over 1,500 pages of documents relating to his political activities over a twenty-year period. These findings were summarized into a truly ludicrous portrait of the aged physicist based on hearsay evidence and the ravings of right-wing zealots such as Elizabeth Dilling, author of *The Red Network* (Jeansonne, 10–28). In fact, the first document in the FBI's file on Einstein was the memo submitted in 1932 to the State Department by the Woman Patriot Corporation. According to this source, Einstein was the leader of a relativity movement that aimed not only to undermine Christian values but also to "shatter" America's "military machinery" as a preliminary step toward a Communist world revolution.

Hoover was particularly curious about the alleged link with Fuchs, so the agency stepped up its investigation. By 1952 the FBI was working hand in hand with the Immigration and Naturalization Service, which was considering possible denaturalization and deportation proceedings against Einstein under the provisions of the 1940 Alien Registration Act (Jerome, 220–221). Two years later, however, the trail of evidence that supposedly led from Fuchs to Einstein went cold, leaving Hoover with nothing more than a thick pile of papers.

Tucked away in Princeton, Einstein surely had no inkling of the powerful forces that had gathered in the expectation, at the very least, of embarrassing him politically. He was never subpoenaed to testify before a government committee, nor did he feel any need to defend himself on those few occasions when he was openly criticized. If he was a target, he knew very well that others stood in front of him on the firing line; and so he spoke out in their defense, just as he had earlier on behalf of the two prominent pacifists, Gumbel and Ossietzky.

In June 1953 Einstein placed himself at the center of a blazing controversy when he authorized the *New York Times* to publish a short letter he had written to a Brooklyn schoolteacher, William Frauenglass (chapter 10). Frauenglass had refused to cooperate when called before the Senate Internal Security Subcommittee, thereby jeopardizing his career. Einstein not only hailed him for his moral courage but also commended his action as the only appropriate response to such an "inquisition [which] violates the spirit of the Constitution." Citing the teachings of Gandhi, he appealed to "every intellectual . . . called before one of the committees . . . to refuse to testify," even if this meant going to jail or economic ruin. Indeed, the situation demanded "the sacrifice of his personal welfare in the interest of the cultural welfare of his country." Only through a spirit of solidarity, however, could these tactics prevail: "If enough people are ready to take this grave step they will be successful. If not, then the intellectuals of this country deserve nothing better than the slavery which is intended for them."

These were brave words, indeed, but the message was true to form. In some respects it was reminiscent of his "Two Percent Speech" of twenty years earlier in which he pleaded for civil disobedience on an even larger scale. Reactions this time were mixed. Writing to Carl Seelig, his Swiss biographer, Einstein noted that "all the important newspapers have commented in a more or less politely negative tone" (N & N 1960, 547). Predictably, he was flooded with mail, most of it positive. A *New York Times* editorial questioned whether using "the unnatural and illegal forces of civil disobedience" was not merely replacing one evil by another (N & N 1960, 550). Einstein was pleased when Bertrand Russell sent a letter to the editor offering a strong but humorous rebuttal (chapter 10).

In the last years of his life, Einstein drew a strong parallel between the political climate in the United States and the one he had lived through during the death throes of the Weimar Republic. Characteristically, he saw the challenge of the former and the failure of the latter in essentially moral terms. Writing to Norman Thomas, a leading American socialist, he belittled the "Communist conspiracy" as essentially "a slogan used in order to put those who have no judgment and who are cowards into a condition which makes them entirely defenseless." Recalling how liberal democracy had come under a similar attack in Germany, thereby enabling Hitler "to deal it the death-blow with ease," he predicted that the same would happen in the United States "unless men with vision and willingness to sacrifice come to the defense" (chapter 10).

Two peace initiatives occupied Einstein during the final months of his life. In February 1955 Einstein received a letter from Russell outlining a new plan for mobilizing leading scientists around the world, independent of their nation's place within the postwar political system, against the use of nuclear weapons, in particular the hydrogen bomb. This undertaking came to fruition soon thereafter as the Russell-Einstein Manifesto, which eventually paved the way for the Pugwash conferences on nuclear arms control (see Doty). Though Einstein had died two years earlier, one might allow that he was represented in spirit at the first Pugwash meeting by his friend, Leo Szilard, that most political of all nuclear physicists (Lanouette, 369–373).

Einstein's second initiative had a more personal character: he planned to deliver an address to the Israeli people offering a "somewhat critical analysis of the policies of the Western nations with regard to Israel and the Arab states" (letter to Reuven Dafni, 4 April 1955). Unfortunately, his text remained a mere torso at the time of his death. Nevertheless, what he did manage to write elegantly summarizes the central political themes that dominated his attention for so long.

The conflict between Israel and Egypt that would break out into open warfare the following year was nothing new, merely the latest instance of an "old-style struggle for power, once again presented to mankind in semireligious trappings." The only difference, he noted, was the very real possibility that the next major war would be the last one should the parties involved resort to the use of nuclear weapons. In spite of the risks at stake, responsible leaders in the Communist and so-called Free Worlds continued "seeking to intimidate and demoralize the opponent by marshaling superior military strength." None "has dared to pursue the only course that holds out any promise of peace, the course of supranational security." Politicians appeared willing to take the world to the brink of disaster, and Einstein held both sides fully accountable. All the same, he pointed no fingers, adopting a dispassionate tone and measured language. The message—surely written not just for the people of Israel but for the whole world—was familiar, too. And it is this last communication to the world—ending with the warning that "Political passions, once . . . fanned into flames, exact their victims"—that may well serve as Einstein's political epitaph ("A Final Undelivered Message . . .," chapter 10).

CHAPTER 1

The First World War and Its Impact, 1914–1921

Einstein arrived in Berlin in 1914 at the age of 35, a scientific prodigy but innocent of the world of politics. The outbreak of the war was a rude awakening for him and led to his first tentative grappling with political issues. What he retained was the sensibility and vocabulary of a citizen of the republic of letters, not of someone prepared to engage in public debate or political action at close quarters. In the very first months of the war, moral outrage triggered his collaboration in a countermanifesto protesting the solidarity of German intellectuals with the army in its violation of Belgian neutrality. He continued, however, to devote most of his time to his scientific passions for which he sought out and embraced isolation. To a colleague he wrote in January 1915 that "in spite of the troubling, disgusting war I work quietly in my room" (CPAE 8A, Doc. 44). A police report in early 1916 offers corroboration. After mentioning his membership in the pacifist New Fatherland Association, the report noted that Einstein had not yet made his mark politically (Gülzow, 234). Perhaps his harshest critic in this regard was Einstein himself. In the last year of the war he revealed to a friend that he was unsure whether or not to rebuke himself for his own passivity (CPAE 8B, Doc. 537). The friend was Georg Nicolai. A physiology professor in Berlin, confidant of Einstein's cousin and future wife Elsa, Nicolai had drawn up the first draft of the countermanifesto which Einstein signed at the outbreak of the war. In the summer of 1915, he was working on a manuscript to be published two years later as *The Biology of War* (Nicolai; for commentary, see Zuelzer, 52–57). Many of its salient points are strikingly similar to the themes in Einstein's essay on the psychological roots of war, which was written in autumn 1915. Both authors emphasize that the survival of the human race depends on cultivating social impulses and channeling aggressive instincts to ends that benefit the entire community.

The fact that Einstein composed his essay on the origins of war at the same time that he was struggling with the final stages of his general theory of relativity suggests that the recluse was engaging, however hesitantly, with a world beyond his single-minded devotion to science. The crystallizing event that brought him at least partially out of his seclusion was his induction into the New Fatherland Association in summer 1915, presumably at Nicolai's prompting. Membership in this pacifist organization opened to him in Berlin a network of like-minded intellectuals, their politics stretching from the center to the left, whose selfless dedication and refusal to accept blind authority struck a resonant chord (Goenner 2005, 75–88). Only months later he wrote that "these times show us that everyone must do his part for the organization of the whole" (Vol. 8A, Doc. 152). In this endeavor, he also found colleagues farther afield in neutral Europe. They included the forensic physician Heinrich Zangger and the prominent pacifist Romain Rolland in Switzerland, as well as the physicists Hendrik A. Lorentz and Paul Ehrenfest in the Netherlands. Though the New Fatherland was closed down by the authorities in early 1916, Einstein retained an affinity for the camaraderie he had experienced there. Still cautious, he was nevertheless concerned enough about the sad state of wartime politics a year later to lay out concrete steps for an international pacifist union of sovereign states.

Five years after his arrival in Berlin, the German Empire lay in ruins and a hopeful new Weimar Republic struggled to find its footing. Meanwhile Einstein had vaulted to international fame. His cautious testing of the political waters during the war gave way to an increasingly urgent engagement with social and political issues. Two factors determined his greater access to and interest in the political realm. One was thrust on him in late 1919 after a British solar expedition confirmed his general theory of relativity. Now a world figure, pronouncements on public affairs came to be expected of him. The other was a conscious redefinition for himself of the role of the intellectual who has access to the media in the mass society of the twenties.

Dearest to his heart in the first phase of his political involvement were the issues of a Jewish homeland—a theme taken up at length in chapter 3—and the need for international reconciliation, as well as for revitalizing scientific research and cooperation across national borders. In dealing with these matters he confronted the need to find a balance between empathy for specific constituencies, such as his fellow scientists in Central Europe and his Jewish brethren, as opposed to his overarching commitment to internationalism. This was firmly rooted in his faith in the transnational character of science as well as his instinctive distaste for parochial nationalism. Particularly offensive in this regard was the enmity between Germany and France, which could only be overcome, he thought, by a passionate commitment to the principle of human solidarity.

After he became ever more engaged in political concerns, Einstein remained without partisan political affiliations. Though he called for the founding of the liberal German Democratic Party, he took pains to deny publicly that he was a member. Eager to educate the German public about the events of the war and to counter feelings of revenge against the Allies, he joined a nonpartisan private commission to evaluate German war guilt. He proved equally evenhanded in assessing blame for the turmoil of the early Weimar years. More significantly, the political novice was developing a discerning eye for domestic and foreign affairs.

On the home front he expressed his distrust for the extremes of right and left, though he saw the greater danger from the former, particularly after the military putsch of March 1920. Still, he

defended the Germans as a whole against charges of innate belli-
cosity, asserting that "on average, the moral qualities of a people
do not differ very much from country to country" (CPAE 9, Doc.
80). Most worrisome for him were the ravages of the rapidly
spreading economic malaise and its consequences for maintaining
the high standards of German science.

Turning his attention abroad, he took the Allies to task for the
punitive Versailles Treaty imposed on the defeated Germany,
damning them with the faint praise that they were the lesser of
two evils. Alternating between optimism and pessimism about the
prospects for the newly formed League of Nations, he calculated
coolly that the venture was doomed without the participation of
an internationally minded America.

Initial Moral Outrage

Einstein experienced "a mixture of pity and disgust" at the outbreak of
the First World War, an instinctive recoiling from "Europe in its madness" (CPAE 8A, Doc. 34). Two months later, insult was added to injury when ninety-three German intellectuals published a "Manifesto to
the Civilized World" (also known as the "Manifesto of the 93") declaring that "the German army and the German people are one" (Böhme,
47–49). Einstein was outraged by this appeal to a narrow nationalism,
a declaration he found even more offensive because it was proclaimed
from the ranks of a cultural elite, to which he himself had recently been
recruited. In response and in collaboration with Nicolai and two others, he drew up a countermanifesto reaching out to all Europeans.

Manifesto to the Europeans, mid-October 1914
Nicolai, 9–11; CPAE 6, Doc. 8

While technology and commerce clearly compel us to recognize
the bond between all nations, and thus a common world culture,
no war has ever so intensively disrupted cultural cooperation as

the present one. Perhaps our acute awareness of the disruption that we now sense so painfully is due to the numerous common bonds we once shared.

Even should this state of affairs not surprise us, those for whom a common world culture is the least bit precious should redouble their efforts to uphold these principles. Those, however, of whom one should expect such conviction—in particular scientists and artists—have thus far only uttered things which suggest that their desire for maintaining relations has vanished simultaneously with their disruption. They have spoken with an understandable hostility—but least of all of peace.

Such a mood cannot be excused by any national passion; it is unworthy of what the entire world has until now come to understand by the name of culture. It would be a disaster should this mood pervade the educated classes.

Not only would it be a disaster for civilization, but—and we are firmly convinced of this—a disaster for the national survival of individual states—in the final analysis, the very cause in the name of which all this barbarity has been unleashed.

Through technology the world has become *smaller*; the *states* of the large peninsula of Europe today move in the orbit of one another much as did the *cities* of each small Mediterranean peninsula in ancient times. Through a complex of interrelationships, Europe—one could almost say the world—now displays a unity based on the needs and experience of every individual.

Thus it would appear to be the duty of educated and well-meaning Europeans at the very least to attempt to prevent Europe—as a result of an imperfect organization of the whole—from suffering the same tragic fate which befell ancient Greece. Should Europe too gradually exhaust itself and collapse in fratricidal war?

The struggle raging today will likely produce no victor; it will probably leave only the vanquished behind. Therefore, it seems not only *good*, but rather bitterly *necessary, that intellectuals of all nations* marshal their influence such that—whatever the still uncertain end of the war may be—the *terms of peace shall not become the cause of future wars.* The fact that through this war European relationships have to some extent become *volatile and malleable* should rather be used to make of Europe an organic entity. The technological and intellectual prerequisites are given.

How this European order is to be brought about should not be discussed here. We wish merely to emphasize as a matter of principle that we are firmly convinced that the time has come when *Europe must act as one in order to protect her soil, her inhabitants, and her culture.*

We believe that the will to do this is latently present in many. In expressing this will collectively we hope that it gathers force.

To this end, it seems for the time being necessary that all those who hold European civilization dear, in other words, those who in Goethe's prescient words can be called *"good Europeans"* join together. After all, we must not give up the hope that their collective voice—even in the din of arms—will not trail off entirely unheard, especially, if among these "good Europeans of tomorrow," we find all those who enjoy esteem and authority among their educated peers.

First it is necessary, however, that Europeans get together, and if—as we hope—enough *Europeans in Europe* can be found, that is to say, people for whom Europe is not merely a geographical concept, but rather a worthy object of affection, then we shall try to call together a union of Europeans. Such a union shall then speak and decide.

We wish only to urge and appeal; and if you feel as we do, if you are similarly determined to *lend the most far-reaching resonance to the European will,* then we ask that you sign.

The countermanifesto was circulated among a large number of academics, but aside from its three authors, only one graduate student was prepared to sign. Wartime censorship in Germany consigned it to three-year oblivion, from which it only emerged in 1917. In that year the countermanifesto was published in neutral Switzerland in the preface to Nicolai's *Biology of War.*

Nicolai was only one of a number of colleagues with whom Einstein felt a growing kinship of the like-minded. Another was his former colleague at the University of Zurich, Heinrich Zangger, a professor of forensic medicine with excellent political ties to the Allied camp. A third was Romain Rolland, the most prominent pacifist of his generation, who, though a French national, was living in exile in Switzerland.

Letter to Heinrich Zangger, ca. 10 April 1915
Einstein Archives 39-662; CPAE 8A, Doc. 73

You have the patience of an angel not to be cross with me because of my silence. But I console myself with the fact that your memory is not adequate to determine reliably the degree of my negligence. I'm now beginning to feel comfortable in the mad turmoil of the present, in conscious detachment from all things that occupy the deranged public. Why should one not live enjoyably as a member of the madhouse staff? All madmen are respected as those for whom the building in which one lives is constructed. To a certain degree the institution can be freely selected—but the difference between them is less than what one in younger years expects. Romain Rolland, who currently lives in Geneva, recently sent me a proposal, which—to continue the metaphor—leads to an organization of the sane staff of all madhouses for the purpose of not becoming deranged as well. Moreover, he has hopes that such an organization might more or less cure the inmates. The optimist! If you have the opportunity, look after him; he is being persecuted for his internationalism. . . .

The "organization of the sane staff" to which Einstein referred was not just an idle metaphor. Established recently in the Netherlands as the Central Organization for a Durable Peace, it was dedicated to working toward a stable postwar Europe. Einstein joined its executive council a half year later.

In an attempt to build bridges to more conservative intellectuals, Einstein put aside resentment against colleagues, including the eminent physicist Max Planck, who had signed the Manifesto of the 93 in October 1914. At the same time, he appealed to the internationally respected Dutch physicist Hendrik A. Lorentz to initiate an effort to restore the bonds of cooperation among intellectuals from the belligerent states. In the letter that follows, Einstein bemoaned his lack of political contacts in Berlin, though he had, only a month earlier, become a member of the pacifist New Fatherland Association.

LETTER TO HENDRIK A. LORENTZ, 21 JULY 1915
EINSTEIN ARCHIVES 83-432; CPAE 8A, DOC. 98

Recently I spoke with Planck, and we both gloomily recalled the bitter division that has arisen between us and our highly esteemed foreign colleagues as a result of the unfortunate war. Whatever errors in deplorable political agitation may have been committed on either side, it is never too late to change. It is certain that we academics are all innocent of the war and that the present miserable circumstances ought to induce us even more to solidarity; whatever occurred before has to be regarded simply as not having happened.

What to do? If I did not live in Berlin, I would write personally to our closest colleagues in France and England with the request that they pull back from the general misfortune, in order that earlier friendly relations within our community can be restored. I would ask them to assemble, completely voluntarily and unofficially, at an appropriate location (Holland or Switzerland) now during this vacation, primarily to nurture personal contacts.

But I live in Berlin, have few contacts, and also have little skill in communicating with people. That is why I am confiding in you in the hope that you will be able to transform all this, which I can only dream of, into reality. Would you not enjoy devoting some time to this worthwhile mission? Planck encouraged me very much to do everything that was in my power; he also would do everything to restore the good relations. This is all the more important as there are signs that the official relations among the learned societies and academies could rupture; for there is a great surge of nationalistic blindness. But I note here that especially the most highly regarded are fighting against it with all their might and this will surely be the same in other countries as well . . .

Though Planck and Emil Fischer, another colleague in the Prussian Academy, had signed the Manifesto of the 93, Einstein sought to reassure Lorentz that they and many other German intellectuals had given their assent hastily and regretted it. One example of Planck making partial amends was his signing in July of a petition opposing annexationist war aims. On the other hand, he was not prepared to retract publicly his signing of the Manifesto, a rebuff of Lorentz's suggestion in spring 1915 that German intellectuals might take the sting out of that document if they were to declare publicly that they recognized the equality of other cultures. Lorentz, in turn, was not interested in playing host to the gathering suggested by Einstein in the letter above.

Letter to Hendrik A. Lorentz, 2 August 1915
Einstein Archives 16-438; CPAE 8A, Doc. 103

Your negative reply did not come as a surprise, as I already had indications of the mood of our colleagues abroad. It is strange in Berlin. Professionally, scientists and mathematicians are strictly internationally minded and guard carefully against any unfriendly measures taken against colleagues living in hostile foreign countries. Historians and philologists, on the other hand, are mostly chauvinistic hotheads. The well-known and notorious "Manifesto to the Civilized World" is deplored by all levelheaded people here. The signatures had been given irresponsibly, some without prior reading of the text. That is how it was for Planck and Fischer, for example, who have supported upholding international ties in a very resolute manner. I am going to talk to Planck about your suggestion. But I believe that these persons cannot be prompted to retract their words.

I must admit that the narrow nationalistic sentiment even of people of high standing is bitterly disappointing to me. Moreover, I must say that my respect for the politically more advanced states has diminished significantly on perceiving that they are all in the hands of oligarchies that own the press and wield the power and can do what they like. A malicious person has altered a fine proverb thus

"vox populi, vox ox."

Add to this that the perceptive and powerful have no feelings for the many; there you have the sad picture of what is revered as the "fatherland" by those who belong to it. This does not change on the other side of a frontier, but is everywhere essentially the same. And must relations between persons who have come to respect one another privately and professionally pale before this threadbare ideal? It is beyond belief and unacceptable. It seems that people constantly need a fantasy for the sake of which they can hate one another; earlier it was religious faith, now it is the state. . . .

During a visit to his estranged family in Zurich, Einstein was invited to meet with Romain Rolland in Switzerland. Assuming that he would not find the time, Einstein put his thoughts to paper, writing "Confidential" at the head of the letter.

Reiterating the regrets of many of his colleagues in signing the Manifesto of the 93, Einstein commented on the fate of the New Fatherland Association, a pacifist group he had joined some months earlier. A campaign of official intimidation against it had recently been signaled by an article in a newspaper associated with the government accusing pacifist organizations of delivering "bullets in the back." In part the authorities were emboldened by the collapse of the Russian army on the Eastern Front. Right-wing allies of the government in the Prussian Academy of Sciences were similarly heartened. In reaction to a French academician's call to resist the German dream of "making Germany the center of a world that is organized like a battleship" (Déclarations, 5–7), the Prussian Academy deliberated on retaliatory measures against French intellectuals. Planck, with the assistance of Fischer, was able to win approval in the Prussian Academy to postpone any action against foreign institutions until after the war. All but three scientists voted for Planck's compromise, whereas more than half of the humanists were against it. The speech to the Academy of Sciences that Einstein mentions and attributes to the physicist Jules Violle was presumably given by the mathematician Paul Appell, a member of the Academy.

Letter to Romain Rolland, 15 September 1915
Einstein Archives 33-006; CPAE 8A, Doc. 118

Your cordial invitation to visit you in Vevey [on the shore of Lake Geneva] has whetted my appetite for making the acquaintance of one of the rare conciliatory Europeans. Well, my various obligations do not leave enough time for this trip; so I am going to use my time in Switzerland at least to send you an uncensored letter. The New Fatherland Association is going through quite difficult times; it is being harassed by the authorities and being condemned (on the whole) by the press. It looks as if the military successes in Russia have given the pro-military party and the pan-Germans increased influence with the government. On the other hand, those among my acquaintances who are most discerning in economic matters are not particularly optimistic; this seems to be due to a shortage in certain raw materials. Strangely enough, beside a quaint egotism, one finds in Germany a love for France and its population, whereas great animosity against England is quite universal. Among the uncritical masses there is general confidence in victory and an equally prevalent greed for annexation. It is strange how the man on the street can feel compensated for his heavy sacrifices by the seizure of territory, from which he certainly benefits little personally. I hope it does not come to that! A decisive victory for Germany would be a misfortune for the whole of Europe, but especially for itself.

One of the most disheartening phenomena of this terrible time is that in many cases intellectuals have completely lost their perspective. I regret to say that the unfortunate and ridiculous war of words in Berlin has already begun. You have certainly been amazed that so many men, who in times of peace were justifiably regarded as sensible, signed the notorious "Manifesto to the Civilized World." Condemnation of this step is now quite universal in Berlin as well. Incidentally, the best among them had in fact given their approval *over the telephone*, without having read the appeal! Recently there was a great furor at the Prussian Academy, because someone responded to Violle's Academy speech by submitting a petition to sever all ties with French academies. Then a curious thing happened: almost all the historians, philologists, etc., supported the petition, while most scientists and mathematicians worked avidly to maintain international ties. Thank God, the latter won, if only by a small majority; Planck (a physicist) and Fischer (a chemist) deserve particular credit for their great resolve and firmness. . . .

One day after writing Rolland, Einstein did meet with him. As Rolland's diary entry for that date indicates, most of the themes raised in the letter were discussed (Rolland, 510–515).

Only five weeks later Einstein was asked to contribute an essay to a patriotic commemorative volume entitled "The Land of Goethe 1914/1916." The heirs of the greatest German poet were called upon to defend German culture. Instead, Einstein seized the opportunity to display his visceral feelings on the phenomenon of war in general. When the editors of the volume deemed portions of the essay too unpatriotic and requested that Einstein delete the fourth and fifth paragraphs, he agreed. Yet he could not resist pointing out that he was only reiterating Tolstoy's comparison of patriotism to a mental disorder. The article was published the following year without the offending paragraphs (Einstein 1916, 30). Presented here in its entirety, it makes public the thoughts Einstein had penned at the end of the previous year to his friend Zangger: "What drives people to kill and maim each other so savagely? I think, in the end, it is the sexual aggressiveness of the male that leads to such wild explosions from time to time. . . . The special calamity of our times, however, is that bestial instincts together with the available technologies are leading to genuine destruction" (CPAE 10: Vol. 8, Doc. 41a).

The timing of the essay's composition is more than of passing interest: Einstein wrote it just as he was struggling with the final versions of his gravitational field equations submitted on four consecutive Thursdays to the Prussian Academy beginning 4 November. The final form of the paper, submitted on 25 November, provides the capstone of what is known as the general theory of relativity.

MY OPINION ON THE WAR, LATE OCTOBER, EARLY NOVEMBER 1915
EINSTEIN 1916, 30; CPAE 6, DOC. 20

The psychological roots of war are—in my opinion—biologically grounded in the aggressive nature of the male. We "lords of creation" are not the only ones who sport this crown: some animals—the bull and the rooster—surpass us in this regard. This aggressive tendency comes to the fore whenever individual males are placed side by side and even more so when relatively close-knit societies have to deal with each other. Almost without fail they will end up in disputes that escalate into quarrels and murder unless special precautions are taken to prevent such occurrences. I will never forget what honest hatred my classmates felt for years for the first-graders of a school in a neighboring street. Countless brawls ensued, resulting in many a gash in the heads of the boys. Who could doubt that vendetta and duelling spring from such feelings? I even believe that the *honor* that is so carefully culti-vated by us derives its major nourishment from such sources.

Understandably, the more modern organized states have had vigorously to push these expressions of primitive virile traits into the background. But wherever two nation states are neighbors and do not belong to a supranational organization, those traits from time to time generate tensions that lead to the catastrophes of war. By the way, I consider so-called aims and causes of war rather meaningless, because they are always to be found when passion requires them.

Subtle intellects of all times have agreed that war is one of the worst enemies of human development, and that everything must be done to prevent it. Notwithstanding the unspeakably sad con-ditions at present, I am convinced that it is possible, in the near future, to form a supranational organization in Europe that prevents European wars, just as now war between Bavaria and Württemberg is impossible in the German Reich. No friend of in-tellectual advancement should fail to vouch for this most impor-tant political goal of our time.

One can ponder the question: Why doesn't an individual in peacetime—when the state suppresses almost every expression of manly rowdiness—lose the capabilities and motivations that en-able him to commit mass murder in wartime? The reasons seem to me as follows. When I peer into the mind of a decent average citi-zen, I see a dimly lit comfortable space. In one corner stands the pride of the man of the house, a lovingly cared for shrine, to the presence of which every visitor is loudly alerted and upon which is written in huge letters the word "patriotism." It is usually a taboo to open this cabinet. In fact, the man of the house scarcely knows, or not at all, that this shrine contains the moral requisites of bestial hatred and mass murder, which when war is declared he dutifully takes out and uses. You will not find this shrine in my parlor, dear reader. I would be happy if you would consider placing a piano or a bookshelf in that same corner of *your* parlor. Either would be a more fitting piece of furniture than the one you tolerate only because you have grown accustomed to it since childhood.

I have no intention of making a secret of my internationalist sentiments. How close I feel to a human being or a human organization depends only on how I judge their intentions and capabilities. The state, to which I as a citizen belong, has no place at all in my emotional life; I consider affiliation with a state to be a business matter, somewhat akin to one's relationship to life insurance. From what I have said above, there can be no doubt that I must seek the citizenship of a country that will in all likelihood not force me to take part in a war.

How can a powerless individual creature contribute to reaching this goal? Should everyone perhaps devote a considerable portion of his abilities to politics? I really believe that the intellectually more mature people in Europe have sinned in their neglect of general political questions; yet I do not regard the pursuit of politics as the path to an individual's greatest effectiveness in this area. I rather believe that everyone should act privately in such a way that those feelings, which I have discussed in detail earlier, can no longer represent a curse to society.

Every individual who is conscious of acting to the best of his knowledge and ability should feel a sense of honor, without consideration of words and deeds. The words and actions of others or of other groups cannot *offend* one's personal honor or that of one's group. The thirst for power and greed should, as in the past, be treated as despicable vices; the same applies to hatred and contentiousness. I do not suffer from an overvaluation of the past, but in my opinion we have not made progress on this important point; on the contrary, we have declined. Every well-meaning individual should work hard at improving himself and his personal surroundings in this regard. Then the grave afflictions which plague us in such terrible fashion today will vanish.

But then again, why so many words when I can say all in one sentence and in a sentence very appropriate for a Jew: Honor your Master Jesus Christ not only in words and hymns, but above all by your deeds.

After the German authorities disbanded the New Fatherland Association in February 1916, Einstein lacked a public venue for expressing antiwar sentiments. Privately, as he confessed to Zangger some months later, "I shut my eyes as best I can to the insane goings-on in the world at large" (CPAE 10: Vol. 8, Doc. 232a). Though he was increasingly wary of publishing his views, he did not hesitate to make his feelings known within the German physics community. In an obituary for Ernst Mach written the following month, he praised the humanitarian impulse in Mach's works, which "made him immune to another disease of our time from which today few are spared—national fanaticism" (CPAE 6, Doc. 29). In the last two years of the war he restricted his political musings almost exclusively to private correspondence, but the care and seriousness with which he grappled with such concerns are evident in the very concreteness of the proposal that he set out below. His deliberations also make clear that his earlier contacts with confidants Heinrich Zangger, Romain Rolland, and H. A. Lorentz had borne fruit. In the following letter, Einstein suggested a strategic plan for peace that anticipated many of the features of Wilson's League of Nations.

LETTER TO HEINRICH ZANGGER, 21 AUGUST 1917
EINSTEIN ARCHIVES 89-523; CPAE 10: VOL. 8, DOC. 372A

... I have given much thought in recent days to the political situation and have arrived at a more optimistic view. I have something in mind that I would like to submit to you here for critical consideration.

While the war continues, that is, at this very moment, a pacifist organization consisting of as many states of the Entente as possible and maybe drawn also from neutral states should be founded on the following principles:

1) a court of arbitration to resolve disputes between these treaty states;

2) a common institution to decide to what extent these states shall and may apply the principle of universal military conscription. Collective deployment of troops outside the home countries. Reduction of the standing army in keeping with the possibilities afforded by the external relations of the treaty states;

3) in tariff policy, a most-favored-nation principle among the treaty states, linked with the tendency eventually to abolish tariff walls between them;

4) any state can become a member of the union if it fulfils the following conditions:

a) a parliament elected according to democratic principles;

b) ministers who are dependent on a parliamentary majority (such ministers naturally to be in complete control of the executive authority);

5) military alliances with states not belonging to the union are impermissible and will result in the loss of membership in the union;

6) the union guarantees the territorial defense of each treaty state against external aggression.

Essentially, the advantage to this proposal is that even a union that does not begin to comprise *all* states can be very useful, in that it guarantees the territory of its members at the price of renouncing territorial ambition. The more states join, the greater the reduction in the individual military burden. Should the Entente bring about such a union, which encompasses the United States, England, France, and Russia, it can without concern make pacts with Germany, which would be compelled economically to seek admission to the union without it being said that its "national honor" had been slighted. . . .

———————————— ✹ ————————————

In the following letter to his pacifist colleague Rolland, written hard on the heels of the one above, Einstein reiterated in more general terms his proposal for an international alliance of democratic states (omitted here). His interest in current events was clearly conditioned by sensitivity to Germany's recent past. He placed much of the blame for its great-power vainglory at the feet of propagandists like the historian Heinrich von Treitschke, whose account of successes in the Franco-Prussian War had instilled a misguided sense of triumphalism in German intellectuals.

Letter to Romain Rolland, 22 August 1917
Einstein Archives 84-166; CPAE 8A, Doc. 374

I am touched by the cordial interest you have in a person whom you have met but once. I would definitely not neglect to visit you if my health were a bit more stable; but the smallest undertaking often takes its toll afterwards. The bad experiences that we have had in the meantime with the actions of others have actually not made me *more* pessimistic than I was two years ago. I even find that the surge of an imperialistic mentality, which dominates influential sectors in Germany, has subsided somewhat. Yet I still find that it would be extremely dangerous to form a pact with Germany as it is today.

Due to military victory in 1870 and successes in the fields of commerce and industry, this country has arrived at a kind of religion of power, which has found fitting and by no means exaggerated expression in Treitschke. This religion holds almost all intellectuals in its sway; it has eradicated almost completely the ideals of Goethe and Schiller's time. I know people in Germany whose private lives are guided by virtually unbounded altruism, but who were awaiting the declaration of unlimited submarine warfare with the greatest impatience. I am firmly convinced that this aberration can only be curbed by hard facts. These people must be shown that it is necessary to have consideration for non-Germans as worthy equals, that it is essential to earn the *trust* of foreign countries, in order to be able to exist, that the goals that one sets for oneself cannot be achieved through force and treachery. Even combating the goal with intellectual weapons seems hopeless to me; people like Nicolai are characterized with genuine conviction as "utopians." Only facts can dissuade the majority of the misled from their delusion that we live for the state, and that its intrinsic purpose is to accumulate at any price the greatest power possible. . . .

Confined to his apartment with an abdominal ulcer for three months beginning Christmas 1917, Einstein was unable even to attend the weekly meetings of the Prussian Academy. Soon after returning to his normal routine in early April 1918, he helped organize another appeal by German intellectuals of an internationalist stripe. As with the countermanifesto of October 1914, the idea was Friedrich Nicolai's brainchild, though it was Ilse Einstein, his stepdaughter and secretary, who broached the suggestion to Einstein directly. In a letter to the eminent mathematician David Hilbert written about the same time as the following one, Einstein asked for the names of others who might be interested in the project, saying that Hilbert was the only one among the mathematicians and physicists of Germany whom he could approach.

Letter to David Hilbert, before 27 April 1918
Einstein Archives 13-115; CPAE 8B, Doc. 521

In these sorry times of widespread infatuation with nationalism, men of science and the arts have on countless occasions published declarations, which have damaged incalculably the solidarity cultivated so hopefully before the war among those dedicated to nobler and freer ends. The clamor of hidebound priests and vassals of the odious principle of might rises to such a level, and public opinion is so misled by the purposeful gagging of all publications, that well-meaning individuals, feeling their bleak isolation, dare not raise their voices. Daily the danger grows that those who until now have clung with conviction to the ethical ideals of a happier phase of human development gradually lose heart and fall victim to the general malaise of the spirit. This grave situation places a responsibility on those who by virtue of their intellectual accomplishments have achieved great prestige among their colleagues in the entire civilized world. This responsibility, from which they may not shrink, consists of making an open declaration that may serve as support and consolation for those who in their lonely existence have not yet abandoned their faith in moral progress.

I suggest the following. In a brief essay—no more than ten printed pages—each of us publicly acknowledges that which is intended to have the effect discussed above. These essays, which may serve as a testimonial to internationalism, are to appear in the book trade as a small volume, perhaps first in the neutral countries. In order to emphasize their international character we might try to convince individuals from the countries presently at war with one another and from neutral countries to contribute essays as well. . . .

Hilbert's initial reaction was positive. On closer reflection, however, in apparent acknowledgment of the indifferent if not hostile reception accorded the countermanifesto almost four years earlier, he advised Einstein to let the matter drop. Not only would the word "international" raise hackles, publication of such an appeal would be counterproductive until such a time as the "hurricane of madness had blown over" (CPAE 8B, Doc. 530). Einstein agreed, though he said that he had never conceived of the enterprise as a political one. It merely allowed each of the signatories to declare that "I am first of all a civilized individual [*Kulturmensch*] and secondarily a German or a Frenchman" (CPAE 8B, Doc. 548).

Until the end of the war Einstein would take an even more conservative stance. In response to a request to sign an appeal shortly after the emperor's abdication on 9 November, he counseled academics to "Keep your trap shut!" ("Maul halten!"; CPAE 8B, Doc. 653), advice he applied to himself as well.

Into the Fray

The collapse of the empire also spelled the end of Einstein's self-enforced isolation. His convergence with the new world of politics occurred almost immediately. On the same day that the emperor abdicated, 9 November 1918, Einstein and two friends attempted to free a number of professors of the University of Berlin who had been placed under arrest by students. Einstein delivered a speech to their revolutionary council, which had convened in the Reichstag. He described the situation some twenty-five years later to Max Born, who had accompanied him: "Do you recall . . . when we took a tram to the Reichstag building, convinced that we could really help turn those fellows into honest democrats? How naive we were, for all our forty years! I have to laugh when I think of it. Neither of us realized how much more powerful is instinct compared to intelligence. We would do well to bear this in mind or the tragic errors of those days may be repeated" (letter to Max Born, 7 September 1944).

Less than a week later Einstein sought to defend the idea of a parliamentary democracy against calls by the left for a government of revolutionary workers', soldiers', and students' councils. The following speech, one of a number, was made before the newly reconstituted New Fatherland Association. More than a thousand people attended.

Einstein deleted the last two paragraphs in the original manuscript, as well as a clause in the middle of the text. All are included below. (A facsimile of the first page of the address can be seen in Plate 1.)

On the Need for a Legislative Assembly, 13 November 1918
Einstein Archives 28-001; CPAE 7, Doc. 14

Comrades!

As an old-time believer in democracy, one who is not a recent convert, may I be permitted a few words.

Our common goal is democracy, the rule of the people. This is only possible if the individual holds two things sacred: faith in the salutary judgment and healthy will of the people, and a willing subordination to the will of the people, as expressed in the electoral process, even when this popular will is at odds with one's own personal will or judgment.

How can we attain this goal? What has been achieved so far? What must still be done?

The old society of caste rule has been abolished. It collapsed of its own sins and by the liberating acts of the soldiers. For the time being, we must accept as the organs of the popular will their swiftly elected Soldiers' Council, acting in concert with the Workers' Council. In this critical hour we owe these public authorities our unconditional obedience and must support them with all our might, whether or not we approve of their decisions in detail.

On the other hand, all true democrats must be vigilant lest the old class tyranny of the right be replaced by a new class tyranny of the left. Do not be tempted by feelings of vengeance to the fateful view that violence must be fought with violence, that a temporary dictatorship of the proletariat is necessary in order to hammer the concept of freedom into the heads of our fellow countrymen. Force breeds only bitterness, hatred, and reactionary activity.

We must, therefore, unconditionally demand of the present dictatorial authority, whose directives we must willingly follow, that, irrespective of party interests, it immediately prepare for the election of a legislative assembly, so that all fears of a new tyranny may be dispelled as soon as possible. Only after such an assembly has been convened and has satisfactorily completed its task, only then can the German people state with satisfaction that they have achieved freedom for themselves.

Our current Social Democratic leaders deserve our unqualified support. Full of confidence in the appeal of their program, they have already decided in favor of convening a legislative assembly. Thus they have shown that they respect democratic ideals. May they succeed in leading us out of the grave difficulties in which the sins and half measures of their predecessors have mired us.

A legislative-representative system did triumph, and elections to the National Assembly in mid-January 1919 legitimized the political ascendancy of the Social Democrats and their moderate allies, the Catholic Center and the German Democratic Party. Einstein's own inclination was toward the latter grouping, a liberal, middle-class party whose founding manifesto he had signed immediately after the armistice though he never joined its ranks.

Hopes that a government elected by parliament might control the political situation were, however, dashed. When successive left-wing revolts in early 1919 threatened to turn some cities, including Berlin, into bastions of the proletarian dictatorship, the Social Democratic government called in right-wing militias to crush them. Einstein wrote to his close friend Paul Ehrenfest in late March 1919, describing the domestic scene as dominated by "reactionary activity, with all its vile deeds decked out in disgusting revolutionary disguise" (CPAE 9, Doc. 10).

After the confirmation of general relativity in November 1919 and his elevation almost overnight to international prominence, his views on a variety of issues became the subject of intense scrutiny. In mid-December 1919 he was forced in an interview to defend himself against wild exaggerations of his political leanings: "In various newspapers I am portrayed as an emphatic Communist and anarchist, obviously due to confusion with someone who has a similar name. Nothing is farther from my mind than anarchist ideas" (*Neues Wiener Journal*, 25 December 1919).

Meanwhile a new element had crept into the political equation in Germany. Accusations of treason to the old regime were now combined with an upwelling of anti-Semitism. One of the favorite targets of right-wing groups in Berlin was Georg Nicolai, Einstein's collaborator on the countermanifesto, who fled Germany in the last year of the war. Returning to Berlin, he accepted a position as professor of medicine at Berlin University, where he was vilified as a traitor and a Polish Jew by the right-wing press and like-minded students. In addition, his timing could not have been worse. As a result of the ratification of the Versailles Treaty on 10 January, Germany ceded extensive territory, a loss that extreme nationalists attributed in part to wartime betrayal by pacifists like Nicolai. Twice in the week of 12 January Nicolai's lecture was interrupted by extreme nationalists in the audience. In his defense Einstein circulated the following statement to a few Berlin colleagues:

IN SUPPORT OF GEORG NICOLAI, 26 JANUARY 1920
EINSTEIN ARCHIVES 78-124; CPAE 7, DOC. 32

In recent days a systematic hate campaign has flared up in newspapers against the pacifist writer and courageous fighter for his convictions, Professor Nicolai, who is well known here and abroad. Before that, "pan-German students," with their unruly riots at the university, already had made it impossible for him to deliver his lectures. The undersigned academic teachers consider it their duty to express how deeply they deplore these events, which, in their opinion, are a symptom of narrow-minded intolerance and can only damage the reputation of the University of Berlin.

We who know Nicolai's work and actions deny quite emphatically that he did anything to harm Germany. To the contrary, his actions only helped to raise sympathies for Germany.

But even if one has a different opinion about the effects of Nicolai's actions, one should not attempt to attack him with blatant untruths and slander.

Einstein's efforts were to no avail; even the Prussian Social Democratic minister of education, Konrad Haenisch, could not protect Nicolai against the combined opposition of students and faculty. In 1922 he left Berlin to take up a position in Argentina (Zuelzer, 280–326).

The Primacy of Reconciliation and the Rescue of Science

Throughout the war years and after, one constant in Einstein's attitude toward public life was his unwavering commitment to the international character of all intellectual activity. A centerpiece of this concern was his efforts to ease the traditional enmity between France and Germany. Rejoining the reconstituted New Fatherland Association after the war, he devoted much thought to bridging the gulf between the two neighbors. On one particular occasion, he welcomed Paul Colin, a member of the board of directors of the pacifist Clarté movement, to Berlin. Following Einstein's introduction, Colin sought to reassure his German audience that he and his friends in Paris had never ceased loving Germany and were outraged by the harsh terms of the Versailles Treaty.

WELCOMING ADDRESS TO PAUL COLIN, 16 DECEMBER 1919
EINSTEIN ARCHIVES 28-005; CPAE 7, DOC. 27

In the name of the New Fatherland Association, which even during the dreadful war held high the ideals of humanity and of the reconciliation of nations, I am moved to greet you with all my heart. You are the first Frenchman to come to us after the war in the service of the sacred aim of reconciliation.

Our duty is grave and the hour difficult. It is difficult to say whether your victory or our defeat has more disastrously ignited the nationalist passions which threaten to perpetuate the state of blood revenge between our neighboring countries. The root of the calamity lies not in the present historical moment, but in the traditions, which have been passed down through the educated classes of the European states from family to family, in spite of the lip service paid to the teachings of Christian morality: rape and oppression bring honor and fame; to endure injustice brings shame and dishonor. We wish to confront these old, bad traditions, which threaten to destroy our continent completely with a passionate commitment to that feeling of human solidarity, without which no individuals or states can bear to live together.

. . .

The growing plight of German and Central European intellectuals, victims both personally and professionally of a continued economic blockade by the Allies and an increasingly destructive inflation, was a special area of concern for Einstein. His own Kaiser Wilhelm Institute of Physics had to abandon plans to support large-scale scientific research. To counter such developments, a number of German academies and professional associations banded together in October 1920 to stave off complete collapse. Einstein became a member of this association, the Emergency Society for German Science and Scholarship, directing much of his effort to attracting American donations.

Recognizing the importance of strengthening the bonds between German science and the German-American community, Einstein accepted a solicitation by the German Social and Scientific Society of New York to write a brief essay for a memorial volume. Proceeds from sales of the volume were to flow to the Schiller Foundation in Weimar to assist needy intellectuals in Germany and Austria.

On the Contribution of Intellectuals to International Reconciliation, after 29 September 1920
Einstein 1920, 10–11; CPAE 7, Doc. 47

In my opinion the most valuable contribution of intellectuals to international reconciliation and to the lasting fraternity of man lies in their scientific and artistic creations, because these elevate the human spirit above personal and selfish nationalistic aims. Concentrating energy on questions and goals that unite all intellectuals quite naturally generates a feeling of camaraderie, which must inevitably bring together true scholars and artists of all countries, though it is unavoidable that the less magnanimous and less independent among them will from time to time as a result of political and other passions fall out with one another. Intellectuals should never weary of emphasizing the international character of mankind's most treasured possessions, nor should their organizations lend themselves to public declarations or other steps that inflame political passions. Finally, I believe that international reconciliation would be advanced if young students and artists, in greater numbers than before, were to study in former enemy countries. Direct experience most effectively counteracts those disastrous ideologies which under the influence of the World War have been planted in many heads.

In spring 1921 Einstein embarked on a journey to the United States, which will receive more coverage in chapters 2 and 3. Two days before his departure, he granted an interview to Elias Tobenkin of the *New York Evening Post*, conducted in the study of Einstein's Berlin apartment. The reporter wanted to know Einstein's opinion about the plight of German science, but he found that the physicist was preoccupied with larger concerns.

Without mentioning by name the League of Nations, established in January of the previous year, Einstein touched on the general framework necessary for peace. He remained optimistic, in spite of the fact that President Wilson's barnstorming campaign to win support for joining the League had been thwarted by a final vote in the U. S. Senate in March 1920.

The interview appeared under the title "How Einstein, Thinking in Terms of the Universe, Lives from Day to Day." The version presented here follows the text in the *New York Evening Post*, which differs somewhat from that in Einstein 1933.

On Internationalism
New York Evening Post, 26 March 1921; Einstein 1933, 9–10

Of course, science is suffering from the terrible effects of the war, but it is humanity that should be given first consideration. Humanity is suffering in Germany, everywhere in Eastern Europe, as it has not suffered in centuries. Humanity is suffering from too much and too narrow a conception of nationalism. The present wave of nationalism, which at the slightest provocation or without provocation passes over into chauvinism, is a sickness.

The internationalism that existed before the war, before 1914, the internationalism of culture, the cosmopolitanism of commerce and industry, the broad tolerance of ideas—this internationalism was essentially right. There will be no peace on earth, the wounds inflicted by the war will not heal, until this internationalism is restored.

[Does this imply that you oppose the formation of small nations?]

Not in the least. Internationalism as I conceive it implies a rational relationship between countries, a sane union and understanding between nations, mutual cooperation, mutual advancement without interference with a country's customs or inner life.

[And how would you proceed to bring back this internationalism that existed prior to 1914?]

Here is where science, scientists, and especially the scientists of America, can be of great service to humanity. Scientists, and the scientists of America in the first place, must be pioneers in this work of restoring internationalism.

America is already in advance of all other nations in the matter of internationalism. It has what might be called an international "psyche." The extent of America's leaning to internationalism was shown by the initial success of Wilson's ideas of internationalism, the popular acclaim they met with the American people. That Wilson failed to carry out his ideas is beside the point. The enthusiasm with which the preaching of these ideas by Wilson was received shows the state of mind of the American public. It shows it to be internationally inclined.

American scientists should be among the first to attempt to develop these ideas of internationalism and to help carry them forward. For the world, and that means America also, needs a return to international friendship. The work of peace cannot go forward in your own country, in any country, so long as your Government or any Government is uneasy about its international relations. Suspicion and bitterness are not a good soil for progress. They should vanish. The intellectuals should be among the first to cast them off.

... I am a convinced pacifist. ... I believe that the world has had enough of war. Some sort of an international agreement must be reached among nations preventing the recurrence of another war, as another war will ruin our civilization completely. Continental civilization, European civilization, has been badly damaged and set back by this war, but the loss is not irreparable. Another war may prove fatal to Europe.

The price of continued wrangling was too horrible to contemplate. In December 1921, he reiterated his hope that "when this development [war as a matter of international significance] has entered the consciousness of human beings, after sufficient calamitous experiences, then men will also find the energy and good will to create organizations that have the power to prevent wars" (CPAE 7, Doc. 69).

The same month, requested by an Austrian daily to write an article that might further the "scientific community of interest" between Germany and Austria, Einstein directed the following comments to a general public but made no bones about focusing his appeal on wealthy members of his audience. The urgency in his words was due to the ever-worsening situation of the economy in general and specifically of the degenerating state of scientific research. By the end of 1921 the mark had lost half its value since the beginning of 1920.

THE PLIGHT OF GERMAN SCIENCE: A DANGER FOR THE NATION, 21 DECEMBER 1921
NEUE FREIE PRESSE [VIENNA], 25 DECEMBER 1921; CPAE 7, DOC. 70

*The great creator of the theory of relativity
appeals to the Austrian public to rescue science.*

The German-speaking countries are threatened by a danger that must be pointed out emphatically by those who recognize it. The economic distress that accompanies the heavy blows of political fate does not fall equally hard on everybody. Especially hard hit are institutions and persons whose material existence depends directly on the state, among them the scientific institutions and researchers upon whose work rests not only the economic prosperity but also, to a large part, the cultural standing of Germany and Austria.

To recognize the full gravity of the situation, one has to be aware that in time of distress people care only for the needs of the moment.

One pays for the labor that creates material value *immediately*. Yet science cannot focus on immediate practical results if it is not to wither away. *The insights and methods developed by science serve practical purposes only indirectly and often only for future generations;* but if we neglect science we will later lack the scientific workers who, by virtue of their broad vision and judgment, are able to create new niches in the economy or adapt to new challenges.

If scientific research crumbles, the intellectual life of the nation shuts down and, with it, numerous possibilities for future advancement. This must be prevented. As the powers of the state decline due to the turn of events in the world, it becomes an obligation of financially well-situated citizens to lend a hand in preventing scientific life from fading away.

Men of clear vision have judged the prevalent conditions correctly and created institutions that support all research in Germany and Austria. Please help in making these efforts a splendid success. During my teaching activities I have seen with admiration that financial cares have not been able to suffocate the desire and love for scientific research. To the contrary! *It seems the heavy shocks have even heightened the love of ideals.* Everywhere people work with enthusiasm under difficult circumstances. Take care that the ambitions and talent of our youth today do not waste away—it would be a heavy loss for all.

By late 1921 few Viennese readers would have been unfamiliar with the name Einstein. For those who might have forgotten, the reference in the *Neue Freie Presse* to the "great creator of the theory of relativity" surely would have sufficed to remind them. By this time no other scientific figure enjoyed comparable recognition and fame, a circumstance that was due in part to the proliferation of mass-circulation newspapers after the war.

Only two years earlier, few outside the world of science had any inkling of who Einstein was, much less what he thought about the affairs of the world. Thus his emergence in the political arena was an integral part of the stardom he attained during the immediate postwar era when the media first became fascinated with him and his revolutionary new theory. In the following chapter, we step back in time to document briefly the earlier controversies surrounding the theory of relativity as background to the explosive events that followed Einstein's emergence as one of the most famous cultural figures of his day.

CHAPTER 2

Science Meets Politics: The Relativity Revolution, 1918–1923

Einstein's rise to fame as a scientist took place during a period of tumultuous political events throughout Europe, but particularly within Germany. With the abdication of the Kaiser, the parallels between impending scientific and political revolution began to color public discourse about the theory of relativity. Soon after the signing of the armistice that ended the fighting in November 1918, Einstein published his "Dialogue about Objections to the Theory of Relativity." Using this literary genre in a manner reminiscent of Galileo's famous *Dialogues Concerning the Two Chief World Systems,* he publicly responded to two of his harshest critics, Ernst Gehrcke and Philipp Lenard.

One year later, on 6 November 1919, British scientists enthusiastically embraced Einstein's revolution at a joint meeting of the Royal Society of London and the Royal Astronomical Society. After considerable discussion—for the data obtained from the two expeditions sent to Sobral and Principe to prove the general theory of relativity were by no means consistent—the experts decided it was high time to dethrone Newton and crown Einstein, or so at least was the impression conveyed in the press reports that followed. In a brief account of his theory of relativity for the *Times* of London, Einstein stressed its foundational character and its compatibility with Newton's theory of gravitation. He also noted with amusement that his national identity—was he a

German man of science or a Swiss Jew?—seemed to be subject to a political law of relativity. Two weeks later, his face was emblazoned on the title page of Berlin's most popular illustrated periodical, the *Berliner Illustrirte Zeitung* (see Plate 2).

For anti-relativists like Gehrcke and Lenard, the matter was plain enough: they were convinced that Einstein's theory was both unsound and un-German. Moreover, they saw his sudden rise to fame as a fabrication of the popular press, which hailed him not only as *the* scientific genius of his day but also as a spokesman for humanity. Gehrcke's activism brought the anti-relativity movement to the attention of Paul Weyland, an obscure right-wing demagogue who tried to suggest that relativity theory was not even legitimate science but rather akin to Dadaism, described by its proponents as an anti-art movement of those who had lost confidence in European culture. As if to emphasize the connection, Einstein was portrayed in a prominent Dada collage of the period (see Plate 3).

Backed by Gehrcke and a handful of other conservative scientists, Weyland launched a campaign against Einstein and his theory at a meeting held in the Berlin Philharmonic Hall. Einstein personally witnessed Weyland's tirade, which naturally aroused his curiosity about the man at the podium (Kleinert 1993). In his "Reply to the Anti-Relativity Co." he wondered aloud about Weyland: Was he "a doctor? engineer? politician?" only to add, "I could not find out." Nor did it matter, as the publicity generated by Einstein's response in the *Berliner Tageblatt* led to a whirlwind of controversy, particularly after newspapers reported that he was contemplating leaving Germany. Letters of support from colleagues like Arnold Sommerfeld helped convince him that he should stay.

These events culminated in a direct confrontation between Einstein and Lenard at a highly publicized scientific conference in Bad Nauheim. There they rehashed the thought experiments set forth in Einstein's "Dialogue," as Lenard sought to brand relativity theory as "counterintuitive" (*unanschaulich*). Later, he would attribute this lack of physical common sense to Einstein's ethnic background, making him into the prototype of what the Nazis would dub "Jewish physics" (Beyerchen, 85–91).

Einstein's travels to the United States in 1921 and to France one year later were both closely followed by the press, and occasionally his wit and sarcasm got him into trouble. An amusing example is illustrated by remarks he made to a Dutch reporter about his impressions of American life, views he tried to "relativize" in an effort to repair the damage. No similar faux pas occurred during his trip to Paris, which was hailed by the liberal media in both Germany and France. According to the Ullstein press, which again put Einstein on the cover of its *Berliner Illustrirte Zeitung* (*BIZ*), he undertook this trip mainly for scientific purposes, scoring a stunning triumph during his stay. "Einstein," so read the caption below a photo showing him calmly smoking a pipe, "participated as the guest of honor in disputations over his theory of relativity and received accolades acknowledging that his theory is the most significant accomplishment in modern science" (*BIZ*, 23 April 1922). Yet the political import of this journey was evident to all, as underscored by Charles Nordmann's article in the French news magazine *L'Illustration*.

The deep political significance of Einstein's trip to France in the spring of 1922 was also borne out by a conversation he had beforehand with foreign minister Walther Rathenau, who advised him to accept the French invitation despite tense relations between the two countries. Einstein's warm personal relations with the physicist Paul Langevin helped ensure the success of this venture into foreign affairs, which came just weeks before Rathenau's diplomatic triumph with the signing of the Rapallo Treaty. Two months later, Germany's leading diplomat was assassinated by right-wing radicals, the most shocking of the many acts of political violence that shook the early Weimar Republic. Einstein's eulogy to Rathenau put the blame for his death squarely on the shoulders of Germany's conservative elite. Warned of threats to his own life, he took the opportunity to visit Japan, the crowning stop on a tour that did nothing to diminish his fame.

As an unofficial emissary for the much-maligned German Republic, Einstein knew that his political activities were bound to arouse passions. Yet what he found particularly puzzling were the deep ideological currents that made his theory of relativity a political target throughout much of his life. In discussions with his friend Philipp Frank, he encouraged the latter to explore this theme in hopes of shedding light on the social-psychological ramifications of the relativity revolution. Einstein's preface to Frank's *Einstein: His Life and Times* (written around 1942, but first published in the 1979 reprint of the German edition) reveals his ongoing fascination with the political ramifications of the relativity revolution while inviting comparison with Galileo's role in promoting the Copernican revolution. Indeed, Einstein's impassioned foreword to the new English edition of Galileo's *Dialogue*, written at the height of the McCarthy era, suggests a strong identification with the embattled Italian physicist.

Confronting Critics

Ernst Gehrcke had been crusading against relativity theory since 1911. The following year he referred to it as a case of "mass suggestion" in physics. After Einstein elaborated on his final version of the general theory in May 1916, Gehrcke accused him of plagiarizing the work of one Paul Gerber, who had earlier obtained precisely the same formula Einstein had found for deriving the perihelion of Mercury. Philipp Lenard sided with Gehrcke in this controversy, though he did not charge Einstein with plagiarism. Rather than responding to these criticisms directly, Einstein chose to answer both men by way of a playful Galilean dialogue between a relativist and a polite critic of the theory. The flavor of his response, if not its full substance, is conveyed in the following excerpt.

DIALOGUE ABOUT OBJECTIONS TO THE
THEORY OF RELATIVITY
DIE NATURWISSENSCHAFTEN 6 (29 NOVEMBER 1918), NO. 48, 697–702;
CPAE 7, DOC. 13

KRITIKUS: People like me have quite often expressed doubts of the most varied kind about the theory of relativity in journals; but rarely has one of you relativists responded. We do not want to examine the reasons for this neglect, whether it was out of arrogance, a feeling of weakness, or laziness—perhaps it was an especially effective mix of all these psychological forces—or maybe the criticism also revealed, often clearly, that the critic really had too little knowledge of the subject matter. As I have said, let's not discuss this; but one thing I want to state right away: I have come to you today personally in order to make it impossible for you to dodge the issue effortlessly once again. Because, I assure you, I will not budge from the spot until you have answered all my questions.

In order not to give you too much of a shock, and perhaps even to give you a certain pleasure in the task (which you cannot in any case escape), I will say something consoling. I am not, like some of my colleagues, so steeped in the dignity of my guild as to come on like a superior being with otherworldly insight and self-assurance (like a science writer or, worse, a theatre critic). I will rather talk like a mortal, the more so as I know that the source of criticism can quite often be found in a paucity of one's own ideas. I also do not intend—as one of my colleagues recently did—to pester you as would a prosecutor, and charge you with theft of intellectual property or other dishonorable acts. My attack was only motivated by a desire to clarify several points on which opinions still diverge too much. But I must also ask you to allow the publication of our dialogue, not least because the shortage of paper is not the only deficiency which robs my friend, the editor Berolinensis [Arnold Berliner], of sleep.

Since I can see your willingness, I come right to the matter at hand. Since the special theory of relativity has been formulated, its result of the delaying influence of motion upon the rate of clocks has elicited protest and, as it seems to me, with good reason. This result seems necessarily to lead to a contradiction with the very foundations of the theory. To make things perfectly clear between us, let this result of the theory be phrased first and precisely enough. [Kritikus proceeds to expound the famous clock paradox.] . . . Do you agree with this conclusion?

RELATIVIST: I agree, absolutely. It saddened me to see that some authors, who otherwise stand on the ground of the theory of relativity, seek to avoid this inescapable result.

KRITIKUS: But now comes the snag. According to the principle of relativity, the entire process must occur in exactly the same way . . . [when the reference systems for the two clocks are reversed and the one formerly in motion is now regarded as at rest]. . . . Even the most devout adherents of the theory cannot claim that of two clocks, resting side by side, each is late relative to the other.

RELATIVIST: Your last assertion is, of course, incontestable. But the entire line of reasoning is not legitimate because the special theory of relativity . . . claims only the equivalence of all Galilean (non-accelerated) systems, i.e., coordinate systems relative to which sufficiently isolated material points move uniformly in straight lines. . . .

KRITIKUS: I admit that you have defused my objection, but I must tell you that your argument nudges rather than convinces me. Besides, my objection is immediately resurrected from the dead if one accepts the general theory of relativity. According to that theory, coordinate systems of *arbitrary* states of motion are equivalent, and I can describe the previous process [in this setting as well]. . . .

RELATIVIST: [Proceeds to explain that even in the general theory of relativity the time measurements in two arbitrary coordinate systems are by no means equivalent.] . . .

KRITIKUS: I see you have extricated yourself very skillfully indeed, but I would be lying if I declared myself completely satisfied. The bone of contention has not been removed, it has only shifted.

Your consideration only shows me the connection between the difficulty just discussed and another difficulty, which has repeatedly cropped up. You have solved the paradox by taking into account the influence of the gravitational field on clocks. . . . But isn't this gravitational field only a fiction? Its existence is, I should say, only simulated by the choice of coordinates. After all, real gravitational fields are always generated by masses and cannot be made to vanish by a suitable choice of coordinates. How could one believe that a merely fictitious field could influence the rate of clocks?

RELATIVIST: First I have to point out that the distinction between real and non-real is not very productive. . . . [RELATIVIST argues further that the distinction between "real" and "non-real" quantities has no evident meaning, whereas one can distinguish between quantities that are inherent in a physical system and those that depend upon the coordinate system chosen.]

. . . The components of the gravitational field in a space-time point, for example, have no equivalent quantity that is independent of the choice of coordinates; the gravitational field *at a certain location* represents nothing "physically real," but the gravitational field together with other data does. Therefore, one can neither say the gravitational field at a location is "real," nor that it is "only fictitious."

[Regarding KRITIKUS's claim that gravitational fields can only be generated by masses] . . . there are two points to consider. First, it is not an a priori necessary requirement that the Newtonian concept of every gravitational field being generated by masses should be retained in the general theory of relativity. . . . Second, one cannot say there are no masses to which the generation of the field could be attributed. It is true that accelerated coordinate systems cannot be employed as the real causes of the field—though a humorous critic once thought he could attribute that idea to me. But all stars in the universe can be thought of as taking part in the generation of the gravitational field. . . . From these considerations it is clear that a complete clarification of the questions raised by you can only be obtained if one forms a picture of the geometric-mechanical constitution of the universe as a whole that is compatible with the theory. This I attempted last year and I found—as it seems to me—a completely satisfying model; but going into it would lead us too far afield [Einstein refers to his cosmological model of a finite but unbounded universe].

KRITIKUS: After your last explanations it seems indeed that the clock paradox is not suited to deduce internal contradictions in the theory of relativity. In fact, it now seems not improbable to me that the theory may have no internal contradictions; but this is not sufficient to take it into serious consideration. *I just cannot see why one should be willing to take on such horrible complications and mathematical difficulties merely for the sake of an intellectual preference, namely for the idea of relativity.* In your last answer you yourself showed clearly enough that they are not minor. Will it, for example, occur to someone to make use of the possibility offered by the theory of relativity to relate the motion of the celestial bodies of our solar system to a geocentric system of coordinates, a system which itself takes part in the rotational motion of the earth? Could such a coordinate system really be considered "at rest" and equivalent to others, when relative to it the fixed stars race around the earth at enormous velocities? Would this idea not run counter to all common sense, and to the postulate of the economy of thought? I cannot refrain from repeating some harsh words which [Philipp] Lenard recently uttered on this subject. After he discussed special relativity, where he characterized the "moving" coordinate system by a railroad train in motion, he said: Now, let the imagined railroad train make a clearly non-uniform motion. When then, due to inertial effects, everything in the train goes to pieces while outside nothing is damaged, then—I believe—*no sound reason* can escape the conclusion that it was the train and not the surroundings which changed its motion with a jolt. Now, the generalized principle of relativity demands, in its simple elementary sense, to admit even in this case that it might possibly have been the surroundings which suffered a change in velocity, so that the whole disaster in the train would only be a consequence of the jolt of the outside world transmitted as a "gravitational effect" of the outside world upon the interior of the train. To the obvious question why the church steeple next to the train did not collapse when *it* was jolted together with the surroundings—why the consequences of such a jolt appear *so unilaterally* only in the train, while *nevertheless* no such unequivocal conclusion about the reference frame for the change in motion is possible—to this question the principle apparently has no answer that could satisfy simple reason.

RELATIVIST: For several reasons we must willingly accept the complications to which the theory leads us. For one, it is a great satisfaction for a person who thinks consistently to understand that the concept of absolute motion—which kinematically makes no sense anyway—does not need to be introduced into physics. It cannot be denied that by the avoidance of this concept the foundation of physics benefits in its logical structure. Furthermore, the fact that the inertia and gravity of bodies are equal stands in need of clarification. Aside from this, physics needs a method of getting to a local field theory for gravitation. Theoreticians could not attack this problem without an effective limiting principle because one could postulate *a large number of theories* which all agree with the rather limited experience in this field. *Embarras de richesse* is one of the most malicious foes making the life of a theoretician difficult. The postulate of relativity limited these possibilities such that the road which the theory *had* to go was marked. Finally, the secular motion of the perihelion of the planet Mercury had to be explained. Its existence had been established with certainty by the astronomers, but the Newtonian theory could not find a satisfactory explanation for it. The postulate of the *principal* equivalence of coordinate systems does not claim that every coordinate system is equally *convenient* for the investigation of a specific physical system; this is already the case in classical mechanics. Strictly speaking, one should, for example, not say the earth moves in an ellipse around the sun, because this statement assumes a coordinate system in which the sun is at rest, whereas classical mechanics also allows for systems relative to which the sun *moves* uniformly in a straight line. But nobody would seriously want to use the latter coordinate system for an investigation of the motion of the earth. Similarly, nobody would conclude from this example that coordinate systems in which the center of gravity of the system under consideration remains at rest at the origin are in principle privileged over other coordinate systems. The same applies to the example you mentioned. Nobody would investigate our solar system in a coordinate system in which the earth is at rest—because this would be impractical. But *in principle*, such a coordinate system would still be equivalent to any other one in the general theory of relativity. The phenomenon that in such coordinate systems fixed stars would race around at tremendous velocities is no argument against its *admissibility*, but merely against the usefulness of this choice of coordinates. The same is true of the complicated structure of the gravitational field relative to such a coordinate system, which, for example, would also have components corresponding to the centrifugal forces. A similar situation prevails in Herr Lenard's example. In the theory of relativity one cannot interpret the case in the sense that "*possibly* it was the environment (of the train) which suffered the change in velocity." We do not have two mutually exclusive hypotheses about the location of motion but rather two, in principle equivalent, ways to describe the same factual phenomenon.

Only utilitarian reasons can decide which representation has to be chosen; not arguments about principles. The following counterexample will show how inadvisable it is to appeal to so-called common sense as an arbiter in such things. Lenard himself says: so far no pertinent objections have been found to the validity of the *special* principle of relativity (i.e., the principle of relativity between uniformly translational motions of coordinate systems). The uniformly moving train could as well be seen "at rest" and the tracks, including the landscape, as "uniformly moving." Will the "common sense" of the locomotive engineer allow this? He will object that he does not go on to heat and grease *the landscape* but rather the locomotive, and that consequently it must be the latter whose motion shows the effect of his labor.

KRITIKUS: I have to admit after this discussion that it is not as simple to disprove your point of view as I previously thought. Still, I have several objections in mind, but I don't want to bother you with them until I have thoroughly thought about the discourse today. But before we part, I have one more question—not an objection—just pure curiosity: What is the present situation with that sick man of theoretical physics, the ether, whom some of you have declared to be permanently dead?

RELATIVIST: He has had a fluctuating fate, but it would certainly be wrong to say that he is now dead. [Einstein then explains how the new conception of ether associated with his field theory of gravitation differs from earlier models.]

Einstein's "dialogue" subsequently prompted more brief skirmishes, but these exchanges were soon forgotten after the sensational confirmation of general relativity in November 1919. Predictably, neither Gehrcke nor Lenard felt that Einstein had met their objections. At Bad Nauheim in September 1920, Lenard resuscitated his thought experiment with train and church steeple in order to argue that the principle of general relativity was counterintuitive.

In the meantime, Einstein had become suddenly famous, but not through the efforts of German scientists. It was their British rivals who performed the rites of "beatification and canonization" (Pais, 305), while English and American newspapers trumpeted the news in banner headlines: "Revolution in Science / New Theory of the Universe / Newtonian Ideas Overthrown" (*The Times* [London], 7 November 1919) and "The Lights of the Heavens Askew" (*New York Times*, 10 November 1919).

Asked to contribute an article on his theory to *The Times*, Einstein took the opportunity not only to praise his English colleagues but to bow before the genius of Isaac Newton.

TIME, SPACE, AND GRAVITATION
THE TIMES (LONDON), 28 NOVEMBER 1919; CPAE 7, DOC. 26

I respond with pleasure to your Correspondent's request that I should write something for *The Times* on the Theory of Relativity.

After the lamentable breach in the former international relations existing among men of science, it is with joy and gratefulness that I accept this opportunity of communication with English astronomers and physicists. It was in accordance with the high and proud tradition of English science that English scientific men should have given their time and labour, and that English institutions should have provided the material means, to test a theory that had been completed and published in the country of their enemies in the midst of war. Although investigation of the influence of the solar gravitational field on rays of light is a purely objective matter, I am none the less very glad to express my personal thanks to my English colleagues in this branch of science; for without their aid I should not have obtained proof of the most vital deduction from my theory.

... The new theory of gravitation diverges widely from that of Newton with respect to its basic principle. But in practical application the two agree so closely that it has been difficult to find cases in which the actual differences could be subjected to observation. ...

The great attraction of the theory is its logical consistency. If any deduction from it should prove untenable, it must be given up. A modification of it seems impossible without destruction of the whole.

No one must think that Newton's great creation can be overthrown in any real sense by this or by any other theory. His clear and wide ideas will for ever retain their significance as the foundation on which our modern conceptions of physics have been built.

A final comment. The description of me and my circumstances in *The Times* shows an amusing feat of imagination on the part of the writer. By an application of the theory of relativity to the taste of readers, to-day in Germany I am called a German man of science, and in England I am represented as a Swiss Jew. If I come to be regarded as a bête noire, the description will be reversed, and I shall become a Swiss Jew for the Germans and a German man of science for the English!

Einstein's little joke with regard to the ambiguities of his nationality was received with amusement among his friends in Holland, but in Germany some of Einstein's long-standing critics found nothing to laugh about. By the following summer, the anti-relativists, led by Ernst Gehrcke, were preparing to mount a counteroffensive. After stirring up attention for this cause in the right-wing press, Paul Weyland launched what he advertised as a series of anti-relativity lectures in the main auditorium of the Berlin Philharmonic Hall. The first of these took place on 24 August 1920 when Weyland and Gehrcke stepped to the podium before a large crowd comprised mainly of curiosity seekers, including Einstein, who responded three days later.

MY RESPONSE: ON THE ANTI-RELATIVITY COMPANY
BERLINER TAGEBLATT, 27 AUGUST 1920; CPAE 7, DOC. 45

A motley group has joined together to form a society under the pretentious name "Syndicate of German Scientists" with the single purpose of denigrating in the eyes of non-scientists the theory of relativity and me as its author. Recently Messrs. Weyland and

Gehrcke held a lecture with this intent at Philharmonic Hall, which I attended personally. I am fully aware of the fact that both speakers are unworthy of a reply from my pen; for I have good reason to believe that there are other motives behind this undertaking than the search for truth (were I a German nationalist, whether bearing a swastika or not, rather than a Jew of liberal international bent . . .). I only respond because I have received repeated requests from well-meaning quarters to have my views made known.

First of all I observe that to my knowledge there is hardly a scientist among those who have made substantial contributions to theoretical physics who would not concede that the entire theory of relativity is logically and consistently structured and that it agrees with the experimental data now available. Prominent theoretical physicists—I name H. A. Lorentz, M. Planck, Sommerfeld, Laue, Born, Larmor, Eddington, Debye, Langevin, Levi-Civita—support the theory, and most have made valuable contributions to it. Among physicists of international renown I can only name Lenard as an outspoken critic of relativity theory. I admire Lenard as a master of experimental physics; however, he has yet to accomplish anything of importance in theoretical physics, and his objections to the general theory of relativity are so superficial that I had not deemed it necessary until now to reply to them in detail. I intend to make up for this.

I have been accused of conducting a tasteless advertising campaign for the theory of relativity. Yet I can say that all my life I have been a friend of well-considered, tempered words and concise presentation. High-flown phrases and slogans give me goose bumps, whether they deal with the theory of relativity or anything else. I have often been amused by effusions for which I am now being called to account. Besides, I gladly grant this pleasure to the gentlemen of the Company.

Now to the talks. Herr Weyland, who does not seem to be an expert at all (is he a doctor? engineer? politician? I could not find out), has presented nothing of pertinence. He indulged himself in coarse abuse and base accusations. The second speaker, Herr Gehrcke, in part has presented direct falsehoods and in part attempted to create a false impression among uninformed laymen through biased data selection and distortion of the facts. The following examples show this:

Herr Gehrcke alleges that the theory of relativity would lead to solipsism, an allegation that any specialist would take for a joke. He supports this with the familiar example of the two clocks (or twins), where *one* takes a round trip in relation to the inertial system while the other does not. He asserts—despite having been repeatedly corrected both orally and in writing by the most qualified experts on the theory—that in this case the theory leads to the truly nonsensical result that each of two clocks, side by side at rest, will be slowed in relation to the other. I can only interpret this as a deliberate attempt to mislead the lay public.

Furthermore, Herr Gehrcke alludes to Herr Lenard's objections, which many relate to examples from mechanics in everyday life. These are already invalidated by my general proof that the statements of the general theory of relativity, in first approximation, conform with those of classical mechanics.

What Herr Gehrcke has said about the experimental confirmation of the theory is to me the most striking proof that he was not concerned with revealing the true facts.

Herr Gehrcke wants to give credence to the idea that the motion of the perihelion of Mercury could also be explained without the theory of relativity. There are two options. Either one can postulate suitable interplanetary masses that are large enough and distributed in such a way so as to cause the amount of motion observed at Mercury's perihelion; this is naturally a highly unsatisfactory alternative to the one offered by the theory of relativity, which explains Mercury's perihelion motion without making any special assumptions. The other alternative is to cite a paper by Gerber, who found the correct formula for Mercury's perihelion motion before me. Yet experts in the field not only agree that Gerber's derivation is totally incorrect, but also that the formula cannot even be obtained as a consequence of Gerber's initial assumptions. Herr Gerber's paper is therefore utterly worthless, a failed and irreparable attempt at a theory. I claim that the general theory of relativity has provided the first realistic explanation for the motion of the perihelion of Mercury. I did not mention Gerber's article originally, because I did not know about it when I wrote my paper on the motion of Mercury's perihelion; but even if I had been acquainted with it, there would have been no reason to mention it. The personal attacks which Messrs. Gehrcke and Lenard have directed against me in this matter have been generally regarded as unfair by real experts in the field; I have until now considered it beneath my dignity to waste even one word on it.

In his lecture Herr Gehrcke has placed the reliability of the masterfully conducted English measurements on the sun's deflection of passing light rays in a distorted light when he mentioned only *one* of the *three* independent groups of exposures, namely the one where faulty results were produced as a result of the warping of the heliostat mirror. He concealed the fact that in their official report the English astronomers themselves called their results a brilliant confirmation of the general theory of relativity.

With regard to the problem of the redshift of spectral lines, Herr Gehrcke has concealed the fact that current measurements still contradict one another, and that a final decision on this matter has not yet been reached. He only quoted evidence *against* the existence of the line shifts predicted by relativity theory, but omitted mentioning that the conclusiveness of these earlier results has been shaken by more recent investigations by Grebe and Bachem and also by Perot.

Finally, I would like to note that on my initiative arrangements are being made for a discussion to be held on relativity theory at the scientific conference in Nauheim. Anyone daring to face a professional forum can present his objections there.

Seeing how the theory and its creator are slandered in such a manner in Germany will make a strange impression in foreign countries, especially with my Dutch and British colleagues H. A. Lorentz and Eddington, gentlemen who have worked intensively in the field of relativity and repeatedly given lectures on the subject.

Einstein's heated reply to his critics lacked the playful quality found in his earlier "Dialogue," a difference his friends noted immediately. Replying to his good friend Paul Ehrenfest in early September 1920, Einstein explained that he simply had to defend himself against the steady stream of charges publicly leveled against him, including dishonest self-promotion, literary theft, and outright plagiarism: "I had to do this if I wanted to stay in Berlin, where every child knows me from photographs. If one is a democrat, then one must grant the public this much right as well" (CPAE 10, Doc. 139).

Shortly after Einstein's response appeared in the *Berliner Tageblatt*, he received a letter from Arnold Sommerfeld, the presiding officer of the German Physical Society. Sommerfeld was intent on damage control, but he was also anxious to show Einstein that he sympathized fully with his plight. He therefore suggested to Einstein that he consider answering his critics in the *Süddeutsche Monatshefte*, a conservative periodical. The *Berliner Tageblatt*, a liberal newspaper with a large Jewish readership, he added, "does not appear to me the right place to settle accounts with the brawling anti-Semites" (CPAE 10, Doc. 131). Einstein appreciated such expressions of solidarity, but he also thought that self-respecting Jews should not stoop to answering anti-Semitic critics, a view that he had arrived at by the spring of 1920 (see following chapter).

Letter to Arnold Sommerfeld, 6 September 1920
Einstein Archives 21-395; CPAE 10, Doc. 134

In fact I did ascribe too much importance to that venture directed against me, believing that many of our physicists were involved. For two days I really thought about "desertion," as you call it.

Upon reflection, though, I soon came to the realization that it would be wrong to leave the circle of my proven friends. Perhaps I should not have written the article. But I wanted to ensure that no one would interpret my persistent silence in the face of these systematically repeated objections and accusations as a concession. It's too bad that every comment of mine is marketed by the journalists. I really must seclude myself completely.

I cannot possibly write an article for the *Süddeutsche Monatshefte*. I'd be happy, in fact, if I could just catch up with my correspondence. Such a declaration at Nauheim might well be appropriate for setting the record straight for foreigners. This should by no means occur for my sake, however. I am already once again happy and content and read nothing that is written about me except what is objective. . . .

Max Planck and Arnold Sommerfeld were especially concerned that the earlier attacks leveled against Einstein at the Berlin Philharmonic Hall might spill over into the Bad Nauheim meeting. They therefore persuaded the society's presiding officer, Friedrich von Müller, to take preemptive action against the anti-relativists in his opening address. Müller kindly obliged. Referring directly to the special session on relativity theory, he emphatically stated that:

> . . . it will be treated in an entirely different spirit than that of the tumultuous gatherings in Berlin. Scientific questions of such difficulty and great significance as the theory of relativity cannot be brought to a vote in popular assemblies with demagogic slogans, nor can they be decided by personal attacks in the political press. They will receive here the objective appreciation that their brilliant creator deserves. (*Berliner Tageblatt*, 20 September 1920)

This counteroffensive evidently produced the desired effect, as Müller's statement was greeted with thunderous applause.

American Hoopla: A Twice-Told Tale

The following year Einstein and his wife traveled to the United States as part of a small delegation that aimed to raise funds for founding a new Jewish university in Jerusalem. In an interview aboard the S.S. *Rotterdam* before disembarking in New York on 3 April 1921, Einstein was asked about those who opposed his theory. "No man of culture, of knowledge, has any animosity toward my theories. Even the physicists opposed to the theory are animated by political motives" (*New York Tribune*, 3 April 1921). When asked what he meant, he said he referred to anti-Semitic feeling. He would not elaborate on this subject, but said the attacks in Berlin were entirely anti-Semitic.

After returning from his first visit to the United States, Einstein was interviewed in his Berlin apartment by a Dutch reporter for the *Nieuwe Rotterdamsche Courant*. The account she filed led to a minor public relations fiasco, causing Einstein to distance himself from the version he read in the *Berliner Tageblatt*. Since the latter was an accurate synopsis of the original Dutch article, Einstein was not in a position to challenge its accuracy. This probably explains why he chose to express his "shock" not to the readers of the *Berliner Tageblatt*, but in a rival paper, the *Vossische Zeitung*. Therein he tried to evade a central issue— was the "Einstein craze" in the United States mainly an effort to escape the boredom that dominated most Americans' lives?—by referring to irrational currents in all cultures that were mistakenly associated with the theory of relativity.

An Interview with Professor Einstein
Nieuwe Rotterdamsche Courant, 4 July 1921; CPAE 7, Appendix D

... [The] vast enthusiasm for me in America appears to be typically American, though, and as far as I can judge I rather understand it: *the people are so uncommonly bored*, yes honestly much more so than is the case with us. And there is so little for them there anyhow. Of course places like New York, Boston, and Chicago have their theaters and concerts, but other than that?

Places with a population of a million—ah, what poverty, what spiritual poverty. So folks are happy when they are given something to play with and which they can revere, and that they then do with exceptional intensity. Most of all it is the women, by the way, who dominate all of American life. The men are interested in nothing at all; they work, work as I haven't seen anyone work anywhere else.

For the rest, they are toy dogs for their wives, who spend the money in the most excessive fashion and who shroud themselves in a veil of extravagance. They will do anything that's in vogue and in fashion, and, as it happens, have thrown themselves among the throngs of the "Einstein-craze." Does it make an outlandish impression upon me, the crowd's excitement here and there about my beliefs and theories, about which it doesn't understand anything? I find it amusing and also interesting to watch the game. I certainly believe that it is the magic of non-comprehension that attracts them . . . one tells them about something tremendous that will influence all future life, and of a theory that is within the realm of comprehension of only a select group of the very learned, and famous names are mentioned of predecessors who also made discoveries, of which the crowd couldn't understand a thing. It overly impresses them, it assumes the colors and the charm of mystery . . . and people begin to get enthusiastic and excited. . . .

EINSTEIN'S IMPRESSIONS OF AMERICA: WHAT HE REALLY SAW
VOSSISCHE ZEITUNG, 10 JULY 1921; CPAE 7, APPENDIX E

. . . I was shocked as I read this newspaper. I must assume that this report, which naturally I had not read beforehand, came about as follows: a young Dutch woman, who introduced herself to me by mentioning some mutual acquaintances, struck up a casual conversation about my trip to America. She afterward published the substance of this conversation in the *Nieuwe Rotterdamsche Courant* of 4 July 1921 and in a thoroughly friendly, sympathetic, and humorous tone, which also marked the conversation. Of course there were a few places in the Dutch account with which I would take issue had I seen it earlier. The damage is slight, however, and one won't hold me responsible for the extemporaneous account that the Dutch journalist prepared from memory. The words I am supposed to have uttered cannot correspond exactly to what I actually said. On the other hand, the excerpt from the Berlin paper gives a completely false impression of my feelings and also of what the Dutch journalist wrote. Under these circumstances I regard it as my duty to say something about my true experiences even though I am reluctant to appear before the public *in this manner.*

That which satisfies me most when I think back on America is the feeling of gratitude for the warm and heartfelt reception given me by all of the colleagues, public authorities, and private individuals. I delivered lectures at the universities in Princeton and Chicago, and at two New York universities, Columbia and the City College of New York, and everywhere found great interest and deep knowledge among the colleagues there. I must particularly acknowledge that no one objected that I spoke *German.* Those who spoke with me did so *also in German* and with efforts of the most touching kind. All who had connections with Germany mentioned these as well as the scientific ties that bound them with Germany.

I was very enraged by the statement attributed to me regarding the American public. It is surely true that the sensational interest in relativity theory shown by the public at large is mainly based on a kind of misunderstanding. But this applies not only to the American public, it is also just as true of our German one as well. . . . There is a peculiar irony in this, as many people believe that the anti-rational tendencies of our time find support in relativity theory. Yet the latter is a strictly objective theory that is not merely accessible to a "circle of the chosen" but rather to any intellectually capable person who possesses the necessary prior knowledge and who has the diligence to study it. . . .

Crossing the Rhine

By 1921 Einstein's political views began to disturb even sympathizers within his own camp. Thus, when Arnold Sommerfeld read an interview that appeared on the front page of the Parisian periodical *Figaro*, he was convinced that it must be "a lie from start to finish" (Einstein/Sommerfeld, Doc. 40). Einstein was quoted as saying that the Germans had played a disastrous role in the war and richly deserved the defeat that it had brought. After writing to him about this, Sommerfeld was crushed to learn that the views attributed to his colleague in that interview were substantially accurate. A series of invitations to visit France soon followed, the most tempting coming from Einstein's friend, Paul Langevin, one of the leading experts on relativity in France. It took considerable soul-searching before Einstein agreed to accept the offer.

LETTER TO PAUL LANGEVIN, 27 FEBRUARY 1922
EINSTEIN ARCHIVES 15-342

When I received your dear letter of invitation, it gave me a feeling of great and pure joy, and now a week later I pick up my pen feeling hesitant and sad because I cannot now accept the invitation, as much I would personally like to, and despite the warm feelings of friendship for you that I cherish. You know that I am of the opinion that relations between scholars should not suffer because of politics and that consideration of the scientific community should take precedence over all other considerations. You also know that I am unconditionally international in my outlook and that the fact that I receive my salary from the Prussian Academy of Sciences has no effect on my views. But after conscientious deliberation I have come to the conclusion that at this moment of political tensions my visit to Paris would have more unfavorable consequences than favorable ones. My colleagues here are still excluded from all international scientific meetings and are of the opinion that their French colleagues are most to blame. I know and can well appreciate the deep causes that have led to this position. On the other hand, *you* too can imagine that individuals here, whose touchiness has risen to almost pathological levels due to the events and experiences of the last years, would regard my trip to Paris at this moment as an act of faithlessness and be so hurt that very nasty consequences might develop. In Paris too there is the threat of unforeseeable complications. I know of nothing more agreeable than the prospect of chatting with you, Perrin, and Mme. Curie in a pleasant little room as once before and painting a subjective picture of relativity theory for your students. Still, the wider public and politics have long ago seized my theory and person and have tried somehow to customize both to their ends. A considerable number of people would lie in wait for any unguarded words from me in order to serve them up to newspaper readers after suitable preparation. My latest experiences in this regard suggest to me that this danger is very great; the end effect is always hatred and animosity rather than reason and goodwill. I would certainly also be asked my opinion about French-German relations, and since I can only speak honestly my answer would register sympathy neither on this side of the Rhine nor the other.

It is true that I did not shy from visiting North America, England, and Italy last year. But my trip to America was in the first instance devoted to the university in Jerusalem, and as for the other two countries, the psychological conditions were simpler and more favorable than in our case (unfortunately!).

Dear Langevin, it hurts me that I cannot comply, as you are dear to me. I also have the need to express my heartfelt thanks to you and your colleagues at the Collège de France for the friendly accommodation and conciliatory spirit underlying your decision.

LETTER TO PAUL LANGEVIN, 6 MARCH 1922
EINSTEIN ARCHIVES 15-345

Further reflection and a coincidental conversation with Rathenau have convinced me that I should have accepted your invitation despite all the misgivings expressed in my letter. In the effort to overcome the calamity of this war step by step one must not allow oneself to be led astray by petty considerations and—*you and your colleagues did not allow yourselves to be led astray.* I declare myself ready to come, if you have not already chosen some other person. But even if this has already happened, I nevertheless feel the need to document my goodwill and my mettle with this letter. Should something come of this, I would come during the first half of April. The language will admittedly be a problem for me. Still, I prefer to develop the subject matter freely rather than reading something from a written text. The formulas help a lot and a friendly colleague will act as prompter and pull words out of my throat that remain stuck there. Certainly it would have been perhaps nicer and more fruitful if we had held a kind of mini-congress on relativity, at which I just would have answered questions; that way my deficiencies in expressing myself would have proven less disturbing than when giving a more or less exhaustive exposition of the theory. I can imagine, however, that, with regard to the mode of presentation, you are bound by the statutes of the foundation.

Einstein and Langevin subsequently arranged to meet on 28 March at the French frontier. Since the distinguished visitor wished to have as little contact as possible with the public, his accommodations were to remain a secret throughout the trip. He also informed the Frenchman that "I want to have absolutely nothing to do with journalists, though I would quite like to have a word with one or two politicians should the opportunity arise." Perhaps, he hoped, something positive could be done about the ill-will that French academics still harbored toward their German counterparts (letter to Langevin, 23 March 1922, cited in Fölsing, 592). At the same time, Einstein informed the Prussian Academy of plans for this forthcoming diplomatic venture.

His first invitation was extended by the French branch of the League of Human Rights, a sister organization of the reconstituted New Fatherland Association, to which Einstein had belonged during the war and which he rejoined thereafter (see chapter 1).

LETTER TO THE PRUSSIAN ACADEMY OF SCIENCES, 13 MARCH 1922
EINSTEIN ARCHIVES 43-017; GRUNDMANN, 195–196

It is now the third time this year that I have received an invitation of an official character from Paris, the first time from the "Ligue pour les droits des hommes," the second from the French Philosophical Society, and the third from the Collège de France. This last invitation requests that I deliver some guest lectures. The invitation was extended by the faculty of the Collège de France and was passed on to me by my friend and colleague Langevin. In his letter he expressly points out that these lectures should serve to restore relations between German and French scholars. The relevant sentences read: "The interests of science demand that relations between the German scientists and us be reestablished. If you accept the invitation you will contribute to this better than anyone else, and you would render a very great service not only to your colleagues in Germany as well as France but above all to our common ideal."

I answered this letter from my friend at first with a polite refusal, invoking as the main reason considerations of solidarity with my colleagues here.

I could not rid myself of the feeling, however, that in declining I had chosen the path of least resistance rather than following my true duty. A conversation with Minister Rathenau solidified this feeling into a firm conviction. So a few days after my first refusal,

I sent Prof. Langevin a second letter in which I reversed my initial refusal and declared that I wished to accept the invitation, if other arrangements had not already been made with regard to the guest lectures. Langevin and I thus agreed that I will travel to Paris at the end of the month in order to deliver the lectures.

In view of the fact that the Academy is interested in everything that transpires concerning international relations, I regard it as proper to make the above information known to the Academy.

Three days after his arrival in Paris, Einstein began a series of lectures before a select audience at the Collège de France. He was not presented to the Académie des Sciences, however, since thirty of its members had threatened to leave should he enter its hall. Charles Nordmann, an astronomer and popularizer of Einstein's theories who accompanied him throughout his stay, was mesmerized by the latter's modest bearing. The German guest had but one request: he hoped to be taken to some of the former battle sites so as to witness firsthand the ravages of the Great War. At the conclusion of his stay, on a Sunday morning, 9 April, Einstein was whisked away from Paris to see the devastation with his own eyes. Nordmann, who was deeply moved by what he, Langevin, and Einstein's longtime friend Maurice Solovine saw that day, recorded Einstein's reaction. The original article contains four photographs of Einstein surveying the damage to the cathedral at Rheims, the ruins of a village (see Plate 6), and fortified French and German positions.

WITH EINSTEIN IN THE DEVASTATED REGIONS, 9 APRIL 1922
CHARLES NORDMANN, "AVEC EINSTEIN DANS LES RÉGIONS DÉVASTÉES,"
L'ILLUSTRATION 80 (15 APRIL 1922), NO. 4128, 328–331; EINSTEIN 1933, 9

We ought to bring all the students of Germany to this place—all the students of the world—so they can see how ugly war is.

All the peoples have a great many false ideas of one another because their ideas are too bookish. The majority of Germans certainly had an idea of the French that was far too literary—by which I mean an idea that conformed closely to what they had read. A good many men, perhaps, have an idea of war and of the ruin it causes that is also merely literary. How very desirable it would be if they could all go and see!

It is terrible!

A year later, Langevin returned the favor by joining Einstein at a pacifist rally in Berlin (see Plate 7).

The German ambassador in Paris filed a detailed report on Einstein's trip, characterizing it as a personal triumph, but also as a notable event for German science and the German spirit (Grundmann, 212). Only isolated criticisms appeared in the press, and these were quickly countered by the liberal media. Einstein's mission abroad clearly stirred deep emotions on both sides of the Rhine, and the reactions it engendered fell along predictably political lines. His visage appeared on the covers of both the *Berliner Illustrirte Zeitung* and its French counterpart, *L'Illustration*, both of which heaped lavish praise on his genius.

In the report filed by the German ambassador in Paris to the Foreign Ministry, he explicitly cautioned against seeing Einstein's personal triumph as a harbinger of renewed relations between the French and German scientific communities. Two reasons were given for regarding Einstein's visit as entirely exceptional: first, that this event was "a sensation that the intellectual snobbery of the capital could not pass up"; and second, even before his arrival he was made "palatable" to the public through press reports that announced "he had not signed the Manifesto of the 93 but rather wanted to sign a countermanifesto [see Chapter 1], that his opposition to the German government during the war was known, and that, in any case, he was a Swiss citizen and only born in Germany" (Grundmann, 212).

Shattered Hopes: The Rathenau Assassination

Two months after his return, a congress of fifteen German pacifist organizations, including the German League of Human Rights, convened in Berlin along with a French delegation, led by the Sorbonne professor of aesthetics, Victor Basch. The goal of the peace congress was to promote goodwill and build what Basch called a "Bridge over the Chasm." At a meeting on the floor of the Reichstag, its presiding officer addressed the delegates, and Basch drew thunderous applause when he introduced Einstein as living proof that the great men of France and Germany could still work together (N & N 1960, 50). Afterward, Einstein presented his sober reflections.

FROM THE VANTAGE POINT OF THE MOON, 11 JUNE 1922
EINSTEIN 1933, 11–13

I should like to present this situation of ours as soberly as if we were able to watch, from the moon, what happens on this miserable planet of ours. We ask ourselves this: In what respect does the complex of international problems which we face today differ from any we have faced before—and by "today" we mean not merely this present day but rather the last half-century.

The answer is very simple. Through the development of new technical devices, distances between men and between their institutions have suddenly shrunk to one-tenth their former size. In consequence, mankind has organized a production system whose components are spread over the globe. It would not only be reasonable but it is actually necessary that the enlargement of the territories utilized in human production should be followed by the development of an appropriate political organization.

In these terrible years all of us have realized by our own frightful experience that this organization is really a dire necessity. The man in the moon undoubtedly would be surprised if he could see that so few forces are at work to complete this organization after the horrible experiment from which humanity is just emerging. What is the reason for this?

Well, I think men are the victims of their memory of history. This is a very singular phenomenon. The common people register the occurrences in their lives directly as these happenings come along, and they can change with comparative ease their conceptions about the great truths; but the man who has swallowed a lot of printer's ink is in a much more fatal plight. This is where the community of language plays an equally disastrous role, as the nation in its entirety is really made up of men who continuously influence one another by means of the printed word. They do this to such an extent that a uniform bloc is formed, so solid in its strength that those who share the community of language do not even notice the peculiarity and arbitrary nature of their conceptions or their partial view of the world.

When traveling in Holland, France, and the United States, I had occasion to realize with great alarm the rigidity of those language units. These distinctions of separate human groups, cultivated by language and by governments, are so terribly hard to bridge. They are the product of centuries. We should not conceal from ourselves that the fight against these deep-rooted spiritual and linguistic barriers is a dreadfully difficult one, nor that there is any prospect of political solidarity and cooperation among the nations of Europe so long as this battle is not fought out.

Therefore I believe it is of utmost importance that everybody who is aware that necessity—not to speak of idealistic considerations—demands in the present condition of the world a greater unity of material and spiritual cooperation should resolve nevermore to ask, "What can be done for my country?" but much rather, "What must my country do to make it possible for the greater entity to exist?" Without this entity his own country will not be able to exist either.

I believe that only a person who has repeatedly said this to himself with determination and energy, and who meets every situation in life with that thought, can break through the barriers erected against a spiritual and intellectual consolidation. I hold it to be of extreme consequence that wherever the possibility arises, men of different languages, of different political and cultural ideas, should get in touch with one another across their frontiers—not with the feeling that something might be squeezed out of the other for their and their country's benefit, but with a spirit of good will to bridge the gap between the spiritual groups in comparatively independent spheres.

Only thus can we hope to accomplish such a political unity—in Europe, for instance—as will give us assurance of being able to survive economically and safeguard our spiritual existence. Only then will life be worth living.

Two weeks later, German Foreign Minister Walther Rathenau was gunned down by right-wing assassins. The main promoter of Einstein's journey to France, Rathenau was a brilliant intellectual and complex personality. Shortly after his return, Einstein paid him a visit at his villa in the Berlin suburb of Grunewald along with their mutual friend, the Zionist Kurt Blumenfeld. Both visitors hoped to persuade their host that he should resign as foreign minister in view of widespread anti-Semitism and violence. (According to documentation compiled by the statistician Emil Gumbel, the assassination campaigns of the German right had claimed the lives of over three hundred victims since the murders of the far-left Spartacist leaders Karl Liebknecht and Rosa Luxemburg in January 1919.) Blumenfeld and Einstein argued with Rathenau into the wee hours of the morning, but they failed to convince him that his allegiance to Germany and his faith in the German people were misplaced (Blumenfeld, 142). Two months later, on 24 June, he paid the ultimate price.

IN MEMORIAM WALTHER RATHENAU, AFTER 24 JUNE 1922
EINSTEIN ARCHIVES 32-819; *NEUE RUNDSCHAU* 33 (AUGUST 1922), NO. 8,
815–816

My feelings for Rathenau were and are ones of joyful esteem and thanks for the hope and consolation he gave me during Europe's presently bleak situation as well as for the unforgettable hours this visionary and warm human being granted me. His command of the larger economic interconnections, his psychological understanding of the character of nations and the groups within them, and his knowledge of individual people were astonishing. Despite knowing them, he loved them all as a person who has the power to affirm this life. A delightful mixture of sobriety and genuine Berlin humor made it a unique pleasure to listen to him when he chatted with friends at the table. It takes no talent to be an idealist when one lives in cloud-cuckoo-land; but he was an idealist, even though he lived on this earth, whose smells he knew better than almost anyone.

I regretted that he became [Foreign] Minister. Given the attitude held by a great many of the educated class of Germany toward the Jews, it is my conviction that it would be most natural for Jews to keep a proud distance from public affairs. Yet I could not have imagined that hatred, blindness, and ingratitude could go so far. I would like to draw the attention of those, however, who have directed the moral education of the German people for the last fifty years, to the following: by their fruits shall ye know them.

A similarly impassioned statement came from Chancellor Wirth of the Catholic Center Party when he addressed the Reichstag the day after Rathenau's assassination. Pointing to the nearly empty seats of the German National Peoples' Party he exclaimed: "There is the enemy, who drips his poison into the wounds of a people. There is the enemy—and there can be no question, this enemy stands on the right"

(Wirth, 406). Outside, crowds gathered in the streets, and for the next several weeks Berlin was in a state of shock. Labor unions organized massive demonstrations, and rumors spread about other names on the hit lists of right-wing assassins. In a letter to an unidentified colleague Einstein supported the political purpose of a memorial service for Rathenau at the University of Berlin.

On the Memorial Service for Rathenau, 1 July 1922
Einstein Archives 32-816; N & N 1975, 73

You ask my opinion about a memorial service at the university on the occasion of Walther Rathenau's violent death. Here are my views. Naturally, to hold a service for the mourning of a state minister would ordinarily not be justified. It should only be conceived as a protest demonstration against political murder. In general, the intervention by a cultural institution into political affairs is to be condemned. But here it is a question of *affirming a broad moral position*, thereby preserving those values that are above partisan strife. In my opinion, the university should unconditionally protest against political murder (students as well as professors should speak). The university must forthrightly denounce the despicable crime of premeditated murder in the service of politics; it must likewise affirm that any society which does not insist upon respect for all life must necessarily decay.

I am convinced that a public meeting, at which such sentiments were expressed unanimously and outspokenly, could exert considerable influence on the formation of a healthier public opinion. Indeed, silence on the part of the university would *also* be construed as taking a political stance at the present time.

Anti-Relativists on the March

After Bad Nauheim, Einstein had little or no interest in participating at any further meetings of the Society of German Natural Scientists and

Physicians. In September 1922, however, this venerable organization would celebrate its centenary in Leipzig, an event Max Planck sought to adorn with a special session on the theory of relativity at which Einstein by all rights would serve as the main speaker. Partly under pressure from his wife, the world's most celebrated physicist reluctantly agreed. After Rathenau's murder, prudence became a priority, however, especially in light of earlier warnings from the authorities that he too might be a target for assassination. From Kiel, where he had gone into hiding, Einstein wrote to Planck, regretfully withdrawing from participation in the Leipzig meeting.

LETTER TO MAX PLANCK, 6 JULY 1922
EINSTEIN ARCHIVES 19-300; GRUNDMANN, 176

This letter is hard for me to write; but it must be done. I have to inform you that I cannot deliver the lecture I promised for the meeting of the Society of Natural Scientists despite my earlier firm agreement to do so. I have, in fact, been warned against remaining in Berlin by people (several independent sources) who are to be taken seriously. It seems I belong to that group of people whom the radical right plan to assassinate. Naturally, I don't have definitive proof of this, but the prevailing situation at the present time would appear to make this altogether credible. If the matter involved were one of essential significance, I would not let myself be deterred by such motives, but this is merely a formal affair at which someone else (for example Laue) can easily take my place. The whole difficulty stems from the fact that newspapers have printed my name too often and in doing so mobilized the riff-raff against me. Now there is nothing more to be done than to be patient and travel. I only ask this of you: accept this little incident with humor, as do I.

When news of the threat to Einstein's life reached the public, the right-wing press scoffed at the accusation: "Einstein should not have taken such nonsense seriously; then the 'tribute' intended for him on a grand scale [the plenary lecture] would not have eluded him. For it really is not credible that crazy people who toy with murderous intentions actually exist" (*Die Wahrheit*, 23 September 1922, cited in Wazeck, 80). An even more standard response—echoing Gehrcke's long-standing claims that relativity theory was simply the product of a charlatan—came from the *Rheinisch-Westfälische Zeitung*: "the gist of the matter in this affair is that the flight that he [Einstein] staged can be interpreted as a stroke of publicity, calculated to impart new lustre to his by now considerably faded star" (5 August 1922, cited in ibid., 80).

Months before, Einstein had already made plans to visit Asia, having earlier received an invitation from the Japanese publisher *Kaizo*, a progressive firm that had brought Bertrand Russell to Japan in 1921. Before his departure, the leftist journalist Maximilian Harden was shot and nearly killed outside his villa in Grunewald. Harden, a longtime associate of Rathenau, later testified in court that "Professor Einstein went to Japan because he did not consider himself safe in Germany" (Grundmann, 177). This news reached the German Embassy in Tokyo on 15 December and immediately caused a panic lest it leak out to the public, thereby compromising the diplomatic benefits the German Foreign Ministry hoped to reap from Einstein's visit. As he was then traveling in southern Japan, the embassy contacted him via telegraphic cable and asked for permission to publish a rebuttal of Harden's claim. Einstein wired back to the embassy official stating that the matter was too complicated to be answered in a telegram, and that a letter would follow.

Reply to the German Embassy in Tokyo, 20 January 1923
Grundmann, 177

I hasten to send you further information to expand on the answer I gave earlier by telegraph. The statement by Harden is certainly

awkward for me, as it makes my situation in Germany more difficult; it is also not entirely correct, although it is not altogether
false. In fact, people who have a good overview of the conditions
in Germany are of the opinion that my life is in a certain amount
of danger. To be sure, before Rathenau's murder I did not assess
the situation the same as I did afterward. My decision to accept
the invitation to visit Japan was to a large extent the result of
longing for the Far East, as well as of feeling the need to get away
for a certain time from the tense atmosphere in our homeland,
which has so often placed me in difficult situations. After Rathenau's murder, I was indeed glad to have this opportunity to be
absent from Germany for a longer period of time, as it allowed
me to avoid the temporarily heightened danger without requiring
me to do anything that might have proved awkward for my
German friends and colleagues.

In the meantime, Lenard and other anti-relativists launched an attack
that once again rocked the German physics community. At the special session on relativity in Leipzig, where Max von Laue took Einstein's place on the program, flyers were distributed outside the
lecture hall. One of these fell into the hands of a young pupil of Sommerfeld's, Werner Heisenberg, who read: "[The undersigned] deplore
most deeply deceiving public opinion by extolling the theory of relativity as the solution to the riddle of the universe, and by keeping
people in the dark about the fact that many . . . scholars . . . reject
the theory of relativity . . . as a fundamentally misguided and logically untenable fiction. The undersigned regard it as being irreconcilable with the seriousness and dignity of German science that such a
highly disputable theory is conveyed to the layman so prematurely
and with such charlatanism, particularly when the Society of

German Scientists and Physicians is used to promote such efforts" (Schönbeck 2000, 37).

Among the nineteen signatories were Lenard and Gehrcke, whom Einstein once referred to as the leaders of "the Nauheim brigade." The pamphlet they disseminated in Leipzig received considerable coverage in right-wing newspapers, but very little in mainstream or liberal publications. Speaking for the anti-relativists, one writer gave this standard "explanation": "The major publications in Germany are almost exclusively in the hands of Einstein's kinsmen and they protect him. He is their protégé and pet. Public discussion is prevented in the exclusive interest of Einsteinianism" (*Luzerner Neueste Nachrichten*, 28 October 1922, cited in Wazeck, 82).

Looking Back on Scientific Revolutions

Long before Hitler came to power, Einstein had been convinced that the reception of relativity theory in Germany was deeply influenced by political events. On the other hand, his travels abroad brought home to him that the "Einstein craze" had been part of an international postwar phenomenon connected with mankind's hopes for the future. From a purely scientific standpoint, he did not regard the theory of relativity as a revolutionary contribution, a point he stressed in his 1919 article for *The Times* (London) in which he characterized it as a theory of principle. Thus he found many of the parallels between "his" scientific revolution and the earlier one linked with the names of Copernicus and Galileo to be far-fetched.

Still, viewed as a phenomenon of social psychology, the impact of Einstein's theory on twentieth-century thought and culture is undeniable. Not only did Einstein emerge as one of the most famous and controversial figures of the postwar world, but relativity became a veritable symbol of modernity. Looking back on the twentieth century, the distinguished historian Paul Johnson went so far as to assert that the "modern world began on 29 May 1919 when photographs of a solar eclipse . . . confirmed the truth of a new theory of the universe" (Johnson, 1). Enemies saw it as a scourge and linked it with Zionism, Bolshevism, and other "isms" and deleterious influences of the times. Supporters took it to be a cornerstone of modern scientific thought that carried deep-seated philosophical implications (see Hentschel). A leading figure in this latter group was Philipp Frank, who had succeeded Einstein as professor of physics in Prague and later promoted relativity as a philosopher of science at Harvard University. After joining Einstein in the United States, Frank began work on a biography of Einstein that sought to account for the massive public interest in the man and his theory of relativity. Did Frank succeed in solving this historical puzzle? Einstein's answer was cautious.

FOREWORD TO PHILIPP FRANK, *EINSTEIN: HIS LIFE AND TIMES*, CA. 1942
EINSTEIN ARCHIVES 28-581

I must confess that biographies have seldom attracted me or captured my imagination. Autobiographies mostly arise out of narcissism or negative feelings toward others. Biographies from the pen of another person tend in their psychological traits to reflect the intellectual and spiritual nature of the writer more than that of the person portrayed; the bridge of understanding between individuals is usually fragile, so that the projected image is analogous to the image produced by a warped and colored mirror.

If the biographical subject is a personality whose life consists of a series of actions with widely observable effects, then such an effort is hardly a bad thing, and the same goes for cases where personal ties to other people and human organizations play a central role. However, should the subject be a person who spent the most important part of his life struggling with problems of knowledge, then the biographer's task is surely not an enviable one nor is it likely to be crowned with success. For such persons lead an existence in which external experiences, including contacts with other people, generally have only secondary importance, so long as these external circumstances do not impose unusual personal hardships and obstacles. Professional and personal bonds often inhibit and distract rather than having a positive effect on such lives, while the decisive experience takes place within as if nourished by an invisible source. This is how it was in my case, even if my temperament and external circumstances occasionally led me to undertake detours that gave my life, seen superficially, something like a colorful exterior. Should such a person, whose life has consisted of striving to know and comprehend, be the subject of a biography at all? Should such a life be described when its story can adequately be characterized by a few ideas that—after a long and for the most part fruitless struggle—emerged and have stood the test of purifying criticism?

I am inclined to answer this question in the negative so far as my own person is concerned—hence no biography! Nevertheless, I encouraged my old comrade, Professor Frank, to write this book, which has the appearance of a biography, so to speak. For this man has an unusual understanding of the psychological relations that are fundamental not only for the development of scientific ideas but also for the value judgments and the political feelings of various generations.

For me it was always incomprehensible why the theory of relativity, whose concepts and problems are so far removed from practical life, should have found such a lively, even passionate resonance in the widest circles of the population for such a long time. Since the time of Galileo nothing quite like that has happened. Yet then the church's officially sanctioned view of man's place in the cosmos was shaken—an event of enormous significance for cultural and political history—whereas the theory of relativity is concerned with the attempt to refine fundamental physical concepts and to develop a logically complete system of hypotheses for physics. How could this have occasioned such a great and long-lasting psychological reaction?

I have never till now heard a really convincing answer to this question. I knew, though, that Dr. Frank was especially qualified to bring such connections to light by virtue of his philosophical-psychological intuition as well as his deep knowledge of contemporary physics and the history of physical thought. This, as well as my curiosity about what would result, led me to encourage Herr Frank to write this book.

Has he succeeded in grasping the truth behind these connections? When it comes to these social-psychological things, the question asked by Pontius Pilate—what is truth?—has an especially harsh ring. But what I believe I can promise the reader is this: he will find intelligent, interesting, and plausible explanations in this book, at least some of which will be new and surprising.

Einstein surely had the figure of Galileo Galilei in the back of his mind when in 1918 he decided to answer his harshest critics by way of a "Dialogue about Objections to the Theory of Relativity." Galileo's *Dialogue Concerning the Two Chief World Systems* was not only the Italian scientist's most famous work, it also precipitated the conflict with the Catholic Church that strongly colored the subsequent course of the scientific revolution of the seventeenth century. Einstein's understanding of Galileo's career was presumably close to the one set forth in Bertolt Brecht's *Life of Galileo*. According to this fairly standard view, the pivotal issue at stake in Galileo's conflict with the Church concerned man's central place in God's creation rather than theological matters that hinged on the interpretation of Holy Scripture (Remmert 2005). Einstein, like many of his contemporaries, was not immune to hero worship, and Galileo stood high on his list of scientific greats. His foreword to the 1953 edition of the *Dialogue* also clearly echoes the tense political times in which it was written (see chapter 10 for relevant political texts).

FOREWORD TO GALILEO GALILEI, *DIALOGUE CONCERNING
THE TWO CHIEF WORLD SYSTEMS,
PTOLEMAIC AND COPERNICAN*, CA. 1953
EINSTEIN ARCHIVES 1-174; GALILEI 1953, VII–XIX

Galileo's *Dialogue Concerning the Two Chief World Systems* is a
mine of information for anyone interested in the cultural history
of the Western world and its influence upon economic and political development.

A man is here revealed who possesses the passionate will, the
intelligence, and the courage to stand up as the representative of
rational thinking against the host of those who, relying on the
ignorance of the people and the indolence of teachers in priest's
and scholar's garb, maintain and defend their positions of authority. His unusual literary gift enables him to address the educated
men of his age in such clear and impressive language as to overcome the anthropocentric and mythical thinking of his contemporaries and to lead them back to an objective and causal attitude
toward the cosmos, an attitude which had become lost to humanity with the decline of Greek culture.

In speaking this way I notice that I, too, am falling in with the
general weakness of those who, intoxicated with devotion, exag-
gerate the stature of their heroes. It may well be that during the
seventeenth century the paralysis of mind brought about by the
rigid and authoritarian tradition of the Dark Ages had already so
far abated that the fetters of an obsolete intellectual tradition
could not have held much longer—with or without Galileo.

Yet these doubts concern only a particular case of the general
problem concerning the extent to which the course of history can
be decisively influenced by single individuals whose qualities impress us as accidental and unique. As is understandable, our age
takes a more sceptical view of the role of the individual than did
the eighteenth and the first half of the nineteenth century. For the
extensive specialization of the professions and of knowledge lets
the individual appear "replaceable," as it were, like a part of a
mass-produced machine.

Fortunately, our appreciation of the *Dialogue* as a historical document does not depend upon our attitude toward such precarious questions. To begin with, the *Dialogue* gives an extremely lively and persuasive exposition of the then prevailing views on the structure of the cosmos in the large. . . . The conception of the world still prevailing at Galileo's time may be described as follows:

There is space, and within it there is a preferred point, the center of the universe. Matter—at least its denser portion—tends to approach this point as closely as possible. Consequently, matter has assumed approximately spherical shape (earth). Owing to this formation of the earth the center of the terrestrial sphere practically coincides with that of the universe. Sun, moon, and stars are prevented from falling toward the center of the universe by being fastened onto rigid (transparent) spherical shells whose centers are identical with that of the universe (or space). . . . The outer shells with their heavenly bodies represent the "celestial sphere" whose objects are envisaged as eternal, indestructible, and inalterable, in contrast to the "lower, terrestrial sphere" which is enclosed by the lunar shell and contains everything that is transitory, perishable, and "corruptible."

Naturally, this naïve picture cannot be blamed on the Greek astronomers who, in representing the motions of the celestial bodies, used abstract geometrical constructions which grew more and more complicated with the increasing precision of astronomical observations. . . .

Thus, briefly, had the ideas of later Greece been crudely adapted to the barbarian, primitive mentality of the Europeans of that time. Though not causal, those Hellenistic ideas had nevertheless been objective and free from animistic views—a merit which, however, can be only conditionally conceded to Aristotelian cosmology.

In advocating and fighting for the Copernican theory Galileo was not only motivated by a striving to simplify the representation of the celestial motions. His aim was to substitute for a petrified and barren system of ideas the unbiased and strenuous quest for a deeper and more consistent comprehension of the physical and astronomical facts.

The form of dialogue used in his work may be partly due to Plato's shining example; it enabled Galileo to apply his extraordinary literary talent to the sharp and vivid confrontation of opinions. To be sure, he wanted to avoid an open commitment in these controversial questions that would have delivered him to destruction by the Inquisition. Galileo had, in fact, been expressly forbidden to advocate the Copernican theory. Apart from its revolutionary factual content the *Dialogue* represents a downright roguish attempt to comply with this order in appearance and yet in fact to disregard it. Unfortunately, it turned out that the Holy Inquisition was unable to appreciate adequately such subtle humor. . . .

The *leitmotif* which I recognize in Galileo's work is the passionate fight against any kind of dogma based on authority. Only experience and careful reflection are accepted by him as criteria of truth. Nowadays it is hard for us to grasp how sinister and revolutionary such an attitude appeared at Galileo's time, when merely to doubt the truth of opinions which had no basis but authority was considered a capital crime and punished accordingly. Actually we are by no means so far removed from such a situation even today as many of us would like to flatter ourselves; but in theory, at least, the principle of unbiased thought has won out, and most people are willing to pay lip service to this principle. . . .

Taken together, these retrospective reflections on scientific revolutions clearly indicate that while Einstein rejected the notion that his theory of relativity overturned classical physics, he nevertheless deeply identified with Galileo as a champion of truth who opposed blind authority. And while the theme of persecution clearly reflected the atmosphere of the McCarthy era, it also echoed the dramatic events of 1920 that nearly caused Einstein to leave Berlin. Back then he, like everyone else, was baffled by the strange convergence of scientific and political interests that gave the relativity revolution its special allure. Yet when he confronted the German anti-relativists' organized campaign to vilify him and his theory, the "new Copernicus" presumably drew strength from the image of a defiant Galileo, a figure far closer to his heart than Copernicus, Kepler, or even Newton.

CHAPTER 3

Anti-Semitism and Zionism, 1919–1930

After the influx of significant numbers of East European Jews to Germany during and after the war, antagonism in the public toward these immigrants and fear of Bolshevik infiltration from the East found strong resonance in the right-wing press. This backlash prompted Einstein's first open political act: a newspaper article in late 1919 in defense of these most vulnerable members of the Jewish community.

Initially, he accepted the idea of Jewish assimilation. Soon he became disenchanted, however, with all too frequent attempts by German Jews to deflect popular anti-Semitism onto their poorer East European kinfolk. By April 1920 he had come to believe that only Zionism with its emphasis on Jewish solidarity and its positive message of strengthening Jewish self-esteem might overcome the venom of anti-Semitism. Assimilation was an adaptive response he increasingly likened to the "mimicry of butterflies" (*Vossische Zeitung*, 27 March 1921). His solution to the Jewish question of identity took on the form of a yearning for an earlier age of innocence, a romanticism that was rooted in a longing for community bound together by ties of blood and tradition (Stachel, 64–66).

At the heart of his Zionist conviction lay his hopes for a cultural center in Palestine with a Hebrew University, administered by Jews for Jews as its centerpiece. According to his sister, the roots of this commitment lay in a sense of obligation, already in-stilled in him as a boy, to overcome the obstacles to higher education that were strewn in the path of his deserving brethren (CPAE 1, lx). As a first step to the realization of this dream in Palestine, he began in Berlin in the spring of 1920 to offer accredited univer-sity courses to Jewish students unable to register at the University of Berlin because of the quota system (*Jüdische Pressezentrale Zürich*, 23 April 1920). He continued to attend Jewish educa-tional functions for years after (see Plate 5).

One obstacle that Einstein had had to overcome in his embrace of Zionism was resolving the contradiction between its nationalist emphasis and his own deep-seated affinity for internationalism. His explanation at the time was given in the form of a parable: "I am *against* nationalism but *in favor* of Zionism. . . . When a man has both arms and he is always saying I have a right arm, then he is a chauvinist. When the right arm is missing, however, then he must do all in his power to make up for the missing limb. Therefore I am, as a human being, an opponent of nationalism, but as a Jew I support . . . the Jewish-national efforts of the Zionists" (Blumenfeld, 127–128).

An opportunity to demonstrate his newfound commitment was not long in coming. In early 1921 he accepted an invitation to join leaders of the international Zionist Organization, headquartered in London, on a campaign to raise money in America for its immi-gration and colonization fund and for the Hebrew University (see Plate 4). Visits by prominent individuals from former enemy nations so soon after the war were frowned upon by both sides. When his friend and colleague Fritz Haber scolded him for his dis-loyalty, Einstein explained where his loyalties lay. He reiterated his conviction five years later: "He who remains true to his origin, race and tradition will also remain loyal to the State of which he is a subject. He who is faithless to the one will also be faithless to the other" (Einstein 1931a, 51–53).

The Balfour Declaration of 1917 had recognized the Zionist aim of a national home for the Jewish people. In setting up a

British Mandatory Authority there three years later, the League of Nations used the same expression: "national home." Much in the development of Zionism over the next thirty years would hinge on the interpretation of this phrase. For those like Einstein who emphasized the cultural aspects of Zionism, a national home in Palestine implied colonization in the pursuit of a cultural center that might be shared politically with other groups, all under the aegis of the British Mandate. Back in Germany from his American trip, Einstein summed up the position: "The goal which draws on the leaders of Zionism is not a political one, but social and cultural. The community in Palestine should approach the social ideal of our forefathers, much as it was laid down in the Bible, while at the same time becoming a place of modern intellectual life. An intellectual center for Jews of the whole world" (CPAE 7, Doc. 59).

For those, on the other hand, who stressed the political aspects of Zionism, the goal was achieving statehood for the Jewish people in the land of Palestine. A litmus test of attitudes was apparent in the relative significance assigned by each to the need for an understanding with the Arabs—for Einstein it was paramount, for the political Zionists it was, if not unattainable, far less important than maximizing Jewish immigration to Palestine.

Complementing the vision of a cultural center in Palestine was the ideal of creating farmers and laborers out of the largely urban Jewish population of Europe. Devotion to the land would be the new defining characteristic of the Jew in Palestine. To assist in this Herculean task of transformation, Einstein stressed the importance not of book learning, but of a practical education that would provide the training needed to solve pressing problems in the areas of agriculture and public health.

In the mid-twenties an almost imperceptible messianic tone crept into his reasoning. Rather than attempting as earlier to reconcile internationalism with sympathy for the national aspirations of Jewry, his new point of departure was that the moral resurgence of Palestine would redound to the good of all humankind.

The shocking attacks by Arabs on Jewish settlements in 1929 did not shake his faith in the need for a *modus vivendi* with the Arab population, but he seems to have greatly misjudged the depth of Arab alienation. His instinctive distrust of politicians also led him to underestimate the growing intransigence in the positions of Jews, Arabs, and the British Mandatory Authority.

Forging New Bonds of Solidarity

Thirty thousand East European Jews had performed wartime labor service in Germany. One-half of them had been deported from Russia and Russian Poland, the other half were recruited as contract labor. When the fighting ended, resentment of their presence spread. A steady drumbeat of attacks called for halting the continued trickle of immigrants from the East and deporting all undesirable émigrés. After the government decree of November 1919, it became increasingly clear that venom aimed at East European Jews could at a moment's notice be turned against all Jews. A month later, Einstein put pen to paper at the request of an umbrella agency representing the major Jewish interest groups in Germany. Much of his information was drawn from an article with the same title that had appeared in the liberal *Berliner Tageblatt* a week earlier.

IMMIGRATION FROM THE EAST
Berliner Tageblatt, 30 December 1919; CPAE 7, Doc. 29

Among the German public voices are increasingly heard demanding legal measures against East European Jews. It is claimed there are 70,000 Russian, i.e., East European Jews, in Berlin alone; and these East European Jews are alleged to be profiteers, black marketeers, Bolsheviks, or elements that are averse to work. All these arguments call for the most sweeping measures, i.e., herding all immigrants into concentration camps or expelling them.

Measures that would devastate so many individuals must not be triggered by slogan-like assertions, even less so as objective re-examination has shown that we have here a case of agitation by demagogues. It does not reflect the actual situation and is not a suitable means for counteracting existing wrongs. Agitation against East European Jews in particular raises suspicion that calm judgment is being clouded by strong anti-Semitic instincts and, at the same time, that a *specific* method of influencing the mood of the people is chosen which diverts from the true problems and from the real causes of the general calamity.

As far as is known, an official inquiry by the authorities that would undoubtedly reveal the invalidity of the accusations has not been conducted. It may very well be true that 70,000 Russians live in Berlin; but according to competent observers, only a small fraction of them are Jews, while the overwhelming majority are of *German descent.* According to authoritative estimates, not more than 15,000 Jews have immigrated from the East since the signing of the peace treaty. Almost without exception they were forced to flee by the horrible conditions in Poland and to seek refuge here *until they are given an opportunity to emigrate elsewhere.* Let us hope that many of them will find a true homeland as free sons of the Jewish people in the newly established Jewish Palestine.

It is quite likely that there are Bolshevik agents in Germany, but they undoubtedly hold foreign passports, have at their disposal ample funds, and cannot be arrested by any administrative measures. The big profiteers among the East European Jews have certainly, long ago, taken precautions to elude arrest by officials. The only ones affected would be *those poor and unfortunate ones,* who in recent months made their way to Germany under inhumane privations, in order to look for work here. Only these elements, certainly harmless to the German national economy, would fill the concentration camps, and there perish physically and spiritually. Then one will complain about the self-made "parasitic existences" who no longer know how to take their place in a normally functioning economy. The misguided policy of suddenly laying off thousands of East European Jewish laborers—who were coerced into coming to Germany during the war—and thus depriving them of their means of livelihood, leaving them with nothing to eat and systematically denying them job opportunities, has indeed forced people into the black market to keep themselves and their families from starving. The German economy, too, is certainly best served if the public supports the efforts of those who try to channel East European Jewish immigrants into productive work (as, e.g., the often mentioned Jewish Labor Bureau does). Any "order of expulsion"—now so vigorously demanded—would only have the effect that the worst and most harmful elements remain in the country, while those willing to work would be driven into bitter misery and despair.

The public conscience is so dulled toward appeals for humanity that it no longer even senses the horrible injustice which is here being contemplated. I refrain from going into details. But it is disturbing when even leading politicians do not consider how much their proposed treatment of East European Jews will damage Germany's *political and economic position*. Has it already been forgotten how much the deportation of Belgian laborers undermined the moral credibility of Germany? And today, Germany's situation is incomparably more critical. Despite all efforts, it is extremely difficult to re-establish damaged international relations; in all nations only a few intellectuals among the peoples of the world are initiating some first attempts; the hope for new economic connections (e.g., the material help of America) is still very weak today. The expulsion of the East European Jews—resulting in unspeakable misery—would only appear to the whole world as new evidence of "German barbarism," and provide it with a pretext, in the name of humanity, to hamper Germany's reconstruction.

Germany's recovery can really not be accomplished by the use of force against a small and defenseless portion of the population.

The Jewish Labor Bureau directed most of its activities to finding work for Jewish wartime labor conscripts and former prisoners of war, particularly in the mining industry. The program of deporting workers from occupied Belgium to Western German industrial centers to offset the desperate shortage of skilled domestic labor had already been halted in early 1917 after howls of protest were heard in the neutral countries.

During his initiation into Jewish affairs, Einstein took the conventional line of defending Jews against anti-Semitic barbs. An example was his defense of Georg Nicolai in late January 1920 ("In Support of Georg Nicolai," chapter 1). As he began to examine his reasoning, however, he found it wanting. By April 1920 he was mocking the assimilationist position of the Central Association of German Citizens of the Jewish Faith, the leading self-defense organization of assimilated German Jews. Its defense of Jewish civil rights in the courts, he thought, would avail it little. He went on to condemn the divisiveness in the Jewish community, which he argued was a result of the Association's hostile attitude toward its Jewish brethren from Eastern Europe. To match his newfound understanding of the Jews as a coherent community, he employed terms like "race" and "instinct" as he wrestled with the question "What is a Jew?" (see Stachel, 68–70).

ASSIMILATION AND ANTI-SEMITISM, 3 APRIL 1920
EINSTEIN ARCHIVES 36-625; CPAE 7, DOC. 34

When an intimidated individual or a careerist among my brethren feels inclined or forced to identify himself as a son of his forefathers, then he usually describes himself—provided he was not baptized—as a *"German citizen of the Mosaic faith."* There is something comical, even tragic-comical in this designation, and we feel it immediately. Why? It is quite obvious. What is characteristic about this man is not at all his religious belief—which usually is not that great, anyway—but rather his being of *Jewish nationality*. And this is precisely what he does *not* want to reveal in his confession. He talks about religious faith instead of kinship affiliation, of "Mosaic" instead of "Jewish" because the latter term, which is much more familiar to him, would emphasize affiliation to his kith and kin. Besides, the broad designation "German citizen" is ridiculous because practically everybody you can meet in the street here is a "German citizen." Then, if our hero is no fool—and that is rather rare indeed—there must be a certain intention behind it. Yes, of course! Frightened by frequent slander he wants to assert that he is a good and dutiful German citizen, even though all his life he has been bedeviled by "German citizens" because of his "Mosaic faith," and often not just a little.

For brevity's sake, I have used the term "*Jewish nationality*" above sensing that it could meet with resistance. Nationality is one of those slogans that cause vehement reaction in contemporary sensibilities, while reason treats the concept with less confidence. If somebody finds this word inappropriate for our case, he may choose another one, but I can easily circumscribe what it means in our case.

When a Jewish child begins school, he soon discovers that he is different from other children, and that they do not treat him as one of their own. This being different is indeed rooted in heritage; it is in no way based only upon the child's religious affiliation or on certain peculiarities of tradition. Facial features already mark the Jewish child as alien, and classmates are very sensitive to these peculiarities. The feeling of strangeness easily elicits a certain hostility, in particular if there are several Jewish children in the class who, quite naturally, join together and gradually form a small, closely knit community.

With adults it is quite similar as with children. Due to race and temperament as well as traditions (which are only to a small extent of religious origin) they form a community more or less separate from non-Jews. Aside from social difficulties, due to the changing intensity of anti-Semitism over the course of time, a Jew and a non-Jew will not understand each other as easily and completely as two Jews. It is this basic community of race and tradition that I have in mind when I speak of "Jewish nationality."

In my opinion, aversion to Jews is simply based upon the fact that Jews and non-Jews are different. It is the same feeling of aversion that is always found when two nationalities have to deal with one another. This aversion is a consequence of the *existence* of Jews, not of any particular qualities. The reasons given for this aversion are threadbare and changing. Where feelings are sufficiently vivid there is no shortage of reasons; and the feeling of aversion toward people of a foreign race with whom one has, more or less, to share daily life will emerge by necessity.

Herein lies the psychological root of all anti-Semitism, but by no means is it a justification for the agitation of the anti-Semites. A feeling of aversion may be natural, but to follow it unreservedly indicates a low level of moral development. A nobler individual will guide his actions by reason and insight and not by dull instinct.

But how is it with society and with the state? Can it tolerate national minorities without fighting them? There is no state today that does not regard tolerance and the protection of national minorities as one of its duties. Let us hope the state takes these duties seriously. This involves halting its practice of demanding that Jews in many cases abandon principle and abase themselves (baptism) in order to obtain government employment; it would be even more advisable to drop this custom as it creates a rather unfortunate selection.

The methods used by Jews to fight anti-Semitism are quite diverse. I have already characterized the assimilatory one, that is, to overcome anti-Semitism by dropping nearly everything Jewish and appealing to the civil rights of Jews. This method is not calculated to raise the reputation of the Jewish people in the estimation of the non-Jewish world; besides, it is useless and morally questionable. Another method of combating anti-Semitism, occasionally used by Jews who have not yet broken with everything Jewish, is to draw a sharp dividing line between *East European Jews and West European Jews*. Everything evil blamed on Jews as a totality is heaped on the East European

Jews and, thus, of course granted as an actually existing fact. The result of this not merely bad but also foolish procedure is, of course, just the opposite of what was intended. Anti-Semites have no intention of clearly distinguishing between East European and West European Jews as some West European Jews might wish; instead, they interpret this strange kind of defense as an admission and unfairly accuse those West European Jews of betraying their own people. It is not difficult to prove, in both general and individual cases, that most West European Jews are nothing but former East European Jews; and vice versa for all East European Jews. And since the major concern of anti-Semites is to prove that Jewish inadequacies and vices have not been acquired during a few generations, but can allegedly be shown to have existed through the entire history of the Jewish people, the inference from East European Jews to West European Jews appears logically justified. And here we do not even take into consideration that East European Jewry contains a rich potential of the greatest human talents and productive forces that can well bear comparison to the higher civilization of West European Jews—as is often admitted even by those who are by no means philo-Semitic.

It cannot be the task of the Jews to obtain "immunity" from the anti-Semites by accusing any part of their own people. This attitude reveals a severe misconception of both the law and of the significance of anti-Semitism, whose presumption of sitting in judgment on the Jewish people we will never accept. As Jews we know the faults of our people better than others do, and we alone are called upon and able to remedy this. This can only be achieved, however, if we follow our Jewish duty, which dictates that we always view the Jewish people as a living whole and that standing shoulder to shoulder with our brethren we work for a Jewish and human future for our people.

Responding to a request from the Central Association of German Citizens of the Jewish Faith to attend a meeting to discuss anti-Semitism in the universities (CPAE 9, Doc. 363), Einstein fired off the following letter, in which he clearly indicates how far he has moved from the position of the Association and its tactic of self-defense. Not defensiveness but the achievement of self-respect was the key to gaining respect from others.

The original of the letter was recently found in Russian archives and is facsimilized in Dirks & Simon, 23. Two typescript versions of the letter, dated 5 April 1920, are also preserved in Einstein Archives 43-443 and 35-060. Though intended as a private communication, a version of the letter made its way into the Swiss press in September 1920 by a path that remains mysterious.

A LETTER OF CONFESSION, 3 APRIL 1920
ISRAELITISCHES WOCHENBLATT FÜR DIE SCHWEIZ, 24 SEPTEMBER 1920; CPAE 7, DOC. 37

Today I received your invitation to a meeting on the 14th of this month, which will be devoted to combating anti-Semitism in academic circles. I would gladly attend if I believed that such an undertaking might meet with success. First, however, anti-Semitism and servility among us Jews must be combated through education. More dignity and more independence in our own ranks! Only when we have the courage to regard ourselves as a nation, only when we respect ourselves, can we win the respect of others, or put another way, the respect of others will then follow of itself. Anti-Semitism as a psychological phenomenon will always be with us so long as Jews and non-Jews are thrown together. Where is the harm in that? It may be thanks to anti-Semitism that we owe our survival as a race; that at any rate is what I believe.

When I come across the phrase "German Citizens of the Jewish Faith," I cannot suppress a pained smile. What lies behind this highfalutin' description? What is Jewish *faith* after all? Is there a kind of non-faith by virtue of which one ceases being a Jew? In describing it this way, our *beaux esprits* are confessing two things:

1) I wish to have nothing to do with my poor East European Jewish brethren;

2) I don't wish to be regarded as a son of my people, but only as a member of a religious community.

Is this being candid? Can the "Aryan" respect such dissemblers? I am neither a German citizen nor do I believe in anything that might be described as "Jewish faith." But I am a Jew and am glad to belong to the Jewish people, though I do not regard it in any way as chosen. Let us leave anti-Semitism to the "Aryans," and save our love for kith and kin. . . .

In spite of Einstein's protestations, the Central Association did take offense. After the letter's publication, an Association representative met with him and managed to extract a concession. In the March 1921 issue of the Association's house organ, readers were assured "on good authority" that Einstein had come to appreciate the work of the Association since the letter was published (*Im deutschen Reich* 27, 91–92).

Making the One-Armed Man Whole

When Einstein announced his intention of accompanying the London-based Zionist leadership to America on a fund-raising drive for the Hebrew University, his friend Fritz Haber rebuked him for consorting with former enemies. He begged him to postpone his trip, pointing out how it would not only be interpreted as demonstrating Einstein's own disloyalty to Germany, but also that of his fellow Jews (Grundmann, 184–185).

In the following letter Einstein argued for loyalty to a deeper kinship. Among the standard-bearers of the new anti-Semitism mentioned in the letter were his adversaries at a contentious scientific conference in Bad Nauheim, whom he dubbed the "Nauheim brigade" (see chapter 2 for more on the Nauheim conference and Paul Weyland's role in the anti-relativity company). In addition, he singled out two ultranationalist colleagues, Gustav Roethe and Ulrich von Wilamowitz-Moellendorff, both philologists in the Prussian Academy. In a letter to Romain Rolland, 15 September 1915 (chapter 1), Einstein had already pointed out the more conservative sentiment prevailing among the humanists in the Academy.

LETTER TO FRITZ HABER, 9 MARCH 1921
EINSTEIN ARCHIVES 12-332; GRUNDMANN, 186–187

The following happened to me regarding the American trip, plans for which can under no circumstances be changed. A few weeks ago, at a time when no one was thinking of political complications, a trusted Zionist approached me with a telegram from Professor Weizmann. In it the Zionist Organization requested that I accompany several German and English Zionists to America for consultations on educational matters in Palestine. Naturally they don't need me for my abilities but because of my name, whose luster they hope will attract quite a bit of success with the rich kinsmen of Dollar-land. In spite of my emphatic internationalism, I believe that I am always under an obligation insofar as it is in my power to advocate on behalf of my persecuted and morally oppressed kinsmen. For this reason I gladly acquiesced without even giving it five minutes' thought, though I had already written off invitations from all American universities. Thus it is far more an act of loyalty than one of disloyalty. The prospect of establishing a Jewish university is precisely that which fills me with particular joy after seeing recently on numerous occasions how treacherously and uncharitably one deals here with splendid young Jews and seeks to deny them opportunities for education. I could cite other events of the last year which must impel a self-respecting Jew to take Jewish solidarity more seriously than would have appeared appropriate and natural in earlier times. Think of Roethe, Wilamowitz-Moellendorff, and the infamous Nauheim brigade, which in the end only shook off the fool Weyland for opportunistic reasons. . . .

Einstein was hailed as a conquering hero in America. Though the following article was written during his stay there, it only appeared after his return to Germany. Somewhat shamefacedly, the editors of the paper in which it appeared pointed out a week after its publication that they had neglected to mention that the article was not written by Einstein. It was based on an interview that he granted to a representative of the Jewish Correspondence Bureau in New York. On obtaining the text, the editors submitted it to Einstein, who gave his approval with a number of emendations.

How I Became a Zionist
JÜDISCHE RUNDSCHAU 26 (21 JUNE 1921), NO. 49, 351–352;
CPAE 7, DOC. 57

Until a generation ago, Jews in Germany did not regard themselves as part of the Jewish people. They simply thought of themselves as members of a religious community, and many still do today. In fact they are far better assimilated than Russian Jews. They have attended mixed schools and have accommodated themselves to both the everyday and cultural life of the Germans. Yet in spite of the equal rights they enjoy formally, strong social anti-Semitism remains. Especially the educated class supports the anti-Semitic movement. They have even constructed a "science" of anti-Semitism, while the intellectuals of Russia, at least prior to the war, were usually philo-Semitic and made frequent and honest attempts to fight the anti-Semitic movement. This has a number of causes. To some degree, the phenomenon is based on the fact that Jews exert an influence on the intellectual life of the German people altogether out of proportion to their numbers. While in my opinion the economic position of the German Jews is vastly overestimated, Jewish influence on the press, literature, and science in Germany is very pronounced and obvious to even the casual observer. There are many individuals, however, who are not anti-Semites and are honest in their argumentation. They regard Jews as a nationality distinct from Germans and feel that increasing Jewish influence threatens their national character. Although the percentage of Jews in England, for instance, is perhaps not much less significant than in Germany, English Jews certainly do not exercise a comparable influence on English society and culture. Yet the highest civil-service positions are accessible to them there, and a Jew can become Lord Chief Justice or Viceroy of India, something almost unthinkable in Germany.

Anti-Semitism is frequently a question of political calculation. Whether or not somebody admits to his anti-Semitism is often merely a question of which political party he belongs to. A socialist, even if he is a convinced anti-Semite, will not admit to or act on his conviction because it does not fit into the program of his party. Among conservatives, however, anti-Semitism often stems from the desire to exploit instincts that already exist in the population. In a country like England, where Jewish influence is less and the sensitivity of non-Jews is therefore far less, it is the existence of old, deep-rooted liberal traditions that hinders the rapid growth of anti-Semitism. Never having been in England, I say this without knowing the country personally. Nevertheless, the attitude toward my theory adopted by English science and the press has been characteristic. In Germany, for the most part, a newspaper's political orientation dictated its judgment of my theory; the attitude of English scientists, on the other hand, demonstrated that their sense of objectivity is not clouded by a political point of view. I should add that the English have actually influenced the development of our science to a great degree and have gone about testing the theory of relativity with great energy and with remarkable success. While anti-Semitism in America assumes only a social guise, it is political anti-Semitism that is far more common in Germany. The way I see it, the racial particularity of Jews will necessarily influence their social relations with non-Jews. I believe the conclusion which Jews should draw from this is to acknowledge their particular lifestyle and cultural contributions. For the time being they should display a certain dignified restraint and not be so eager to mix socially, which non-Jews desire only a little or not at all. On the other hand, anti-Semitism in Germany also has consequences that, from a Jewish point of view, should be welcome. I believe German Jewry owes its continued existence to anti-Semitism. Religious forms, which in the past hampered Jews from mixing with and integrating into their surroundings, are now in the process of disappearing due to growing affluence and improved education. Thus, nothing which leads to separation in social life remains but this antagonism to the surroundings called anti-Semitism. Without this antagonism, the assimilation of Jews in Germany would proceed quickly and unimpeded.

I have observed this in myself. Until seven years ago I lived in Switzerland, and as long as I lived there I was not aware of my Jewishness, and there was nothing in my life that would have stirred my Jewish sensibility and stimulated it. This changed as soon as I took up residence in Berlin. There I saw the plight of many young Jews. I saw how anti-Semitic surroundings prevented them from pursuing regular studies and how they struggled for a secure existence. This is especially true of East European Jews, who are constantly subject to harassment. I do not believe they constitute a large number in Germany. Only in Berlin are there perhaps a greater number. Yet their presence has become a question that occupies the German public more and more. Meetings, conferences, newspapers press for their quick removal or internment. The housing shortage and economic depression are used as arguments to justify these harsh measures. Facts are assiduously overstated in order to influence public opinion against East European Jewish immigrants. East European Jews are made the scapegoats for the malaise in present-day German economic life, which is in reality a painful after-effect of the war. Opposing these unfortunate refugees, who have escaped the hell that is Eastern Europe today, has become an effective political weapon that is successfully used by demagogues. When the government contemplated measures against East European Jews, I stood up for them in the *Berliner Tageblatt*, where I pointed out the inhumanity and irrationality of these measures.

Together with a few colleagues, Jews and non-Jews, I held university courses for East European Jews, and I would like to add that our activity met with the official recognition and full support of the Ministry of Education.

These and similar experiences have awakened my Jewish national feelings. I am not a Jew in the sense that I call for the preservation of the Jewish or any other nationality as an end in itself. I rather see Jewish nationality as a fact, and I believe every Jew must draw the consequences from this fact. I consider raising Jewish self-esteem essential, also in the interest of a natural coexistence with non-Jews. This was my major motive for joining the Zionist movement. Zionism, to me, is not just a colonizing movement directed toward Palestine. The Jewish nation is a living fact in Palestine as well as in the Diaspora, and Jewish national feelings must flourish everywhere that Jews live. Under today's living conditions members of the same clan or peoples must have a lively awareness of their kinfolk in order not to lose their sense of self and their dignity. It was the unbroken vitality of the masses of American Jewry that first made it clear to me how sickly German Jewry is.

We live in an age of exaggerated nationalism and, as a small nation, must take this fact into account. But my Zionism does not preclude cosmopolitan views. My point of departure is the reality

of Jewish nationality, and I believe that every Jew has an obligation toward his fellow Jews. Zionism has a varied significance. It
opens the prospect for a dignified human existence to many Jews
who presently languish in Ukrainian hell or degenerate economically in Poland. By leading Jews back to Palestine and restoring
a healthy and normal economic existence, Zionism represents a
productive activity that enriches all of society. The main point,
however, is that Zionism strengthens Jewish dignity and self-
esteem, which are critical for existence in the Diaspora. Moreover,
in establishing a Jewish center in Palestine it creates a strong bond
that gives Jews a sense of self. I have always found repulsive the
undignified addiction to conformity of many of my peers.

The founding of a free Jewish community in Palestine will again
put Jewish people in a position where they can bring their creative
abilities to fruition without hindrance. The establishment of the
Hebrew University and similar institutions will not only allow the
Jewish people to bring about its own national renaissance, it will
also give it the opportunity of contributing to the spiritual life of
the world on a freer basis than ever before.

On his return, Einstein delivered a speech on 27 June to a capacity
crowd of Zionists in Berlin. He read from a slightly different draft that
he had prepared with his secretary (CPAE 7, Doc. 59), but the last
paragraph was delivered extemporaneously.

On a Jewish Palestine
Jüdische Rundschau 26 (1 July 1921), no. 52, 371; CPAE 7, Doc. 60

Rebuilding Palestine is for us Jews not merely a matter of charity
or a colonial issue but rather a problem of paramount importance
for the Jewish people. Palestine is not primarily a refuge for East

European Jews but rather the incarnation of a reawakening national feeling of community of all Jews. Is it opportune and necessary to revive and strengthen this feeling of community? I believe I must answer this question with an unqualified "yes," based not only on spontaneous emotions but on sound reason.

Let us briefly cast a glance at the development of German Jews during the last one hundred years. A century ago, our ancestors, with few exceptions, lived in the ghetto. They were poor, politically disenfranchised, separated from non-Jews by a wall of religious traditions, daily lifestyle, and legal restraints. In their intellectual development they were limited to their own literature, and only faintly influenced by the tremendous revival that European intellectual life experienced during the Renaissance. For the most part ignored, these modestly living people had one advantage over us: every one of them belonged with every fiber of his being to a community that completely absorbed him, in which he felt himself a fully-fledged member, and in which no one demanded anything of him that ran counter to his natural way of thinking. Our ancestors were physically and intellectually rather atrophied, but in social respects they enjoyed an enviable spiritual equilibrium. Then came emancipation. Suddenly an individual had undreamed-of opportunities of development. Individuals rapidly established contact with the higher economic and social strata. Eagerly they absorbed the magnificent achievements that the arts and sciences had created in the West. They threw themselves with ardor into this development, making lasting contributions of their own. In the process they appropriated the external forms of life of the non-Jewish world, and increasingly turned a blind eye to their religious and social traditions, and adopted non-Jewish habits, customs, and ways of thinking. It seemed as if they would be absorbed into the numerically larger, politically and culturally better organized host nations, so that after a few generations no visible trace would remain. Complete dissolution of the Jewish nation in Central and Western Europe appeared inevitable.

Things turned out differently. There seem to be instincts in racially distinct nationalities that counterbalance their assimilation. The accommodation of Jews in language, morals, and even religious forms to the European nations among whom they live could not extinguish the alienation that exists between Jews and their European host nations. In the last analysis this instinctive feeling of alienation is the source of anti-Semitism. Well-meaning tracts thus cannot eradicate it. Nationalities do not want to mix; they prefer to go their own way. A satisfactory situation can only be achieved through mutual tolerance and respect.

Toward this end it is especially important that we Jews again become conscious of our nationality and that we regain the self-respect that we need for a prosperous existence. We must learn to rededicate ourselves to our forefathers and to our history, and as a people we must accept those cultural duties that serve to strengthen our feeling of community. It is not enough to take part as mere individuals in the cultural development of mankind, we must also tackle tasks that only national unity can solve. This is the only way that Jewry can become socially sound.

I ask you to view the Zionist movement from this perspective. Today, history has delegated to us the task of actively participating in the economic and cultural rebuilding of the land of our fathers. Enthusiastic and highly gifted men have laid the groundwork, and many admirable kinsmen are prepared to devote themselves completely to this labor. May every one of you fully appreciate the importance of this task and contribute to its success to the best of your abilities!

The private Gray-Hill residence on Mount Scopus was purchased in 1918 and in its incarnation as University House served as the first building of the future Hebrew University. Some three years later, before departing for America, Einstein emphasized that the relevant institutes of the university ought to "investigate the natural conditions of Palestine," giving precedence to an agricultural institute and perhaps also a chemical institute (*Vossische Zeitung*, 27 March 1921). Lectures "would not be urgent at the outset in view of the fact that the Jewish population is still comparatively small, but the university will grow with the growth of the population and its teaching activities will be increased accordingly" (*Yidishes Tagblatt—The Jewish Daily News*, 3 April 1921).

On the Founding of the Hebrew University in Jerusalem
Jüdische Pressezentrale Zürich, 26 August 1921; CPAE 7, Doc. 62

There are two reasons underlying the need to establish a Hebrew university in Palestine. First, a Jewish homeland in Palestine is not even imaginable without a university. *Bacteriological* and other branches of research are necessary for the rehabilitation of the country. Industries for the chemical analysis of the soil and vegetation are necessary for the training as well as the intellectual and material support of physicians, especially of those who are to *gain a familiarity with the country*, and further for the development of all things dealing with cultivation of the soil. All this would constitute the *natural sciences* faculty. In order to give these institutions an independent character, one needs, of course, pure physics and chemistry, and with that we already have an almost self-contained *philosophical* faculty, as far as the natural sciences are concerned. Of course, no less important are institutes for the humanities, especially *history*, *Jewish culture*, and *Near East* topics in general, in particular the Hebrew language—all in order to spread knowledge of the country and to give the people settled there an *intellectual center*. A large-scale Jewish colonization without such an intellectual center is hardly imaginable; and it is obvious that these special goals—if pursued in a scientific manner—require as a base the cultivation of philosophy, archaeology, etc. What I have said briefly summarizes the needs concerning the rebuilding of the country and the modern development of a scientific terminology for the Hebrew language.

The Hebrew University in Palestine shall have a *national* character insofar as the *language* of instruction will generally be *Hebrew*. Difficulties of language during the first years will be overcome by the fact that initially it will be mainly an institute of research and not instruction, especially in the field of the natural sciences.

The second major task of the Hebrew University in Palestine must be to offer a place of study for Jewish youths from Eastern Europe, because many young and gifted Jewish people are today generally barred from university studies. I am convinced that a large number of East European Jews who are otherwise eligible for the university have, in vain, tried to enroll in central European universities, and in my opinion it is an *obligation of honor* of the *whole* to lend help. We even hope to bring the Hebrew University in Palestine to such a level that Jewish students from the Diaspora seek admission, and not just out of need.

Finally, we think it highly significant that Jewish people as a *whole*, and through their *own institutions* and their *own strength*, take part in international scientific life, since the enriching and successful participation of individual Jews in scientific life and the achievements of many of its sons in the Diaspora have demonstrated its love and aptitude for cultivating knowledge.

I believe that by creating a prosperous Hebrew University of international reputation we can also contribute to counteract the shameful inclination of many successful intellectuals among the Jews who deny and fearfully hide their group affiliation. In my opinion this phenomenon is not always due to a lack of character but is caused more by the fact that individuals succumb to the influence of their non-Jewish surroundings—especially if anti-Semitic feelings are prevalent.

All my life I have considered it a sacred duty to contribute, to the best of my ability, to make the Hebrew University in Palestine a success; not just now when its foundations are laid. I know that many Jewish scientists feel as I do about this question. . . .

Einstein welcomed another aspect of Zionist emigration. Participation on the part of East European Jews in Germany's wartime labor market had set a positive precedent for incipient labor needs in Palestine. The diversification of Jewish occupations to include industrial and agricultural labor would have a positive effect in meeting the practical demands of Jewish settlement.

The Scope of Research at the University, ca. August 1921
Einstein 1931a, 60–62

The Hebrew University in Jerusalem is to be organised in such a way as to meet the existing requirements of the country for scientific research institutes. It cannot be compared in its initial stages with a fully equipped University in the West. It must begin with a number of research institutes devoted to the scientific investigation of the natural conditions of Palestine. The first to be considered will be an Institute of Agriculture, and then probably a Chemical Institute. These Institutes must be in the closest touch with existing and future experimental stations and agricultural schools. The next most urgent need is for an Institute of Microbiology, which on the practical side of its activities will help to fight epidemics in Palestine. Then for one of the early foundations we have to consider an Institute of Oriental Studies, which will have for its province the scientific exploration of the country and its historical monuments, and the philological study of its languages, Hebrew and Arabic, and possibly of other Oriental languages as well. These Institutes will lay the foundation of scientific research work in Palestine. For the present there is less need for actual teaching by professors and lecturers. Indeed, it is positively undesirable to encourage the Jewish population of Palestine, which as yet is very small and can grow but slowly, to repeat the old mistake of one-sided devotion to the professions and intellectual pursuits. On the contrary, the thing to aim at is a normal distribution of the Jewish population among the various occupations. The notorious one-sidedness of the occupational distribution of the Jews in the Diaspora must not be reproduced in Palestine. Only with an increase of the population will there be a gradual extension of the University, and a gradual addition of teaching activities to pure research work.

Due consideration must, however, also be given to the possibility that from the outset Jewish students will be attracted to the Hebrew University from all over the world. To what extent this tendency should be encouraged in the early stages is a problem requiring special consideration. But it is at any rate permissible to hope that in the course of time the Jerusalem University will grow into a centre of Jewish intellectual life, which will be of value not for Jews alone.

A Moral and Spiritual Home

On the return from his trip to Japan in early 1923, Einstein made his only visit to Palestine. While in Jerusalem he stayed in the home of the British High Commissioner, Herbert Samuel, with whom he hit it off famously, becoming lifelong friends (see Plate 8). On 7 February in University House on Mount Scopus, Einstein lectured on the theory of relativity. The chairman of the Palestine Zionist Executive, speaking in Hebrew, concluded his introduction with a dramatic pronouncement: "Mount the platform which has been waiting for you for 2,000 years." Einstein prefaced his lecture with a few words in Hebrew, apologized for being unable to continue in that language, and finished his speech in French (Samuel, 253, cited in Clark, 394).

Briefed on "the political situation and some of the intricacies of the Arab question" (Kisch, 30), Einstein was enthusiastic about much of what he saw. His harshest comments were reserved for the religious Jews. On a visit to the holiest site in Judaism, he noted that it was "where dull-witted kinsmen were praying aloud, their faces turned to the [Western] Wall, their bodies rocking to and fro. A pathetic sight—men living in the past, not in the present" (Travel Diary, 3 February 1923, Einstein Archives 29-129). A visit to the Zionist settlement of Rishon Letzion seems to have agreed with him more (see Plate 9). At his departure on 14 February, he was asked whether things that should be done were being left undone. His response in French was short and sweet: "Collect more money" (Kisch, 31).

My Impression of Palestine
New Palestine 4 (11 May 1923), no. 18, 341

I cannot begin these notes without expressing my heartfelt gratitude to those who have shown so much friendship toward me during my stay in Palestine. I do not think I shall ever forget the sincerity and warmth of my reception—for they were to me an indication of the harmony and healthiness which reigns in the Jewish life of Palestine.

No one who has come into contact with the Jews of Palestine can fail to be inspired by their extraordinary will to work, and their determination which no obstacle can withstand. Before that strength and spirit there can be no question of the success of the colonization work.

The Jews of Palestine fall into two classes—the urban workers and the village colonizers. Among the achievements of the former, the city of Tel-Aviv made a singularly profound impression on me. The rapidity and energy which has marked the growth of this town has been so remarkable that Jews refer to it with affectionate irony as "Our Chicago."

A remarkable tribute to the real power of Palestine is the fact that the Jewish elements which have been resident in the country for decades stand distinctly higher, both in the matter of culture and in their display of energy, than those elements which have only recently arrived.

And among the Jewish "sights" of Palestine none struck me more pleasantly than did the school of arts and crafts, Belazel, and the Jewish workingmen's groups. It was amazing to see the work that had been accomplished by young workers who, when they entered the country, could have been classified as "unskilled labor." I noted that beside wood, other building material is being produced in the country. But my pleasure was tempered somewhat when I learned of the fact that the American Jews who lend money for building purposes exact a high rate of interest.

To me there was something wonderful in the spirit of self-sacrifice displayed by our workers on the land. One who has actually seen these men at work must bow before their unbreakable will and before the determination which they show in the face of their difficulties—from debts to malaria. In comparison with these two evils the Arab question becomes as nothing. And in regard to the last I must remark that I have myself seen more than once insurance of friendly relations between Jewish and Arab workers. I believe that most of the difficulty comes from the intellectuals and, at that, not from the Arab intellectuals alone.

The story of the struggle against malaria constitutes a chapter by itself. This is an evil which affects not only the rural, but also the urban population. During my visit to Spain some time ago, we submitted to the Spanish Jews a proposition that they send, at their expense, a specialist on the subject of malaria to Palestine, and that this specialist should carry on his work in connection with the work of the University of Jerusalem. The malaria evil is still so strong that one may say that it weakens our colonization work in Palestine by something like a third.

But the debt question is particularly depressing. Take for instance the workers of the colony of Degania. These splendid people groan under the weight of their debts, and must live in the direst need in order not to contract new ones. One man, even of moderate means, could, if he were large-hearted enough, relieve this group of its heartbreaking burden. The spirit which reigns among the land and building workers is admirable. They take boundless pride in their work and have a feeling of profound love for the country and for the little localities in which they work.

In the matter of architectural taste, as displayed in the buildings, in the town, and on the land, there has been not a little to regret. But in this regard the engineer, Kaufmann, has done a great deal to bring good taste and a love of beauty into the buildings of Palestine.

To the government considerable credit must be accorded for its construction of roads and paths, for its fight against malaria and, in general, for its sanitary work as a whole. Here the government has no light task before it. One can hardly find another country which, being so small, is so complicated by virtue of the divisions among its own population as well as by virtue of the interest taken in it by the outside world.

The greatest need of Palestine today is for skilled labor. No academic forces are needed now. It is hoped that the completion of the Technical College will do a great deal toward meeting the need of the country for trained workmen.

I am convinced that the work in Palestine will succeed in the sense that we shall create in that country a unified community which shall be a moral and spiritual center for the Jewries of the world. Here and not in its economic achievement lies, in my opinion, the significance of this work. Naturally we cannot neglect the question of our economic position in Palestine, but we must at no time forget that all this is but a means to an end. To me it seems of secondary importance that Palestine shall become economically independent with the greatest possible speed. I believe that it is of infinitely greater importance that Palestine shall become a powerful moral and spiritual center for the whole of the Jewish people. In this direction the rebirth of the Hebrew language must be regarded a splendid achievement. Now must follow institutions for the development of art and science. From this point of view we must regard as of primary importance the founding of the university which, thanks largely to the enthusiastic devotion of the Jewish doctors of America, can begin its work in Jerusalem. The university already possesses a journal of science which is produced with the earnest collaboration of Jewish scientists in many fields and in many countries.

Palestine will not solve the Jewish problem, but the revival of Palestine will mean the liberation and the revival of the soul of the Jewish people. I count it among my treasured experiences that I should have been able to see the country during the period of rebirth and reinspiration.

Degania, located south of the Sea of Galilee, was the first Jewish settlement in Palestine based on communal living. Richard Kaufmann was instrumental in designing many of Jerusalem's early garden suburbs. The Technical College (Technion) outside Haifa opened its doors in 1924.

Perhaps under the influence of his recent visit to Palestine, Einstein struck an almost messianic chord in the following essay, conceiving of the Jewish homeland as part of a larger ethical-political vision: Palestine as a model for a worldwide community of purpose based on international solidarity. In speaking of Zionism as having "a significant role to play for all mankind," he came as close as never before to suggesting the chosen character of the Jews (by contrast, see, for example, "A Letter of Confession" in this chapter). The article, which the editors characterized as an example of "refined Zionism," appeared a month earlier in the first issue of the journal *La Revue juive*.

Mission

Jüdische Rundschau 30 (17 February 1925), no. 14, 129

In Europe as elsewhere the presence of different nationalities and the resulting antagonistic nationalisms must in my opinion be considered a misfortune. Is it necessary to repeat the fact that a certain kind of nationalism represents a real danger for peace and an inexhaustible source of injustice and misfortune?

On the other hand, there remains a fact which no one can deny: almost everywhere Jews are treated as members of a group with clearly defined national attributes. This may appear unfortunate for Jews like me who view affiliation with a nationality that embraces all mankind as the only possible ideal, though admittedly a difficult one to reach.

Nietzsche has said that one of the peculiarities of the Jewish people consists in recognizing and implementing in practice "the subtle use of misfortune."

The Jews must also put their nationality to use. May they do it in such a way as to further the welfare of all!

They must develop from within those virtues and that faith which are indispensable for one who wishes to serve all of humankind. Since, for the moment at least, the vanishing of Jewish nationality seems out of the question, Jews must justify their existence. They must therefore, without being ridiculously arrogant about it, regain an awareness of the human values which they embody. Through the study of their past, through a better understanding of the spirit of their race, they must learn anew the mission they can accomplish.

By recalling to memory a past filled with glory and sorrow and by opening their eyes to a healthier, dignified future, Zionism can teach them self-knowledge and instill courage. It restores the moral force, which allows them to live and act in dignity. It frees the soul from the unforgivable feeling of exaggerated modesty, which can only oppress and make them unproductive. Finally it reminds them that the centuries they have lived through in common sorrow enjoins upon them the duty of solidarity.

Revitalized through the mysticism of Zionism they will perhaps finally be able to accomplish the tasks which are their duty and which the honorable exertion and common labor of Israel demands of them. Only at this price will those who believe that it is not possible for people of all nations to live without fraternal bonds be able in useful fashion to spread words of wisdom and humanity, which are more necessary today than ever.

For this reason I cannot regard the Zionist movement as an outgrowth of the tree of poisonous views that destroys the enjoyment of thinking and of life.

A Jew who strives to imbue his spirit with the humanist ideal can without contradiction declare himself a Zionist.

We must thank Zionism for the fact that it is the only movement that has imparted a justifiable pride to many Jews, that it has restored essential faith to a despairing race, and if I may be permitted to express myself in this fashion, that it has given an exhausted people new meat on their bones.

Zionism is in the process of creating in Palestine a center of Jewish spiritual life, and for that reason we will always have to be thankful to its leaders. The existence of this moral homeland will, I hope, succeed in imparting more vitality to a people that does not deserve to die. I have been able to observe the first manifestations of this moral resurgence.

For this reason I believe that I am able to assert that Zionism, which appears to be a nationalistic movement, has, when it comes down to it, a significant role to play for all mankind.

———————————— ◭ ————————————

The opening of the Hebrew University on 1 April 1925 was attended by a number of dignitaries, including Chaim Weizmann and Lord Balfour. The first institutes to be established were in microbiology, chemistry, and Jewish studies, following closely the guidelines set out five years earlier ("Statement on the Present Position of the University

Question," 21 January 1920, Einstein Archives 36-830). The foundation stone for the Einstein Institute of Mathematics was also laid on this occasion. A year later Einstein summarized his hopes: "I do believe that in time this endeavor will grow into something splendid, and, Jewish saint that I am, my heart rejoices" (letter to Paul Ehrenfest, 12 April 1926).

THE MISSION OF OUR UNIVERSITY
NEW PALESTINE 8 (27 MARCH 1925), NO. 13, 294

The opening of the Hebrew University on Mount Scopus, at Jerusalem, is an event which should not only fill us with just pride, but should also inspire us to serious reflection.

A University is a place where the universality of the human spirit manifests itself. Science and investigation recognize as their aim the truth only. It is natural, therefore, that institutions which serve the interests of science should be a factor making for the union of nations and men. Unfortunately, the universities of Europe to-day are for the most part the nurseries of chauvinism and of a blind intolerance of all things foreign to the particular nation or race, of all things bearing the stamp of a different individuality. Under this regime the Jews are the principal sufferers, not only because they are thwarted in their desire for free participation and in their striving for education, but also because most Jews find themselves particularly cramped in this spirit of narrow nationalism. On this occasion of the birth of our University, I should like to express the hope that our University will always be free from this evil, that teachers and students will always preserve the consciousness that they serve their people best when they maintain its union with humanity and with the highest human values.

Jewish nationalism is today a necessity because only through a consolidation of our national life can we eliminate those conflicts from which the Jews suffer today. May the time soon come when this nationalism will have become so thoroughly a matter of course that it will no longer be necessary for us to give it special emphasis. Our affiliation with our past and with the present-day achievements of our people inspires us with assurance and pride *vis-à-vis* the entire world. But our educational institutions in particular must regard it as one of their noblest tasks to keep our people free from nationalistic obscurantism and aggressive intolerance.

Our University is still a modest undertaking. It is quite the correct policy to begin with a number of research institutes, and the University will develop naturally and organically. I am convinced that this development will make rapid progress and that in the course of time this institution will demonstrate with the greatest clearness the achievements of which the Jewish spirit is capable.

A special task devolves upon the University in the spiritual direction and education of the laboring sections of our people in the land. In Palestine it is not our aim to create another people of city dwellers leading the same life as in the European cities and possessing the European bourgeois standards and conceptions. We aim at creating a people of workers, at creating the Jewish village in the first place, and we desire that the treasures of culture should be accessible to our laboring class, especially since, as we know, Jews, in all circumstances, place education above all things. In this connection it devolves upon the University to create something unique in order to serve the specific needs of the forms of life developed by our people in Palestine.

All of us desire to cooperate in order that the University may accomplish its mission. May the realization of the significance of this cause penetrate among the large masses of Jewry. Then our University will develop speedily into a great spiritual center which will evoke the respect of cultured mankind the world over.

Einstein's enthusiasm for the Hebrew University was complicated for a decade after 1925 by his "guerrilla battle" with its management, particularly the chancellor, Judah Leib Magnes. Einstein's main objection was to what he considered the overweening influence of financiers in directing an academic institution. After the breach became public in 1933, Einstein succeeded in getting Magnes replaced as academic director two years later and made his peace with the institution. For an account of Einstein's involvement in the matter, see Clark, 397–401 and 480–483.

The reciprocal spiritual relationship between Palestine and the Diaspora emphasized in the following article is reminiscent of the semiofficial Program of Zionism, adopted by the Zionist Association of Germany. It reads in part: "Just as the Diaspora will provide the men and means for the upbuilding of Palestine, so on the other side will Palestine be the source of energy for the Diaspora. Palestine will rekindle the spiritual life of the Diaspora" (Poppel, 97–98).

INTERNAL AND EXTERNAL EFFECTS, 1927
EINSTEIN 1931A, 57–60

The Palestine problem, as I see it, is twofold. There is first the business of settling the Jews in the country. This demands external assistance on a large scale; it cannot be successfully accomplished unless the national resources of Jewry are laid under contribution. The second task is that of stimulating private initiative, especially in the commercial and industrial spheres.

The deepest impression left on me by Zionist work in Palestine is that of the self-sacrifice of the young men and women workers. Gathered here from all sorts of different environments, they have succeeded, under the influence of a common ideal, in forming themselves into closely-knit communities and in working together on lines of systematic co-operation. I was also most favorably impressed by the spirit of initiative shown in the urban development. There is something here that almost suggests an avalanche. One feels that the work is being borne along on the wings of a strong national sentiment. Nothing else could explain the extraordinarily rapid advance, especially on the seacoast near Tel-Aviv.

At no time did I get the impression that the Arab problem might threaten the development of the Palestine project. I believe rather that, among the working classes especially, Jew and Arab on the whole get on excellently together. The difficulties which are as it were inherent in the situation do not rise above the threshold of consciousness when one is on the spot. The problem of the rehabilitation and sanitation of the country seems incomparably more difficult.

It is a common thing for Jews to miss the significance of the Palestine question: they do not see what it has to do with them. It is indeed easy to ask what it matters to a scattered nation of so many millions whether a million or a million and a half of them are settled in Palestine. But for me the importance of all this Zionist work lies precisely in the effect that it will have on those Jews who will not themselves live in Palestine. We must distinguish in this connection between internal and external effects. The internal effect, in my opinion, will be a healthier Jewry: that is to say, the Jews will acquire that happiness in feeling themselves at one, that sense of being self-sufficient, which a common ideal cannot fail to evoke. This is already evident in the younger generation of our day—not among the young Zionists only—and distinguishes it, greatly to its advantage, from earlier generations, whose endeavours to be absorbed in non-Jewish society produced an almost tragic emptiness. That is the internal effect. The external effect I see in the status which a human group can attain only by collective and productive work. I believe that the existence of a Jewish cultural centre will strengthen the moral and political position of the Jews all over the world, by virtue of the very fact that there will be in existence a kind of embodiment of the interests of the whole Jewish people.

The social psychologist Willy Hellpach served as Minister of Education for the German state of Baden from 1922 to 1924 and ran for the presidency of Germany in 1925 as the candidate of the German Democratic Party. In 1929 he published an article in the *Vossische Zeitung* critical of the Zionist movement, a position he continued to hold many years later. In the 1950s, he was still arguing that Zionism undermined one of the critical missions of Jews in the Diaspora: the retention of "cosmopolitan feeling and thought, which appears all the more important to me as a 'vitamin' of Western culture as this culture is compartmentalized and broken up in nationalist pieces" (Hellpach to Carl Seelig, 22 March 1953, Einstein 2001, 221–222). In his response, Einstein stressed "communal purpose" as the bedrock of a nationalism that seeks dignity and health, not power.

LETTER TO WILLY HELLPACH, SUMMER 1929
EINSTEIN 1954, 171–172

I have read your article on Zionism and the Zurich Congress and feel, as a strong devotee of the Zionist idea, that I must answer you, even if only shortly.

The Jews are a community bound together by ties of blood and tradition, and not of religion only: the attitude of the rest of the world toward them is sufficient proof of this. When I came to Germany fifteen years ago I discovered for the first time that I was a Jew, and I owe this discovery more to Gentiles than Jews.

The tragedy of the Jews is that they are people of a definite historical type, who lack the support of a community to keep them together. The result is a want of solid foundations in the individual which amounts in its extremer forms to moral instability. I realized that salvation was only possible for the race if every Jew in the world should become attached to a living society to which he as an individual might rejoice to belong and which might enable him to bear the hatred and the humiliations that he has to put up with from the rest of the world.

I saw worthy Jews basely caricatured, and the sight made my heart bleed. I saw how schools, comic papers, and innumerable other forces of the Gentile majority undermined the confidence even of the best of my fellow-Jews, and felt that this could not be allowed to continue.

Then I realized that only a common enterprise dear to the heart of Jews all over the world could restore this people to health. It was a great achievement of Herzl's to have realized and proclaimed at the top of his voice that, the traditional attitude of the Jews being what it was, the establishment of a national home or, more accurately, a center in Palestine, was a suitable object on which to concentrate our efforts.

All this you call nationalism, and there is something in the accusation. But a communal purpose without which we can neither live nor die in this hostile world can always be called by that ugly name. In any case it is a nationalism whose aim is not power but dignity and health. If we did not have to live among intolerant, narrow-minded, and violent people, I should be the first to throw over all nationalism in favor of universal humanity.

The objection that we Jews cannot be proper citizens of the German state, for example, if we want to be a "nation," is based on a misunderstanding of the nature of the state which springs from the intolerance of national majorities. Against that intolerance we shall never be safe, whether we call ourselves a people (or nation) or not.

I have put all this with brutal frankness for the sake of brevity, but I know from your writings that you are a man who stands to the sense, not the form.

In Theodor Herzl, the founder of modern Zionism, Einstein saw someone who, like himself, had initially approached the Jewish question as an assimilated outsider only later to recognize the power of solidarity. In a tribute written in April 1929, Einstein wrote: "Living apart from Jewish tradition and a sense of Jewish kinship [*Volkstum*], Herzl recognized intuitively how Jews in their moral despair and fragmentation must be helped" (Einstein Archives 28-078).

The Shaken Rock

Arab riots had shaken Palestine at the beginning of the 1920s, but the relationship between Jews and Arabs remained relatively peaceful from 1922 until 1928, when a new phase in the violence was initiated. Hostilities lasted more than a year, culminating in a week-long blood bath at the end of August 1929. In spite of British assurances to the Zionist Executive that they were in control of the situation, 133 Jews were killed, including 67 in Hebron and 18 in Safed while British soldiers looked on and did little or nothing to stop the violence. For three days Arabs rampaged through houses in Hebron, looting and destroying Jewish property.

One of the consequences of the 1929 riots was a growing rift between those, like Einstein, who continued to draw attention to the moral underpinnings of the movement, and those who stressed the need for military and political self-sufficiency.

CREATING A *MODUS VIVENDI*, AUGUST 1929
EINSTEIN 1931A, 69–71

Shaken to its depths by the tragic catastrophe in Palestine, Jewry must now show that it is truly equal to the great task it has undertaken. It goes without saying that our devotion to the cause and our determination to continue the work of peaceful construction will not be weakened in the slightest by any such set-back. But what has to be done to obviate any possibility of a recurrence of such horrors?

The first and most important necessity is the creation of a *modus vivendi* with the Arab people. Friction is perhaps inevitable, but its evil consequences must be overcome by organised co-operation, so that the inflammable material may not be piled up to the point of danger. The absence of normal contact in every-day life is bound to produce an atmosphere of mutual fear and distrust, which is favourable to such lamentable outbursts of passion as we have witnessed. We Jews must show above all that our own history of suffering has given us sufficient understanding and psychological insight to know how to cope with this problem of psychology and organisation: the more so as no irreconcilable differences stand in the way of peace between Jews and Arabs in Palestine. Let us therefore above all be on our guard against blind chauvinism of any kind, and let us not imagine that reason and common-sense can be replaced by British bayonets.

But one demand we must certainly make of the Mandatory Power, which is responsible for the well-being of the country. Adequate protection must be afforded to those who are engaged in peaceful work. The measures devised for their protection must have regard on the one hand to the scattered position of the Jewish settlements, and on the other hand to the need for helping to smooth over national differences. It goes without saying that there must be adequate participation of Jews in the police force. The Mandatory Power cannot escape the reproach that this duty has not been fully carried out, quite apart from the fact that the responsible authorities misjudged the true state of affairs in the country.

The greatest danger in the present situation is that blind chauvinism may gain ground in our ranks. However firm the stand we make for the defence of our lives and property, we must not forget for a single moment that our national task is in its essence a supranational matter, and that the strength of our whole movement rests in its moral justification, with which it must stand or fall.

The spark for the conflict between Jews and Arabs in the late 1920s was, as at the beginning of the decade, the right of Jews to pray at the Western Wall in Jerusalem. Haj Amin al-Husseini, whom the British appointed religious and political leader of the Arabs of Palestine as the Grand Mufti of Jerusalem in spring 1921, proved unwilling or unable to control his followers.

Furious at the inability of the British to protect Jewish-populated areas from Arab attack, Einstein counseled Chaim Weizmann as the head of the international Zionist Organization in late November to avoid "leaning too much on the English. If we fail to reach real co-operation with the leading Arabs, we will be dropped by the English, not perhaps formally but *de facto*" (letter to Chaim Weizmann, 25 November 1929, cited in Sayen, 106). On the other hand, he recognized the fragility of the Zionist experiment. Swallowing some of his anger, he made a spirited appeal to the goodwill of the British public and its government. The following letter was published under the subtitle "Einstein's Protest."

The Palestine Troubles, 7 October 1929
Manchester Guardian, 12 October 1929; Einstein 1931a, 71–85

I have been following with anxious concern the comments in the British press on the recent events in Palestine. What I have read has so deeply affected me that, despite my general reluctance to enter the political arena, I feel impelled to ask for the hospitality of your columns for the following observations.

It was with a wonderful enthusiasm and a deep sense of gratitude that the Jews, afflicted more than any other people by the chaos and horror of the war, obtained from Great Britain a pledge to support the re-establishment of the Jewish national home in Palestine. The Jewish people, beset with a thousand physical wrongs and moral degradations, saw in the British promise the sure rock on which it could recreate a Jewish national life in Palestine, which, by its very existence as well as by its material and intellectual achievements, would imbue the Jewish masses, dispersed all over the world, with a new sense of hope, dignity, and pride. Jews of all lands gave of their best in man-power and in material wealth in order to fulfil the inspiration that had kept the race alive through a martyrdom of centuries. Within a brief decade some £10,000,000 were raised by voluntary contributions, and 100,000 picked Jews entered Palestine to redeem by their physical labour the almost derelict land. Deserts were irrigated, forests planted, swamps drained, and their crippling diseases subdued. A work of peace was created which, although still perhaps small in size, compelled the admiration of every observer.

Has the rock on which we have built begun to shake? A considerable section of the British press now meets our aspirations with lack of understanding, with coldness, and with disfavour. What has happened?

Arab mobs, organised and fanaticised by political intriguers working on the religious fury of the ignorant, attacked scattered Jewish settlements and murdered and plundered wherever no resistance was offered. In Hebron, the inmates of a rabbinical college, innocent youths who had never handled weapons in their lives, were butchered in cold blood; in Safed the same fate befell aged rabbis and their wives and children. Recently some Arabs raided a Jewish orphan settlement where the pathetic remnants of the great Russian pogroms had found a haven of refuge. Is it not then amazing that an orgy of such primitive brutality upon a peaceful population has been utilised by a certain section of the British press for a campaign of propaganda directed, not against the authors and instigators of these brutalities, but against their victims?

No less disappointing is the amazing degree of ignorance of the character and the achievement of Jewish reconstruction in Palestine displayed in many organs of the press. A decade has elapsed since the policy of the establishment of a Jewish national home in Palestine was officially endorsed by the British Government with the almost unanimous support of the entire British press and of the leaders of all political parties. On the basis of that official recognition, which was approved by almost every civilised Government, and which found its legal embodiment in the Palestine Mandate, Jews have sent their sons and daughters and have given their voluntary offerings for this great work of peaceful reconstruction. I think it may be stated without fear of exaggeration that, except for the war efforts of the European nations, our generation has seen no national effort of such spiritual intensity and such heroic devotion as that which the Jews have displayed during the last ten years in favour of a work of peace in Palestine. When one travels through the country, as I had the good fortune to do a few years ago, and sees young pioneers, men and women of magnificent intellectual and moral calibre, breaking stones and building roads under the blazing rays of the Palestinian sun; when one sees flourishing agricultural settlements shooting up from the long-deserted soil under the intensive efforts of the Jewish settlers; when one sees the development of water-power and the beginnings of an industry adapted to the needs and possibilities of the country, and, above all, the growth of an educational system ranging from the kindergarten to the university, in the language of the Bible—what observer, whatever his origin or faith, can fail to be seized by the magic of such amazing achievement and of such almost superhuman devotion? Is it not bewildering that, after all this, brutal massacres by a fanaticised mob can destroy all appreciation of the Jewish effort in Palestine and lead to a demand for the repeal of the solemn pledges of official support and protection?

Zionism has a two-fold basis. It arose on the one hand from the fact of Jewish suffering. It is not my intention to paint here a picture of the Jewish martyrdom throughout the ages which has arisen from the homelessness of the Jew. Even to-day there is an intensity of Jewish suffering throughout the world of which the public opinion of the civilised West never obtains a comprehensive view. In the whole of Eastern Europe the danger of physical attack against the individual Jew is constantly present. The degrading disabilities of old have been transformed into restrictions of an economic character, while restrictive measures in the educational sphere, such as the "numerus clausus" at the universities, seek to suppress the Jew in the world of intellectual life. There is, I am sure, no need to stress at this time of the day that there is a Jewish problem in the Western world also. How many non-Jews have any insight into the spiritual suffering and distortion, the degradation and moral disintegration engendered by the mere fact of the homelessness of a gifted and sensitive people? What underlies all these phenomena is the basic fact, which the first Zionists recognised with profound intuition, that the Jewish problem cannot be solved by the assimilation of the individual Jew to his environment. Jewish individuality is too strong to be effaced by such assimilation, and too conscious to be ready for such self-effacement. It is, of course, clear that it will never be possible to transplant to Palestine anything more than a minority of the Jewish people, but it has for a long time been the deep conviction of enlightened students of the problem, Jews and non-Jews alike, that the establishment of a National Home for the Jewish people in Palestine would raise the status and the dignity of those who would remain in their native countries, and would thereby materially assist in improving the relations between non-Jews and Jews in general.

But Zionism springs from an even deeper motive than Jewish suffering. It is rooted in a Jewish spiritual tradition, whose maintenance and development are for Jews the *raison d'être* of their continued existence as a community. In the re-establishment of the Jewish nation in the ancient home of the race, where Jewish spiritual values could again be developed in a Jewish atmosphere, the most enlightened representatives of Jewish individuality see the essential preliminary to the regeneration of the race and the setting free of its spiritual creativeness.

It is by these tendencies and aspirations that the Jewish reconstruction in Palestine is informed. Zionism is not a movement inspired by Chauvinism or by a *sacro egoismo*. I am convinced that the great majority of the Jews would refuse to support a movement of that kind. Nor does Zionism aspire to divest anyone in Palestine of any rights or possessions he may enjoy. On the contrary, we are convinced that we shall be able to establish a friendly and constructive co-operation with the kindred Arab race which will be a blessing to both sections of the population materially and spiritually. During the whole of the work of Jewish colonisation not a single Arab has been dispossessed; every acre of land acquired by the Jews has been bought at a price fixed by buyer and seller. Indeed, every visitor has testified to the enormous improvement in the economic and sanitary standard of the Arab population resulting from the Jewish colonisation. Friendly personal relations between the Jewish settlements and the neighbouring Arab villages have been formed throughout the country. Jewish and Arab workers have associated in the trade unions of the Palestine railways, and the standard of living of the Arabs has been raised. Arab scholars can be found working in the great library of the Hebrew University, while the study of the Arabic language and civilisation forms one of the chief subjects of study at this university. Arab workmen have participated in the evening courses conducted at the Jewish Technical Institute at Haifa. The native population has come to realise in an ever-growing measure the benefits, economic, sanitary and intellectual, which the Jewish work of reconstruction has bestowed on the whole country and all its inhabitants. Indeed, one of the most comforting features in the present crisis has been the reports of personal protection afforded by Arabs to their Jewish fellow-citizens against the attacks of the fanaticised mob.

I submit, therefore, that the Zionist movement is entitled, in the name of its higher objectives and on the strength of the support which has been promised to it most solemnly by the civilised world, to demand that its unprecedented reconstructive effort—carried out in a country which still largely lies fallow, and in which by methods of intensive cultivation such as the Jews have applied room can be found for hundreds of thousands of new settlers without detriment to the native population—shall not be defeated by a small clique of agitators, even if they wear the garb of ministers of the Islamic religion. Does public opinion in Great Britain realise that the Grand Mufti of Jerusalem, who is the centre of all the trouble and speaks so loudly in the name of all the Moslems, is a young political adventurer of not much more, I understand, than thirty years of age, who in 1920 was sentenced to several years' imprisonment for his complicity in the riots of that year, but was pardoned under the terms of an amnesty? The mentality of this man may be gauged from a recent statement he gave to an interviewer accusing me, of all men, of having demanded the rebuilding of the Temple on the site of the Mosque of Omar. Is it tolerable that in a country where ignorant fanaticism can so easily be incited to rapine and murder by interested agitators so utterly irresponsible and unscrupulous a politician should be enabled to continue to exercise his evil influence garbed in all the spiritual sanctity of religion and invested with all the temporal powers that this involves in an Eastern country?

The realisation of the great aims embodied in the Mandate for Palestine depends to a very large degree on the public opinion of Great Britain, on its press, and on its statesmen. The Jewish people is entitled to expect that its work of peace shall receive the active and benevolent support of the Mandatory Power. It is entitled to demand that those found guilty in the recent riots shall be adequately punished, and that the men in whose hands is laid the responsible task of the administration of a country of such a unique past and such unique potentialities for the future shall be so instructed as to ensure that this great trust, bestowed by the civilised world on the Mandatory Power, is carried out with vision and courage in the daily tasks of routine administration. Jews do not wish to live in the land of their fathers under the protection of British bayonets; they come as friends of the kindred Arab nation. What they expect of Great Britain is that it shall promote the growth of friendly relations between Jews and Arabs, that it shall not tolerate poisonous propaganda, and that it shall create such organs of security in the country as will afford adequate protection to life and peaceful labour.

The Jews will never abandon the work of reconstruction which they have undertaken. The reaction of all Jews, Zionist and non-Zionist alike, to the events of the last few weeks has shown this clearly enough. But it lies in the hands of the Mandatory Power materially to further or materially to hamper the progress of the work. It is of fundamental importance that British public opinion and the Governments of Great Britain and of Palestine shall feel themselves responsible for this great trust, not because Great Britain once undertook this responsibility in legal form, but because they are deeply convinced of the significance and importance of the task and believe that its realisation will tend to promote the progress and the peace of mankind, and to right a great historic wrong. I cannot believe that the greatest colonial Power in the world will fail when it is faced with the task of placing its unique colonising experience at the service of the reconstruction of the ancient home of the People of the Bible. The task may not be an easy one for the Mandatory Power, but for the success it will attain it is assured of the undying gratitude not only of the Jews but of all that is noblest in mankind.

The editors of the *Guardian* went out of their way to accord Einstein the hospitality he requested: "Dr. Einstein's letter should make the coldest reader understand that the future of the Palestine State is closely and inextricably associated with the happiness of the Jewish race. But it should make the coldest reader understand that it is also inextricably associated with the self-respect of the British people" (*Manchester Guardian*, 12 October 1929).

As always, for Einstein good relations with the Arabs were paramount: "Should we be unable to find a way to honest cooperation and honest pacts with the Arabs, then we have learned absolutely nothing during our 2,000 years of suffering and will deserve our fate" (letter to Chaim Weizmann, 25 November 1929; cited in Sayen, 106).

The following letter to the editor of the Arab newspaper *Falastin* was written in refutation of an article that appeared in it some months earlier. This paper, the self-proclaimed "Arab National Organ," was published in Arabic and English editions in Jaffa. The version of Einstein's letter presented here follows the text in *Falastin*, which differs significantly from that in Einstein 1931a.

LETTER TO THE EDITOR OF *FALASTIN*, 28 JANUARY 1930
FALASTIN, 1 FEBRUARY 1930; EINSTEIN 1931A, 85–87

My attention has been drawn to an article entitled "Relativity and Propaganda" which was published in the English Edition of "Falastin" dated Oct. 19th 1929.

That it should contain among flattering comments about my person others which are less flattering is of little consequence; neither do I take the scathing remarks about the Jewish people too seriously. Unfortunately it has become customary that when disputes amongst nations arise they must need malign each other.

It is not my intention to enter on the subject of what has happened in the past . . . [y]et your article gives me the opportunity to say a few words about the future as I see it.

You doubt my words when I say that the Jews desire friendly relations with the Arab population of Palestine. But anyone like myself, whose conviction it has been for many number of years, that the future of mankind must rest upon the foundation of mutual understanding amongst all nations and that aggressive nationalism must be overcome, can only envision a Palestine of the future in which peaceful cooperation unites the two peoples, who call it their home.

I had therefore anticipated that the great Arab nation would appreciate more fully the needs of the Jews to restore their national home in the land of ancient Judaism and that by joint effort a larger Jewish settlement could be made possible in that country.

I am convinced that the deep interest the Jewry of the world takes in Palestine can but benefit the entire population of the country not alone materially but also with regard to its cultural and national questions. I feel sure that the Arab Renascence which is taking place throughout the great territory inhabited by the Arab people cannot but gain by enlisting Jewish sympathy.

Ways and means ought surely to be found to discuss these possibilities freely and frankly, for it is my belief that these two great Semitic peoples who have made each in their own way lasting contributions to the civilization of the Western world could have a great future in common.

And instead of being filled with fruitless hostility and mutual distrust they should support each other's national and cultural ideals, and should enlist the sympathetic cooperation of one another. To my mind especially all those who are not politically active should help to create this atmosphere of trust.

I deplore the sad events of last August not alone for revealing human nature in its lowest form, but also for the reason that they alienated the two peoples from each other and for a time, rendered difficult the access to one another which nevertheless must be found.

―――――――――――――― ◈ ――――――――――――――

In a note in the same issue, the editor responded that Einstein was "drawing a heavy draft on our credulity when he asks of us to take his ideal as that of the Zionist in Palestine. While believing to the full in his peaceful intentions and his beautiful ideal, we can not judge the Zionist by Dr. Einstein."

The suggestion that conflicts between Jew and Arab could be resolved by a council of wise men without the intrusion of partisan politics indicates just how much Einstein underestimated the power calculus that prevailed in Palestine after the brutal Hebron riots. Still he clung to the hope that moderation might prevail.

LETTER TO THE EDITOR OF *FALASTIN*, 15 MARCH 1930
EINSTEIN ARCHIVES 46-154; EINSTEIN 1954, 172–174

Your letter has given me great pleasure. It shows me that there is good will available on your side, too, for solving the present difficulties in a manner worthy of both our nations. I believe that these difficulties are more psychological than real, and that they can be got over if both sides bring honesty and good will to the task.

What makes the present position so bad is the fact that Jews and Arabs confront each other as opponents before the Mandatory power. This state of affairs is unworthy of both nations and can only be altered by our finding a *via media* on which both sides agree.

I will now tell you how I think that the present difficulties might be remedied; at the same time I must add that this is only my personal opinion, which I have discussed with nobody. I am writing this letter in German because I am not capable of writing it in English myself and because I want to bear the entire responsibility for it myself. You will, I am sure, be able to get some Jewish friend of conciliation to translate it.

A Privy Council is to be formed to which the Jews and Arabs shall each send four representatives, who must be independent of all political parties. Each group is composed as follows:

A doctor, elected by the Medical Association;
A lawyer, elected by the lawyers;
A working men's representative, elected by the trade unions;
An ecclesiastic, elected by the ecclesiastics.

These eight people are to meet once a week. They undertake not to espouse the sectional interests of their profession or nation but conscientiously and to the best of their power to aim at the welfare of the whole population of the country. Their deliberations shall be secret and they are strictly forbidden to give any information about them, even in private. When a decision has been reached on any subject in which not less than three members on each side concur, it may be published, but only in the name of the whole Council. If a member dissents he may retire from the Council, but he is not thereby released from the obligation to secrecy. If one of the elective bodies above specified is dissatisfied with a resolution of the Council, it may replace its representative by another.

Even if this "Privy Council" has no definite powers, it may nevertheless bring about the gradual composition of differences, and secure a united representation of the common interests of the country before the Mandatory power, clear of the dust of ephemeral politics.

After the 1929 disturbances, the British issued the Passfield White Paper on 21 October 1930. It stated that if Jewish immigration prevented Arab residents from obtaining work, the Mandatory government should curtail such immigration or even terminate it. Incensed, Einstein gave a speech eight days later in London in which he described the position of the Jews in the Diaspora as a "moral barometer for the political world." The barometer was low, he said, which only served to confirm to him that "it is our duty to preserve and consolidate our community. Embedded in the tradition of the Jewish people there is a love of justice and reason which must continue to work for the good of all nations now and in the future. In modern times this tradition has produced Spinoza and Karl Marx." The speech, delivered at the Savoy Hotel, was attended by G. B. Shaw, who introduced Einstein; by H. G. Wells; and by Lord Rothschild ("The Jewish Community," 29 October 1930, Einstein 1954, 174–175).

Einstein need not have been too concerned by the White Paper. It "never went into effect [and is] notable only because the Zionist movement was able to get it revoked" (Segev, 336). His distrust of British policies remained strong, however. On the eve of departure for America to take up residence for the winter semester at Caltech in Pasadena, Einstein issued the following statement in a cable to the Zionist Organization of America. The message was intended as a greeting to American Jews gathered at the Maccabean Festival held at Madison Square Garden on 13 December. He reiterated his belief some days later in arguing that "Jews should seek to reach a direct understanding with the Arab masses and thus diminish the Palestine government's function as the arbiter of Jewish-Arab interests" (*New York Times*, 7 December 1930).

The use of the term "upbuilding" in this text and in the following one is an anachronistic rendering of the German word "Aufbau," today more conventionally translated as "reconstruction." The following article appeared under the headline "Einstein Attacks British Zion Policy."

Redoubling Efforts
New York Times, 3 December 1930

I cannot conceal my own deep disappointment at the attitude taken to Palestine by the British Government and its bureaucratic officials. However, my disappointment is tempered by the remarkable demonstration of Jewish courage and solidarity. The reaction to the British White Paper by the Jews the world over is proof of their indissoluble bond with the upbuilding of Palestine.

With all the nations involved in the present crisis, the Jewish problem is again felt in all its acuteness. We Jews are everywhere subject to attacks and humiliations that result from the exaggeration of nationalism and racial vanity, which, in most European countries, expresses itself in the form of aggressive anti-Semitism. In such a time we Jews must strengthen our self-consciousness and solidarity. This can be attained in no other way than united participation in the rebuilding of the Jewish national home in Palestine.

The value of the undertaking is enhanced by the inevitable handicaps and setbacks to which we are subject. The surmounting of these obstacles can best develop Jewish idealism and Jewish creative energies. Palestine is a country of unexplored possibilities, and through the influx of Jewish energy, perseverance, and capital it will assume the role of a pioneering country for the entire Near East.

By means of modern methods of reconstruction, Palestine affords ample room for both Jews and Arabs, who can live side by side in peace and harmony in a common country. I believe that the setbacks of last year must strengthen within us the recognition of our duty to improve, through patience and continued efforts, our relations with the Arab people and to convince them of the advantages Zionism creates for them.

The success of the rebuilding of Palestine depends primarily upon the resolute efforts of the Jews themselves. The Jewish national home is not a luxury but an absolute necessity for the Jewish people. Therefore, the reply of the Jews to the present difficulties must be a determination to redouble their efforts in Palestine.

In spite of the recent political upheavals, or perhaps because of them, Einstein reiterated once again forcefully his allegiance to the cultural Zionist vision of a Jewish unity based on a moral tradition that was antithetical to the rule of force. It was in this spirit that he cabled the British High Commissioner in Palestine calling for the commutation of the death sentences passed against twenty-five Arabs who had participated in the riots of August 1929 (*New York Times*, 1 June 1930).

The Einsteins arrived in New York aboard the S.S. *Belgenland* on 11 December 1930 on the first leg of their trip to Pasadena and Einstein's winter appointment at Caltech. The ship dropped anchor in the outer harbor for several days as the couple sought in vain to avoid a persistent crowd of reporters, though Einstein apparently found some of the encounter enjoyable (see Plate 12). Two days after his arrival he spoke from the NBC studios in Rockefeller Center in an address to Zionist youth all over the world. The speech was organized by Avukah, the American Student Zionist Federation, which shared Einstein's cultural Zionist and socialist hopes for Palestine.

The Jewish Mission in Palestine, 13 December 1930
Einstein Archives 28-121; Einstein 1933, 33–34

... Undoubtedly certain statements and measures taken and pronounced by British officials have been just subjects for criticism. We cannot, however, be satisfied with this but must learn the lesson of what has recently happened.

In the first place, we must pay great attention to our relations with the Arab people. By cultivating these relations we shall be able to avoid a development in the future of those dangerous tensions which can be exploited to provoke hostile action against us. We can very well attain this end because our upbuilding of Palestine has been so conducted that it serves also the real interests of the Arab population.

In the second place, in doing this we shall be able to avoid the unfortunate necessity—unfortunate for Arabs and Jews alike—of being obliged to call in the Mandatory power to act as judge and umpire between us.

In this way we are not merely following the bidding of wisdom but we also remain faithful to our traditions which, above all else, give substance and meaning to the unity of Israel. For, indeed, this unity of Jews the world over is in no wise a political unity and should never become such. It rests exclusively on a moral tradition. Out of this alone can the Jewish people maintain its creative powers, and on this alone it claims its basis for existence.

One day later, on 14 December, Einstein delivered his "Two Percent Speech" (chapter 5) to a group of pacifists assembled at the Ritz-Carlton Hotel. Its stirring message signaled his emergence as a leading spokesman for militant pacifism, the cause that dominated his attention for the next two years. With Europe engulfed in a state of crisis, he shifted the focus of his immediate attention from the problems facing the Zionist movement to the threat of military aggression.

In early 1933, soon after the Nazis assumed power, Einstein abandoned his pacifist stance. The fate of those persecuted by Hitler's racial and antidemocratic policies once again became his foremost concern. Now himself a refugee from Nazi barbarism, he assumed an active role behind the scenes after he arrived in his new homeland as an ally of Rabbi Stephen Wise, a leading spokesman for Jewish affairs in the United States. The broader context of their relationship is one of the themes presented in chapter 6.

CHAPTER 4

Internationalism and European Security, 1922–1932

Apart from his ties with the New Fatherland Association—reconstituted as the German chapter of the International League of Human Rights soon after the war—Einstein confined his political activity during the mid-1920s mainly to working for the International Committee on Intellectual Cooperation affiliated with the League of Nations. His on-again off-again relationship with this elite body reflects not only the vicissitudes of European politics but also his own ambiguous status as a quasi-official representative of the Weimar Republic, a role that became official after Germany's admission into the League of Nations in 1926.

Those who initially supported Einstein's nomination to the committee faced objections from certain French representatives, who argued that it was premature to admit a German, particularly at a time of severe strain in the relations between their two countries. Reluctance was also expressed by some of his own countrymen, who unwittingly echoed his quip about the "relativity of his political identity" when they claimed that he was "not a German at all but a Swiss Jew" (N & N 1960, 59). By now this was no laughing matter. In words reminiscent of his unheeded advice to Rathenau, he described the political atmosphere in Germany as such "that a Jew would do well to exercise restraint as regards his participation in political affairs," noting further that "I have no desire to represent people who certainly would not choose me as their representative, and with whom I find myself in disagreement on the questions to be dealt with" (N & N 1960, 59). Owing to a variety of such factors, he remained on the sidelines until 1924.

Einstein strongly favored the idea that intellectual elites should have a platform for voicing their views on human affairs (see the section "Intellectual Elitism and Political Idealism" in chapter 9). With the passage of time, however, he came to feel that the International Committee on Intellectual Cooperation could not live up to this goal because of various national constraints that hampered its efforts. The committee's ongoing work was conducted by the Paris-based Institute for Intellectual Cooperation, which was funded exclusively by the French government. This arrangement reflected the dominant influence of the committee's French contingent, a circumstance Einstein found objectionable on the grounds that it undermined the committee's internationalist mission. On several occasions he voiced misgivings regarding the intrusion of national interests in the committee's work, a recurrent theme in his private correspondence with other leading intellectuals, including Marie Curie, H. A. Lorentz, and Paul Painlevé. After attending a meeting in July 1930, he criticized the committee for lacking "the determination necessary to achieve real progress toward improving international relations" (N & N 1960, 110) and resigned as a member.

The texts in the section entitled "On World Affairs" give an impression of Einstein's assessment of the political landscape during the mid-1920s. Two years after touring the Far East, he gave a cautious reading of the problems facing Asian nations, particularly Japan and Russia. Although he wrote very little about European politics during this period, the views he expressed in several newspaper interviews reflect his general optimism at the time. The stabilization of the German economy in 1924 ushered in a brief era of tranquillity marked by the leadership of Foreign Minister Gustav Stresemann. Together with his French counterpart, Aristide Briand, Stresemann pursued a policy of rapprochement that took the wind out of the sails of Germany's radical right. In 1927

Einstein granted several interviews for newspapers both inside and outside Germany in which he stressed that conditions were ripe for further progress toward a lasting peace. He was quoted in the *Berliner Tageblatt* as saying that "the desire for *rapprochement* with France was widespread in every stratum of the [German] population." Einstein sounded an equally upbeat note in the *Neue Zürcher Zeitung*. When asked about those in Germany who still sought revenge and the dismantling of the Versailles Treaty, he answered: "Objective observers will have to concede that the more the republic is consolidated in Germany, the more sincere and widespread the renunciation of nationalist phrases."

Already aligned with mainstream pacifism, by 1929 Einstein began to adopt an uncompromising, though hardly dogmatic position. He gave a series of essentially moral arguments in support of this stance in a letter to the mathematician Jacques Hadamard, a leftist intellectual who normally saw eye to eye with his German friend on political matters. After the Nazis rose to power, Einstein occasionally referred to the arguments in this letter in order to placate critics who attacked him for abandoning pacifism (see chapter 6).

As Einstein probably sensed, Stresemann's death in 1929 came at the worst possible time. His stirring tribute to the leading statesman of the Weimar era serves as a reminder of the hopes he and other liberals still held for the beleaguered republic. Even as late as 1931, Einstein approached the government of Heinrich Brüning to urge renewed cooperation with France in the wake of Japan's invasion of Manchuria in September 1931. By this time tensions across the Rhine were on the rise, whereas the impotence of the League of Nations when faced with the fait accompli of Japanese aggression was obvious to any but the most optimistic observer. Einstein was well aware of the economic conditions undermining European security as well as the fragile foundations of the Weimar Republic. Still, he could not easily have imagined how quickly this house of cards would come tumbling down.

Einstein's last action as an associate of the League of Nations came in 1932 after he had already submitted his resignation.

Asked by the Institute for Intellectual Cooperation to initiate an exchange of views with another leading intellectual about any major problem facing humanity, he decided to ask Sigmund Freud to voice his opinions about the psychological factors that come into play when nations go to war against one another. The choice was fitting in a number of respects, not least the strong sense both men shared of belonging to those select few who constituted the European intellectual elite. Einstein, as noted already in chapter 1, had been stunned by the mass psychosis that swept Germany during the First World War, a phenomenon he recognized as rooted in deep-seated irrational impulses that normally remained latent. Moreover, this brief exchange gave Freud the opportunity to address a topic he had touched upon indirectly in his recent work (Sulloway, 411–412). He essentially affirmed Einstein's views while offering a deeper analysis of the aggressive instincts, which along with erotic impulses, guide human behavior.

Internationalism's Intellectual Elite

In the wake of Rathenau's assassination, Einstein developed deeply ambivalent feelings about his own status as an unofficial spokesman for the Weimar Republic, giving serious thought to declining an invitation to join the Committee on Intellectual Cooperation. Partly at the urging of Marie Curie he overcame his misgivings.

About the same time, he recalled the circumstances that had led to a critical rift of opinion within the Prussian Academy (see letters to H. A. Lorentz, 2 August 1915, and Romain Rolland, 15 September 1915, chapter 1).

INTERNATIONALISM OF SCIENCE, CA. 1922
N & N 1960, 59–60

When nationalism and political passions were reaching a climax during the war, Emil Fischer remarked with emphasis at a session of the Academy: "Whether you like it or not, gentlemen, science is and always will be international."

The great men among scientists have always known this and felt it passionately, even in times of international conflicts when they stood alone amid their narrow-minded colleagues. During the war, the majority of voting members in every country betrayed their sacred trust. The International Association of Academies was destroyed. Congresses were, and still are being, held to which scholars from former enemy countries were not admitted. Political considerations, advanced with pompous solemnity, make it impossible for pure objectivity to prevail, without which great accomplishments cannot mature.

What can be done by well-meaning people who are immune to the emotional temptations of the moment, to restore health to the intellectual community? As long as the majority of intellectual workers remain so embittered, it will not be possible to arrange for an international congress of real significance. Moreover, psychological opposition to the restoration of international associations of scientific workers is still so formidable that the small number of broad-minded people cannot defeat it.

These more enlightened men can make an important contribution to the great task of reviving international societies by keeping in close touch with like-minded men and women the world over, as well as by steadfastly championing the cause of internationalism in their own spheres of influence. Real success will require time, but eventually it will undoubtedly come. I cannot let this opportunity pass without expressing my appreciation to the particularly large number of English colleagues who, throughout these difficult years, have never ceased to manifest a strong desire to preserve the international community of intellectuals.

The attitude of individual citizens is everywhere far superior to official pronouncements. Let men of good will bear this in mind rather than allow themselves to become exasperated or misled: *Senatores boni viri, senatus autem bestia* [the senators are honorable men, but the senate is a monster].

I am extremely hopeful for the progress of a general international organization. My feelings are based not so much on confidence in the intelligence and high-mindedness of scientists as on the inevitable pressure of economic developments. Since these developments are so largely dependent upon the work of even reactionary scientists, they too will have no choice but to assist in the establishment of an international organization.

Einstein was unable to attend the inaugural meeting of the Committee on Intellectual Cooperation, held in Geneva in August 1922. Soon thereafter, the German economy crumbled under the strain of heavy war reparations. In January 1923, the French government of Raymond Poincaré sent troops into the Ruhr industrial area to seize control of production. By the time Einstein returned from his lengthy travels to Asia, Palestine, and Spain, the German mark had become nearly worthless. In March of 1923 he tendered his resignation from the Committee on Intellectual Cooperation, stating that he had become convinced that "the League [of Nations] possesses neither the strength nor the sincere desire which it needs to accomplish its aims" (N & N 1960, 61). Two months later, in a public pronouncement printed in a German pacifist journal, Einstein reiterated his position, while noting that his decision to resign was taken after some soul-searching.

STATEMENT ON RESIGNING FROM THE COMMITTEE ON INTELLECTUAL COOPERATION
DIE FRIEDENS-WARTE 23 (JUNE 1923), 186

On my return from Japan I announced my resignation from the Committee on Intellectual Cooperation of the League of Nations with a heavy heart. I did so because until now the activities of the League of Nations have convinced me that it appears unwilling to stand up to anything perpetrated by the currently powerful group of states, no matter how brutal. I did so because the League of Nations, as it functions at present, fails not only to embody the ideal of an international organization, but actually discredits it. And yet I did so with great reluctance, because hope has not quite been extinguished in me that this husk can eventually be filled with greater substance.... May the League of the future prove my harsh words ... to have been mistaken.

In a letter to Marie Curie, a fellow member of the committee, Einstein elaborated on the reasons behind his resignation as well as his unwillingness to attend the forthcoming Solvay Congress of Physicists. These private reflections, addressed to his "sister in defiance," suggest that he regarded Curie as an important member of internationalism's intellectual elite.

LETTER TO MARIE CURIE, 25 DECEMBER 1923
EINSTEIN ARCHIVES 8-431; N & N 1960, 64-65

I realize that you will be annoyed with me, and justifiably so, for having resigned from the League committee and for having issued a sharply worded statement. After all, scarcely half a year earlier I myself had advised you to participate in the committee's work! I did not resign for base motives, nor because of pro-German sympathies. I had become convinced that the League (unlike the committee to which I belonged) functioned, despite a thin veneer of objectivity, as a willing instrument of power politics. Under the circumstances, I did not want to have anything to do with the League. I felt that a blunt statement to that effect could do no harm. Perhaps I was wrong, but such were my convictions at the time.

I have requested, furthermore, that I not be invited to Brussels [to attend the Solvay Congress]. Although, psychologically speaking,

I can well understand why Frenchmen and Belgians do not like to meet with Germans, it is certainly unworthy of men of true culture to treat one another according to their respective nationalities or by other superficial criteria as would the common mob, which is governed by mass suggestion. If that is the way of the world, I should choose to stay in my study rather than get upset over the conduct of people outside. Do not think for a moment that I consider my own fellow countrymen superior and that I misunderstand the others; that would scarcely be consistent with the theory of relativity. . . . But enough of that. I would not dare grumble to you in this fashion if I did not think of you as a sister in defiance, one who, somewhere in her soul, has always had some understanding of such feelings, and one to whom I have always felt particularly close.

By mid-1924 tensions between France and Germany had subsided, causing Einstein to reconsider his decision. On the initiative of Gilbert Murray, the League of Nations renewed its invitation to him to join the Committee on Intellectual Cooperation, and Einstein accepted the offer on 25 June 1924. The following month he traveled to Geneva to take part for the first time in the committee's work, about which he gave a summary report to the German public. A portrait of one of its meetings is provided in Plate 10.

REPORT ON MEETING OF THE COMMITTEE ON INTELLECTUAL COOPERATION
FRANKFURTER ZEITUNG, 29 AUGUST 1924; N & N 1960, 70–71

I have just returned from the session of the League Committee on Intellectual Co-operation in Geneva and should like to convey some of my impressions to the German public. The object of the committee is to initiate or foster efforts which may promote international co-operation between the scientific and intellectual communities of various countries in the hope that national cultures, heretofore separated by language and tradition, may thereby be brought into closer communication. Rather than entertaining utopian schemes, the committee has initiated several modest but fruitful projects on a small scale, such as the international organization of scientific reporting, the exchange of publications, the protection of literary property, the exchange of professors and students among various countries, etc. Thus far, the greatest progress has been achieved in the sphere of international reporting.

While the specific projects just mentioned may be of little interest to the general public, much consideration should be given to the question of what attitude the German people and the German Government ought to adopt in principle toward the League of Nations. My personal impressions in Geneva are of interest only insofar as they involve this larger question.

All members of the committee were anxious at all times to emphasize the truly international character of the institution. Regardless of the question under discussion, Germany was always considered as though she were, in fact, a member of the League. It is true that the French mentality may unwittingly have dominated the proceedings to some extent, which is not surprising in view of the origin of the League and the nonparticipation of important nations; yet, I was happy to observe an honest desire to be objective. Such a spirit is bound to prove constructive in the future. I am convinced that the League of Nations is an institution which will make a real contribution toward the steady recovery of Europe's material as well as intangible resources. My experience has been that much can be accomplished when reason and unqualified sincerity are allowed to prevail.

While in Geneva, I also had the opportunity to discuss the question of Germany's entry into the League of Nations with various well-informed people, particularly Frenchmen. Without exception everyone was of the opinion that Germany should be treated like the other great powers and should be given a permanent seat on the League Council. It was also felt that Germany should join the League right after the successful termination of the London Conference. I fully concur in this view and believe, further, that Germany ought not attach any conditions or reservations to her entry into the League. Confidence begets confidence, and without confidence fruitful co-operation will not be possible. I hope that all those in Germany who believe in the necessity of international conciliation exert pressure so that the present favorable moment for Germany's entry into the League can be effectively utilized.

The League offered the best prospects for what Einstein considered a necessity: "One need not rack one's brains over the question whether a United States of Europe is realizable or not; it *must* be realized if Europe wishes to maintain to some degree its importance and its human riches" (29 October 1924, Einstein Archives 28-023).

Committee meetings were chaired by the philosopher Henri Bergson. Members included a number of leading physical scientists, including Marie Curie, Robert A. Millikan, and Hendrik A. Lorentz. Already during the war years, Einstein had confided to Lorentz his distress about the political barriers that impeded cooperation between physicists (see chapter 1). He sought his counsel again when he tried to gain support for a German national committee that would be loosely affiliated with the Geneva committee, a proposal to which Max Planck had turned a deaf ear.

LETTER TO HENDRIK A. LORENTZ, 9 JANUARY 1925
EINSTEIN ARCHIVES 16-577; N & N 1960, 73

It is a difficult job to contribute in some way to the life of human beings in our world.

... As you know, I asked Herr Planck ... to organize, or at least keep a watchful eye on, a National Committee on Intellectual Co-operation. I noticed at once that my request made him unhappy, and the poor fellow has been searching his soul ever since. Whenever I inquired he said he had not as yet decided what he would do. Yesterday he himself brought the matter up and said he was unable to do what had been asked of him.

By way of justification he said something like this: "So long as Germans are excluded from international societies and meetings, they must properly hold aloof from all international events, even though they may continue to maintain cordial personal relations with individuals abroad." Of course I protested that international events could not all be lumped together. ...

Despite the illogic of his reasoning, I did not think it wise to argue with him any further. The important thing is his basic attitude. His reasons are secondary and subject to change. If I judge Planck's position correctly, he himself would really like to co-operate, but loyalty to the social group to which he belongs makes this impossible for him. The political events of recent weeks, which have caused great bitterness here, have apparently tipped the scales in favor of his decision. How definite his decision is may be judged from his express request that I ask you not to write him on the matter.

I am quite unhappy about all this. The epidemic that afflicts Europeans is an emotional condition and, hence, cannot be combated with reason. I fear that at the present time we shall be unable to find anyone of sufficient prestige for a National Committee ... someone who would be regarded by the intellectuals as one of their own and who would command their confidence. We shall have to await a resurgence of political awareness. ...

Planck's attitude reflected the widespread intransigence among the German intelligentsia in the face of continued hostility. Tensions gradually subsided after the signing of the Treaty of Locarno in the fall of 1925, an agreement reached by French Foreign Minister Aristide Briand and his German counterpart Gustav Stresemann and which paved the way for Germany's admission into the League of Nations the following year. Einstein was enthusiastic: "The Locarno pact . . . demonstrates that the public's traditional prejudices and sentiments forged in war have diminished to such a degree that governments dare to take such a step" (Einstein Archives 28-032).

Since the Committee on Intellectual Cooperation met only twice each year in Geneva, it was necessary to establish an institute that could pursue the committee's policies. After several years of planning, the new Institute for Intellectual Cooperation opened in Paris in January 1926. Einstein had argued that the institute would better serve its international purpose were it located in a neutral country, but he was outvoted. At a banquet marking the occasion he diplomatically reiterated his misgivings regarding the financial and administrative arrangements that had been undertaken to create the new institute.

THE INSTITUTE FOR INTELLECTUAL COOPERATION, 16 JANUARY 1926

EINSTEIN ARCHIVES 28-037; N & N 1960, 77–78

For the first time during the past year, the leading statesmen of Europe have made a number of important decisions which clearly resulted from the realization that our continent cannot fully recover unless the latent power struggle among the existing countries comes to an end. The political organization of Europe must be strengthened, and an attempt must be made to abolish divisive tariff barriers. But these great objectives cannot be achieved merely by official treaties among the nations. What is particularly necessary is that the minds of the people be receptive to the concept of international co-operation. We must try to awaken in them a sense of solidarity that will not stop at the frontiers, as it has done in the past. It was with this end in mind that the League of Nations created the Committee on Intellectual Co-operation. This committee was meant to be decidedly an international and entirely nonpolitical agency for the purpose of restoring communication among the intellectuals in various countries who had become isolated by the war. It is a difficult task; for alas! it must be admitted that scientists and artists, at least in the countries with which I am familiar, are guided by narrow nationalism to a much greater extent than are men of affairs.

The committee has been meeting twice a year. To make its work more effective, the French Government has now decided to create and maintain a permanent Institute for Intellectual Co-operation. This generous act deserves the gratitude of all of us.

It is always an easy and satisfying job to praise things one approves of and to ignore those which one regrets or disapproves. But honesty is essential if our work is to progress. I shall not hesitate, therefore, to voice some criticism while, at the same time, expressing my good wishes on the establishment of the institute.

Almost every day I have had occasion to notice that the greatest obstacle to the work of our committee is lack of confidence in its political impartiality. Everything must be done to strengthen people's confidence and everything should be avoided that might impair it.

Now, when the French Government organizes and maintains, out of public funds, an institute as a permanent organ of the committee, with a Frenchman as its director, the detached observer can scarcely avoid the impression that French influence predominates in the committee. The impression is deepened by the fact that the chairman of the committee has so far also been a Frenchman. The men in question enjoy the highest reputation. They are esteemed and respected everywhere. Nevertheless, the impression of French predominance remains.

Dixi et salvavi animam meam [I have spoken and saved my soul]. I hope with all my heart that the new institute, by constant interaction with the committee, may succeed in promoting their common goals and may eventually gain the confidence and recognition of intellectual workers throughout the world.

———————— ✴ ————————

Einstein's hopes were not to be realized, as the affairs of the committee and its institute became increasingly subject to the whims of national politics. In 1928 the Weimar Republic created a National Committee on Intellectual Cooperation, appointing Einstein as one of its members, along with Hugo Krüss, director of the Prussian State Library in Berlin. This arrangement soon undermined Einstein's position as an independent voice on the committee, from which he quietly resigned in 1930. These circumstances are clearly reflected in a letter written to his friend Paul Painlevé, an eminent mathematician and former premier of France, who was presiding over the Governing Body of the International Institute for Intellectual Cooperation located in Paris.

Letter to Paul Painlevé, 9 April 1930
Einstein Archives 34-871; N & N 1960, 107–108

I am very glad that your letter provides me with the welcome opportunity to offer my views on the proposals of the German National Committee. To my deep regret and despite the express disapproval of the German Foreign Office, the proposals were adopted by the National Committee during a session at which only about half a dozen members of the board were present. The proposals are the work of Herr Krüss; we were not advised of their content beforehand.

First, ahead of all concrete issues must come the firm desire to foster international co-operation and mutual trust which is essential to fruitful collaboration. The initiative taken by the German National Committee is not at all calculated to promote a spirit of mutual trust. Had they really desired to influence the future work of the institute, they should have sought some initial confidential discussions with the French staff, in clear recognition of the fact that the burden of the institute has so far rested almost exclusively on the shoulders of France. Such a gesture would have been all the more appropriate since, thus far, the contribution of the German group to the work of the International Committee on Intellectual Cooperation has been rather insignificant.

As for the substance, I know that, while I agree with certain critical remarks you voiced regarding the German proposals, I nonetheless feel compelled to *adhere uncompromisingly to the view that the institute must remain an indivisible unit*. Once segments of the institute are relocated in various countries—unfortunately, there is already one precedent—the significance of the institute as a body which is meant to further the development of a true international spirit becomes impaired. If the Committee on Intellectual Cooperation is to remain true to its high mission, we must not make any concessions to national vanity and jealousy, those evil hereditary maladies of European history.

I wish to use this opportunity to make one other remark: I have always regretted the fact that the institute was established in Paris and financed exclusively by French funds. Doubtless, this was done for noble reasons, but, in a time of so much political restlessness, it seems to me it was bound to give rise to strong suspicions. Even Mr. Luchaire, conducting the work of the institute in a truly exemplary spirit of international impartiality, could not altogether dispel these suspicions.

In my view, the cause of international friendship would be greatly served if the French group were to present the following proposal: to move the institute *in toto* to Geneva and have all countries contribute to its financial support under a quota system. In view of the great sacrifices France has made, such a proposal should not come from our side. However, were the French to make it themselves, I believe that such an act of self-abnegation would be gratefully welcomed and viewed by everyone as an important contribution to the cause of internationalism. However, such a proposal is likely to be accepted only in the course of time. Right now, I believe the most urgent task is the preservation of the institute's integrity.

The previous year Einstein had addressed a confidential letter to Foreign Minister Stresemann stressing that because neither England nor Germany helped in financing the institute, French influence dominated, "a fact which is not conducive . . . to international solidarity" (11 July 1929, N & N 1960, 98). He proposed that Germany contribute 40,000 marks toward financing the institute on the assumption that a German assistant director would be promptly appointed. This measure, he felt confident, would cause England to follow suit. No action was apparently taken by the German government on this proposal. Less than three months later, Stresemann died of overwork (see Einstein's obituary for him below).

On World Affairs

Following his tour of Asia in 1922–23, Einstein took a keen interest in economic, social, and political conditions in the Far East. In 1926, he and Fritz Haber helped found a new Kaiser Wilhelm Institute for Asian Languages and Cultures located in Dahlem. For more regarding his views on Soviet Russia, see the section "Early Thoughts on Socialism and the Soviet Experiment" in chapter 9.

The following is part of a series of interviews conducted by reporter Herman Bernstein with prominent European statesmen, diplomats, and intellectuals. It appeared as an article entitled "Europe's Peace Craving Mixed with Fear." Einstein was asked specifically about the Russo-Japanese Convention of January 1925, whereby the two powers agreed to restore diplomatic relations.

ON ASIA
NEW YORK TIMES, 17 MAY 1925

I think the coalition of Russia, Japan and China is quite natural, because they are in a defensive position against the more progressive economic conditions in Western Europe and the United States. They could not help themselves. Circumstances are stronger than political intentions, and I feel certain that there is great danger for the development of the civilized countries if they do not pursue a cautious and far-sighted policy.

The people in the Far East should not be deprived of the possibility of existence. The danger can be averted only through mutual understanding and a bloc of the combined interests of the world. For instance, Japan is now like a great kettle without any safety valves. It has not enough land to enable its population to exist and develop. This should be remedied in some way if a terrible conflict is to be averted.

As for Russia it seems to me that economically she has made very little progress under the present form of government, and she has little to show of a constructive nature. Her production in the industrial and economic fields has been decreasing. But as to the future of Russia, it is as hard and unwise to prophesy about this as about anything else. . . .

As a spokesman for internationalism during the Weimar era, Einstein threw his wholehearted support behind those who sought to improve Franco-German relations. Doing so meant walking a tightrope stretched over an abyss of hostility and mistrust that separated the reasonable voices on both sides of this divide, a reality that no political spokesperson could afford to ignore. How Einstein performed under such circumstances is illustrated by the delicate maneuvering he displayed in the following statement for Berlin's leading liberal newspaper. In it he distanced himself from the views earlier attributed to him in an interview with the French radical socialist newspaper *L'Oeuvre*. What follows is the text of the statement that he gave to the German paper's correspondent in Paris, who had requested an authoritative account of the statements allegedly made to the French paper.

STATEMENT TO *BERLINER TAGEBLATT*, 1927
EINSTEIN ARCHIVES 29-020; N & N 1960, 88–89

The oft-cited "German war spirit," I said, was frequently exaggerated. Contrary to the view widely held abroad, not only the German Left, but even Rightist circles, especially those who are responsible for political and economic decisions, were animated by a sincere desire to come to an understanding with France. True, the term *pacifism* was not well liked in Germany since it was frequently considered synonymous with an unrealistic political philosophy. Nevertheless, the desire for *rapprochement* with France was widespread in every stratum of the population.

Queried on my attitude toward the League of Nations, I said that, with regard to the great problem of world peace, it could not be denied that the League had been a keen disappointment to many and, further, that people everywhere had sensed its failure to act at decisive moments with courage and good will. Yet, I felt that all well-intentioned people should support this first attempt at restoring order in the sphere of international relations; for, despite the existence of the League of Nations, the outmoded and dangerous formula of the European balance of power still exercised a harmful influence.

When I pointed out that France, like other countries, had not yet reached a courageous decision on the disarmament problem, the interviewer contended that European disarmament was inextricably linked with conditions in Russia, a country still outside the European community of nations. In reply, I stated my firm belief that

Russia had no aggressive intentions against any European country and that, further, the European countries could in no way use this argument to account for their failure to promote disarmament.

Shortly after the publication of this interview relativizing the statements attributed to him in the French press, Einstein granted another to a Zurich newspaper in which he expanded on his statement regarding the efficacy of the League of Nations. Particularly noteworthy are the closing comments wherein he minimizes the threat to the Weimar Republic posed by its opponents, in particular the radical right and its youth movements. This assessment reflected his optimistic outlook during the brief interlude of tranquillity that followed the signing of the Locarno treaty.

INTERVIEW IN *NEUE ZÜRCHER ZEITUNG*, 20 NOVEMBER 1927
EINSTEIN ARCHIVES 29-022; N & N 1975, 105–107

[You have recently on occasion also made public your views on political questions. Are you very interested in day-to-day politics and do your scientific labors allow you time to concern yourself with political matters?]

Of course I am not a politician in the conventional sense of the word; there are few scholars who are. All the same I believe that there is a political idea and a political task, which no one who claims to be a contemporary should shirk. By that I mean the task of restoring *the unity between nations* which has been so completely destroyed by the world war and to see to it that a better and more genuine understanding among nations makes a repetition of the dreadful catastrophe which we have lived through impossible. I am convinced that participating in this is a duty which no one can avoid, however great his achievement in any field whatsoever.

[Do you believe that the world has made significant progress since the end of the war toward your ideal? Above all, do you consider the League of Nations a suitable instrument for bringing lasting peace to the world?]

There is no doubt that progress has been made. Yet neither with respect to the scale of progress nor the pace at which it has been achieved can it satisfy leading pacifists or the general population in all countries. No one denies that the League of Nations in its present guise has by no means fulfilled all the expectations which accompanied its founding. Too much of the spirit of prewar politics persists, in which mistrust was considered the foremost political virtue. We can observe the same psychological factors of power politics in the treatment of the *disarmament* question—which I am convinced is the most pressing problem in the international politics of peace—as those that inhibit the League of Nations. Nevertheless it would of course be wrong to abandon the League of Nations and leave it in the lurch. It may be weak and imperfect as an instrument, but it is after all the only sizeable instrument of peace which we have. As much as we should not hold back in our criticism, we are not justified in withholding our participation. For this reason it is most regrettable that two such large and crucial nations as *America* and *Russia* could not yet bring themselves to join the League of Nations. I am familiar with the authoritative reasons and know that at least in these countries they carry great weight. Yet I do not give up hope that in the foreseeable future America and Russia will stand on the foundation of the League of Nations, and I believe that this moment will have a decisive political importance.

[How do you judge the prospects for peace in Europe, which rest chiefly on *German-French rapprochement*?]

I am firmly convinced that by far the majority of both peoples have the earnest and genuine desire for rapprochement. As to my own fatherland I can in any case assure that all reasonable and politically serious people have no greater wish than to achieve a relationship of sincere and complete rapprochement with our French neighbor. Naturally it requires good will on both sides to reach this goal. Let us not conceal the fact that continuing *foreign occupation* of German land remains one of the most difficult psychological *obstacles* to complete German-French understanding.

[In response to a parenthetical remark that this occupation seems necessary to many Frenchmen in light of the revenge desired in many German circles.]

Let us not be misled by the clamor of a few irresponsible individuals and by the fact—completely insignificant in the realm of foreign policy—that a part of the German nation can distance itself only with great difficulty from certain perceptions of the prewar era. Objective observers will have to concede that the more the republic is consolidated in Germany, the more sincere and widespread the renunciation of nationalist phrases, which given the total disarmament of Germany could in any case not be translated into action. You will perhaps counter that it is in the very circles which are closest to me professionally, especially among students, that an aggressive nationalism still dominates. Believe me, however, that the young, who have never witnessed war, today hold views about which they will chuckle in a few years and which are above all not shared by their own parents. No, with mutual distrust we will not advance; that much we should have learned from the past. If we have *trust* in one another, if we believe in the power of the idea of peace, if every one of us at his own station does his duty in the service of our idea, peace will be better protected than with cannon and gas bombs.

Einstein's long-standing commitment to pacifism led him to come out in support of the War Resisters' International in 1928. One year later he elaborated and clarified his views on pacifism in a letter to his friend, the distinguished mathematician Jacques Hadamard. Einstein prefaced his seven-point argument with a remark about the unsuitability of pacifism for African tribes, contrasting such a state of political barbarism with the civilized impulses of European societies. Although he saw the League of Nations as powerless to prevent a future war, he was hopeful that European intellectuals could lead a popular movement against militarism that would rob it of its glorification by Mussolini and other fascists. These remarks clearly indicate the contingent nature of Einstein's pacifism, which aimed to undercut fascism and pave the way for mutual security within Europe, a strategy that Einstein quickly rejected as untenable after 1933.

Letter to Jacques Hadamard, 24 September 1929
Einstein Archives 12-025; N & N 1960, 100–101

I was very glad to receive your letter, first because it came from you, and then because it displays the great earnestness with which you are considering the grave problems of Europe. I reply with some hesitation, because I am well aware that, when it comes to human affairs, my emotions are more decisive than my intellect. However, I shall dare to *justify* my position. But let me first make a qualification. I would not dare preach to a native African tribe in this fashion; for the patient there would have died long before the cure could have been of any help to him. But the situation in Europe is, despite Mussolini, quite different.

The first point I want to make is this: In a Europe which is systematically preparing for war, both morally and materially, an impotent League of Nations will not be able to command even moral authority in the hour of nationalist madness. The people in every country will insist that their own nation is the victim of aggression and will do so in perfectly good faith. . . . You cannot educate a nation for war and, at the same time, make its people believe that war is a shameful crime.

My second point: I admit that the country which decides not to defend itself assumes a great risk. However, this risk is accepted by society as a whole, and in the interest of human progress. Real progress has never been possible without sacrifices.

My third point: While the risks are great, they are not necessarily fatal. Since Germany, after four years of exhausting warfare, did not suffer more permanent damage than she actually has, a European country which does not even engage in war will certainly not suffer more than Germany actually did.

My fourth point: As long as nations systematically continue to prepare for war, fear, distrust and selfish ambitions will again lead to war.

My fifth point: We cannot afford to wait until the governing classes in the various countries decide voluntarily to accept interference with the sovereign power of their nation. Their lust for power will prevent them from doing so.

My sixth point: Public declarations by prominent personalities, who enjoy the respect of the man in the street, to the effect that their country should not engage in any warlike or even military action, will constitute an effective weapon against the war spirit.

My seventh point: To wage war means both to kill the innocent and to allow oneself to be innocently killed . . . How can any decent and self-respecting person participate in such a tragic affair? Would you perjure yourself if your government asked you to do so? Certainly not. How much worse, then, to slaughter innocent men?

To tell the truth, this last argument is, in my opinion, the strongest; at least, this is the way it affects me. As far as I am concerned, the welfare of humanity must take precedence over loyalty to one's own country—in fact, over anything and everything.

In October 1929 Europe lost one of its ablest spokesmen for peace with the death of Germany's foreign minister Gustav Stresemann. An annexationist during the war, Stresemann afterward founded the liberal and moderately nationalist German Peoples' Party. A man of enormous political gifts, Stresemann took his party kicking and screaming into the government, thereby forming a strong front on the right against the monarchists and antirepublicans in the vehemently nationalistic German National Peoples' Party.

The editors of *Nord und Süd* dedicated the entire November issue to tributes to Stresemann. In his eulogy, Einstein emphasized the difficulties Stresemann had faced in the postwar period in which a "massive war psychosis" was still very much alive. Recriminations persisted against dissenting wartime voices such as those of pacifist Friedrich Wilhelm Foerster and the former executive of the Krupp enterprise, Wilhelm Muehlon, who had exposed his firm's complicity in German war plans against France. In contrast to the venom directed at these men, Einstein noted that Leopold Graf Berchtold, the Austrian foreign minister, suffered no such opprobrium, though it was his ultimatum to Serbia after the assassination of Archduke Franz Ferdinand that led to the outbreak of the First World War.

STRESEMANN'S MISSION
NORD UND SÜD 52 (NOVEMBER 1929), NO. 11, 953–954

If we still accept the saying of the Greek sage that no man can be deemed happy before his death, we may now deem Stresemann a happy man. For it was given to him to live successfully in the service of a great idea and to die in that service—an idea whose realization the more progressive spirits of our generation can no longer question.

He was most happily equipped for the fulfillment of his great mission. He was a strong character hardened by a laborious struggle for existence in his youth, which his comparatively humble origin made inevitable. An atmosphere of struggle and conflict was natural and indeed necessary to him, even when his health had been undermined. But nature had endowed him as well with a fine sense for the beauty of language and literature that lent a more delicate quality to his gift of eloquence.

Indeed, his oratory, sustained as it was by his consciousness of a higher task and a certain wholesome optimism, cast a spell that was difficult to resist. With it came a capacity to grasp unfamiliar ways of thought and spheres of feeling that is seldom encountered in our public servants.

His greatest achievement, to my mind, came from his ability to convince a number of large political groups, against their own political instincts, to give their support to a comprehensive campaign of European reconciliation. His success in this matter depended on a subtle psychological phenomenon. Those men who possessed enough insight and courage to resist the massive war psychosis, tell their countrymen the truth, and confront them with the evil consequences of a war that was both intense and prolonged; these spokesmen were regarded with distrust and even hatred not only during the war but even afterward, down to the present day. On the other hand, those who, from indifference, weakness, or even criminal purpose, shared guilt and responsibility for the outbreak of the Great War stood completely absolved. No one casts a stone at Berchtold, but Foerster and Muehlon are hated men.

During the war, Stresemann was warlike enough that he was afterward able to retain the confidence of those whom he wanted to win over to his own wise and noble purposes. He had to contend against the resentment of men defeated in war whose pride had been wounded and who had been robbed of their traditional privileges; and, unlike his counterpart Briand, he did this without the backing of a victorious and exultant military.

Stresemann had one characteristic that is always found in great leaders. He did not exert his influence as the representative of a caste, a profession, or a nation, nor indeed of any type, but directly as a man of intellect and the bearer of an idea. He was as different from politicians of the usual stamp as a genius differs from an expert. Herein lay the magic and the strength of his personality.

Soon after Stresemann passed from the scene, the Catholic Center Party's leader, Heinrich Brüning, became chancellor. Brüning's measures aimed at resolving the financial crisis proved both unsuccessful and unpopular and led to the fall of his government. New elections, however, failed to produce a majority coalition in the Reichstag. President Paul Hindenburg appointed Brüning chancellor by decree, a provision of the Weimar Constitution that would later bring Hitler to power legally. In the meantime, the German army, which had been secretly rearming, began pressuring Brüning's government to seek concessions from the French side on armaments. In December 1930, Brüning told the American ambassador that "a complete understanding with France was the truly decisive goal for a pacification of the politics of the whole world" (Craig, 554).

After the Japanese invasion of Manchuria in September 1931, Einstein thought he saw a way to promote Brüning's stated goal of rapprochement with France while strengthening the League of Nations, which faced a major crisis in world security. In a letter to the German chancellor, he outlined a wide-sweeping plan that went far beyond the immediate problem of easing tensions between France and Germany. In substance, it echoed earlier proposals (see letter to Heinrich Zangger, 21 August 1917, in chapter 1), though Einstein's note of urgency clearly reflects the circumstances of the day.

LETTER TO CHANCELLOR HEINRICH BRÜNING, 3 OCTOBER 1931
FRANKFURTER ALLGEMEINE ZEITUNG, 27 DECEMBER 2002

... Mutual trust and cooperation between France and Germany can only come about if the French requirement of security against military attack is satisfied. If France, however, were to pose such demands, such a step would certainly be regarded with antipathy in Germany, so that an initiative must be taken from the German side.

A procedure like the following seems, however, to be possible. Let the German government of its own free will propose to the French that they should jointly make representations to the League of Nations that it should suggest to all member states to bind themselves to the following:

(1) to submit to every decision of the international court of arbitration;

(2) to proceed with all its economic and military force, in concert with the other members of the League, against any state which breaks the peace or resists an international decision made in the interests of world peace.

Brüning presumably learned about the contents of this letter because Einstein received assurances from his office that this missive would be passed on to him (*Frankfurter Allgemeine Zeitung*, 27 December 2002). Naturally he ignored Einstein's proposal, as it flew in the face of his efforts to achieve parity in armaments with the other leading European powers. Einstein's own assessment of the League of Nations remained grudging at best. As he remarked a year earlier, he was "rarely enthusiastic about what the League of Nations has done or has not done, but I am always thankful that it exists" (*New York Times*, 22 December 1930). By 1932, however, he came to believe that a far stronger international organization would have to be established if there were to be any chance of securing and maintaining world peace (see the section "Debacle in Geneva" in chapter 5).

Einstein and Freud on the Causes of War

In the autumn of 1931 Leon Steinig, an official of the League of Nations, traveled to Berlin in hopes of winning Einstein's support for one last project. The International Institute for Intellectual Cooperation planned to publish a volume of correspondence between leading intellectuals on an issue of major importance for mankind. Steinig, with the backing of the Institute's Director, Henri Bonnet, asked Einstein if he would agree to initiate such a project. Steinig found the physicist more than willing, as he hoped to achieve something of practical value from this undertaking. Writing to Bonnet, Steinig noted that Einstein had a "horror of platonic declarations which did not look forward to an immediately realizable end" (Clark, 363).

Initially, Einstein proposed contacting two men for whom he had the highest respect, Paul Langevin and Sigmund Freud, inviting an exchange of views on education. In the former case, he planned to discuss how educators in Germany and France should go about reforming textbooks on European history in order to remove tendentious errors that promoted national hostility. For the latter, Einstein thought to query Freud about how psychoanalytical theory might serve to promote peace by helping curb the aggressive impulses in children.

Langevin was in China at the time, which probably explains why the first half of this correspondence was never initiated. Steinig agreed to contact Freud, who wrote back on 6 June 1932 that "the words in which you express your hopes and those of Einstein for a future role of psychoanalysis in the life of individuals and nations ring true and of course give me very great pleasure" (Clark, 365). He agreed to participate, calling upon Einstein to suggest the issues and problems that would form the focus of their correspondence, which opened with the following letter. The version presented here follows the text in Einstein/Freud, which differs somewhat from that in N & N 1960.

LETTER TO SIGMUND FREUD, 30 JULY 1932
EINSTEIN/FREUD, 11–20; N & N 1960, 188–191

The proposal of the League of Nations and its International Institute of Intellectual Co-operation at Paris that I should invite a person, to be chosen by myself, to a frank exchange of views on any problem that I might select affords me a very welcome opportunity of conferring with you upon a question which, as things now are, seems the most insistent of all the problems civilisation has to face. This is the problem: Is there any way of delivering mankind from the menace of war? It is common knowledge that, with the advance of modern science, this issue has come to mean a matter of life and death for civilisation as we know it; nevertheless, for all the zeal displayed, every attempt at its solution has ended in a lamentable breakdown.

I believe, moreover, that those whose duty it is to tackle the problem professionally and practically are growing only too aware of their impotence to deal with it, and have now a very lively desire to learn the views of men who, absorbed in the pursuit of science, can see world problems in the perspective distance lends. As for me, the normal objective of my thought affords no insight into the dark places of human will and feeling. Thus, in the enquiry now proposed, I can do little more than to seek to clarify the question at issue and, clearing the ground of the more obvious solutions, enable you to bring the light of your far-reaching knowledge of man's instinctive life to bear upon the problem. There are certain psychological obstacles whose existence a layman in the mental sciences may dimly surmise, but whose interrelations and vagaries he is incompetent to fathom; you, I am convinced, will be able to suggest educative methods, lying more or less outside the scope of politics, which will eliminate these obstacles.

As one immune from nationalist bias, I personally see a simple way of dealing with the superficial (i.e., administrative) aspect of the problem: the setting up, by international consent, of a legislative and judicial body to settle every conflict arising between nations. Each nation would undertake to abide by the orders issued by this legislative body, to invoke its decision in every dispute, to accept its judgments unreservedly and to carry out every measure the tribunal deems necessary for the execution of its decrees. But here, at the outset, I come up against a difficulty; a tribunal is a human institution which, in proportion as the power at its disposal is inadequate to enforce its verdicts, is all the more prone to suffer these to be deflected by extrajudicial pressure. This is a fact with which we have to reckon; law and might inevitably go hand in hand, and juridical decisions approach more nearly the ideal justice demanded by the community (in whose name and interests these verdicts are pronounced) in so far as the community has effective power to compel respect of its juridical ideal. But at present we are far from possessing any supranational organisation competent to render verdicts of incontestable authority and enforce absolute submission to the execution of its verdicts. Thus I am led to my first axiom: The quest of international security involves the unconditional surrender by every nation, in a certain measure, of its liberty of action, its sovereignty that is to say, and it is clear beyond all doubt that no other road can lead to such security.

The ill-success, despite their obvious sincerity, of all the efforts made during the last decade to reach this goal leaves us no room to doubt that strong psychological factors are at work, which paralyse these efforts. Some of these factors are not far to seek. The craving for power which characterises the governing class in every nation is hostile to any limitation of the national sovereignty. This political power-hunger is wont to batten on the activities of another group, whose aspirations are on purely mercenary, economic lines. I have specially in mind that small but determined group, active in every nation, composed of individuals who, indifferent to social considerations and restraints, regard warfare, the manufacture and sale of arms, simply as an occasion to advance their personal interests and enlarge their personal authority.

But recognition of this obvious fact is merely the first step toward an appreciation of the actual state of affairs. Another question follows hard upon it: How is it possible for this small clique to bend the will of the majority, who stand to lose and suffer by a state of war, to the service of their ambitions? (In speaking of the majority, I do not exclude soldiers of every rank who have chosen war as their profession, in the belief that they are serving to defend the highest interests of their race, and that attack is often the best method of defence.) An obvious answer to this question would seem to be that the minority, the ruling class at present, has the schools and press, usually the Church as well, under its thumb. This enables it to organise and sway the emotions of the masses, and make its tool of them.

Yet even this answer does not provide a complete solution. Another question arises from it: How is it these devices succeed so well in rousing men to such wild enthusiasm, even to sacrifice their lives? Only one answer is possible. Because man has within him a lust for hatred and destruction. In normal times this passion exists in a latent state, it emerges only in unusual circumstances; but it is a comparatively easy task to call it into play and raise it to the power of a collective psychosis. Here lies, perhaps, the crux of all the complex of factors we are considering, an enigma that only the expert in the lore of human instincts can resolve.

And so we come to our last question. Is it possible to control man's mental evolution so as to make him proof against the psychoses of hate and destructiveness? Here I am thinking by no means only of the so-called uncultured masses. Experience proves that it is rather the so-called "Intelligentzia" that is most apt to yield to these disastrous collective suggestions, since the intellectual has no direct contact with life in the raw, but encounters it in its easiest, synthetic form—upon the printed page.

To conclude: I have so far been speaking only of wars between nations; what are known as international conflicts. But I am well aware that the aggressive instinct operates under other forms and in other circumstances. (I am thinking of civil wars, for instance, due in earlier days to religious zeal, but nowadays to social factors; or, again, the persecution of racial minorities.) But my insistence on what is the most typical, most cruel and extravagant form of conflict between man and man was deliberate, for here we have the best occasion of discovering ways and means to render all armed conflicts impossible.

I know that in your writings we may find answers, explicit or implied, to all the issues of this urgent and absorbing problem. But it would be of the greatest service to us all were you to present the problem of world peace in the light of your most recent discoveries, for such a presentation well might blaze the trail for new and fruitful modes of action.

Einstein and Freud had met at least once before in 1927 when the psychologist was visiting Berlin (Erikson, 167). It may have been on this occasion that Freud gave him a signed copy of *The Future of an Illusion* (Freud 1927), but Einstein was fairly familiar with Freud's other writings as well. In a diary entry from 6 December 1931, he wrote "I understand Jung's vague, imprecise notions, but I consider them worthless; a lot of talk without any clear direction. If there has to be a psychiatrist, I should prefer Freud. I do not believe in him, but I love very much his concise style and his original, although rather extravagant, mind." (N & N 1960, 185). Later, after a period of doubt, he warmed somewhat to Freud's ideas: "It is always a blessing when a great and beautiful conception is proven to be in harmony with reality" (letter to Freud, 21 April 1936).

In his response to the letter above, Freud expressed surprise. He had not expected that Einstein would ask him to address the problem of war confronting humanity. Still, he rose to the challenge, offering a lengthy, if hardly sanguine account and concluding that "there is no likelihood of our being able to suppress humanity's aggressive tendencies. In some happy corners of the earth, they say, where nature brings forth abundantly whatever man desires, there flourish races whose lives go gently by, unknowing of aggression or constraint. This I can hardly credit; I would like further details about these happy folk. The Bolshevists, too, aspire to do away with human aggressiveness by insuring the satisfaction of material needs and enforcing equality between man and man. To me this hope seems vain" (N & N 1960, 199).

Einstein was genuinely delighted by Freud's reflection on the psychological roots of warfare and thanked him profusely for his contribution.

LETTER TO SIGMUND FREUD, 3 DECEMBER 1932
EINSTEIN ARCHIVES 32-554; N & N 1960, 202

You have made a most gratifying gift to the League of Nations and myself with your truly classic reply. When I wrote you I was thoroughly convinced of the insignificance of my role, which was only meant to document my good will, with me as the bait on the hook to tempt the marvelous fish into nibbling. You have given in return something altogether magnificent. We cannot know what may grow from such seed, as the effect upon man of any action or event is always incalculable. This is not within our power and we do not need to worry about it.

You have earned my gratitude and the gratitude of all men for having devoted all your strength to the search for truth and for having shown the rarest courage in professing your convictions all your life. . . .

This now famous epistolary exchange—Einstein's final contribution to the intellectual efforts of the League of Nations—was published in a limited edition of 2,000 copies by the League of Nations in 1933 under the title *Why War?* (Einstein/Freud). Yet like so many other pacifist efforts from this period, Einstein's correspondence with Freud could do nothing to slow the juggernaut that was about to roll over the fragile democratic institutions of central Europe. Indeed, by the time the letters appeared in print, Hitler had already assumed power in Germany. Not surprisingly, the German edition was banned in Germany, where the Nazis did not even allow the booklet to be advertised.

Einstein had been relatively quiet about politics and world affairs during the late 1920s, but with the advent of the Great Depression and the parallel resurgence of the radical right in Germany he took up where he had left off during the early postwar years. Faced with the prospect of Hitler's virulently anti-Semitic brand of fascism, he spoke out vigorously and often, advocating a policy of civil disobedience in hopes of deterring the threat of escalating militarism. During these crisis years he became internationally known as a spokesman for militant pacifism, a movement that Einstein hoped might avert an impending disaster for Europe and the world.

CHAPTER 5

Articles of Faith, 1930–1933

This chapter begins with two essays that reveal a good deal about the articles of faith that constituted the essence of Einstein's personal philosophy of life. The first, "What I Believe," originally written in 1930 but subsequently reissued in slightly variant forms, did much to enhance his image as a modern-day sage. In the closing paragraph, Einstein deals with matters bearing on the creative life of the individual, offering some novel reflections on the common source of religious and scientific thought: mankind's sense of the mysterious.

American readers were particularly keen to learn more about this controversial topic, which Einstein subsequently elaborated on in his article "Religion and Science" for the *New York Times Magazine*. In tracing the evolutionary history of religious life through stages of primitive fear, submission to authority, and finally to what Einstein calls "cosmic religion," he followed the spirit, if not the substance, of Freud's *The Future of an Illusion* (Freud 1927), a book he knew well. For Einstein, both religion and science shared a common vital core, even if this emotional nexus was seldom appreciated due to the still largely dogmatic character of religious experience (Holton 2002). Not surprisingly, religious conservatives in the United States found these views highly offensive.

Einstein's schedule of political activities intensified during the early 1930s, when he spent three consecutive winters in southern California as Robert Millikan's guest at the California Institute of

Technology. This activity abroad took place just as political conditions in Germany were rapidly deteriorating. Whereas Einstein had downplayed the dangers of Nazi-style fascism in 1927 (see "Interview in *Neue Zürcher Zeitung*," chapter 4), in 1930 he and others on the left were deeply troubled by the sudden ascent of Hitler's Brown Shirts. In the September 1930 election, the National Socialists captured 18.3 percent of the vote and 107 seats in the Reichstag (up from just 12 in 1928).

Shortly afterward, when he returned to the United States for the first time since 1921, Einstein appealed for American help to defeat the fascist specter that was haunting Europe. During this trip abroad he delivered a number of speeches in support of militant pacifism, the position he proclaimed almost from the moment he set foot on American soil. His famous "Two Percent Speech" in New York, advocating civil disobedience in order to paralyze compulsory military service, touched off a maelstrom of controversy that made Einstein a hero among Anglo-American pacifists.

Einstein's three stays in the United States during the period 1930–1933 provided him with ample opportunity to speak out on a variety of political themes from German-American relations to the need for a stronger American role in disarmament negotiations. Along with many others, he closely followed the deliberations leading up to the General Disarmament Conference scheduled to open in Geneva in February 1932. Although far from sanguine about the chances for substantive progress, Einstein felt that much would depend on the role played by the American representatives. When he sensed already in May 1932 that the undertaking was doomed to failure, he joined a small group of peace advocates who held a press conference outside the Geneva conference center in order to denounce the intransigence of the negotiating parties. Einstein characterized the talks as a futile attempt to establish a set of rules for future warfare. "War cannot be humanized," he emphatically proclaimed. "It can only be abolished."

By this time, Einstein had considerable experience dealing with the press, including political journalists who were intent on advancing their own viewpoints. One of these was the Hungarian-born political publicist Imre Révész, who later changed his name to Emery Reves after fleeing the continent of Europe in 1940 and becoming a British citizen. Einstein met him soon after he began the Cooperation Press Service, which distributed syndicated articles by leading political figures in foreign newspapers around the world. After a brief start as a freelance journalist, Révész began this operation in 1930 out of a small office in Berlin after persuading three experts to write monthly articles on current political affairs as seen from their own nation's perspective. These three were Lord Robert Cecil, one of the strongest British supporters of the League of Nations; Louis Loucheur, a kingpin of French politics who had been in charge of German reparations; and the economist Georg Bernhard, editor of Berlin's influential *Vossische Zeitung*. One year later, Albert Einstein joined this elite group of political commentators who wrote occasional articles for Cooperation Press Service. Two of the texts in this chapter—the article on the forthcoming Geneva Disarmament Conference from November 1931 and Einstein's "Statement on the Herriot Plan"—were written at the suggestion of Révész.

During his early visits to the United States, Einstein was overwhelmed by the adulation he received, so much so that he found the American reaction to the "Einstein craze" simply baffling (see chapter 2). Still, he was well aware of deep pockets of resistance to his ideas, both within traditional religious circles but also among self-proclaimed patriots. After learning that a group of such women patriots wanted to bar him from entering the United States, Einstein issued a satirical "Reply to the Women of America" in which he tried to make light of this new form of "Einstein hysteria."

For Caltech's Robert Millikan, this was precisely the kind of publicity he hoped to avoid. Having been severely criticized for hosting a pacifist agitator in past years, the conservative-minded Millikan feared that Einstein's forthcoming visit in the winter of 1932–33 might stir more such controversy. For this visit he had obtained funding from the Oberländer Trust of Philadelphia, an organization that aimed to promote German-American relations. The agreement called for Einstein to deliver a speech on this subject during the course of his stay. Speaking only a week before Hitler was appointed chancellor of Germany, he condemned the role of rhetoric and ideology as obstacles to reaching understanding (*Verständigung*), out of which alone might arise "the will, the readiness, and the ability peacefully to solve ... questions, which are of far-reaching importance in the life of nations."

The Moral Foundations

Einstein conjoined politics and ethics on numerous occasions, but never more poignantly than in "What I Believe." This essay was first published in 1930 as one in a series of "Living Philosophies" in the journal *Forum and Century*. The editors expressed the hope that presenting the "innermost beliefs" of "a man who has always withdrawn himself from social contacts ... may serve to make him somewhat less of a public myth."

The text was reproduced several times under the title "The World as I See It," most notably in *Mein Weltbild* and *Ideas and Opinions*, and in 1932 the German League of Human Rights released a phonograph recording of Einstein reading a slightly variant version entitled "Confession of Belief." The version presented here follows the text in *Forum and Century*, which differs significantly from that in Einstein 1954.

WHAT I BELIEVE
FORUM AND CENTURY 84 (OCTOBER 1930), NO. 4, 193–194;
EINSTEIN 1954, 8–11

Strange is our situation here upon earth. Each of us comes for a short visit, not knowing why, yet sometimes seeming to divine a purpose.

From the standpoint of daily life, however, there is one thing we do know: that man is here for the sake of other men—above all for those upon whose smiles and well-being our own happiness depends, and also for the countless unknown souls with whose fate we are connected by a bond of sympathy. Many times a day I realize how much my own outer and inner life is built upon the labors of my fellow men, both living and dead, and how earnestly I must exert myself in order to give in return as much as I have received. My peace of mind is often troubled by the depressing sense that I have borrowed too heavily from the work of other men.

I do not believe we can have any freedom at all in the philosophical sense, for we act not only under external compulsion but also by inner necessity. Schopenhauer's saying—"A man can surely do what he wills to do, but he cannot determine what he wills"—impressed itself upon me in youth and has always consoled me when I have witnessed or suffered life's hardships. This conviction is a perpetual breeder of tolerance, for it does not allow us to take ourselves or others too seriously; it makes rather for a sense of humor.

To ponder interminably over the reason for one's own existence or the meaning of life in general seems to me, from an objective point of view, to be sheer folly. And yet everyone holds certain ideals by which he guides his aspiration and his judgment. The ideals which have always shone before me and filled me with the joy of living are goodness, beauty, and truth. To make a goal of comfort or happiness has never appealed to me; a system of ethics built on this basis would be sufficient only for a herd of cattle.

Without the sense of collaborating with like-minded beings in the pursuit of the ever unattainable in art and scientific research, my life would have been empty. Ever since childhood I have scorned the commonplace limits so often set upon human ambition. Possessions, outward success, publicity, luxury—to me these have always been contemptible. I believe that a simple and unassuming manner of life is best for everyone, best both for the body and the mind.

My passionate interest in social justice and social responsibility has always stood in curious contrast to a marked lack of desire for direct association with men and women. I am a horse for single harness, not cut out for tandem or team work. I have never belonged wholeheartedly to country or state, to my circle of friends, or even to my own family. These ties have always been accompanied by a vague aloofness, and the wish to withdraw into myself increases with the years.

Such isolation is sometimes bitter, but I do not regret being cut off from the understanding and sympathy of other men. I lose something by it, to be sure, but I am compensated for it in being rendered independent of the customs, opinions, and prejudices of others, and am not tempted to rest my peace of mind upon such shifting foundations.

My political ideal is democracy. Everyone should be respected as an individual, but no one idolized. It is an irony of fate that I should have been showered with so much uncalled-for and unmerited admiration and esteem. Perhaps this adulation springs from the unfulfilled wish of the multitude to comprehend the few ideas which I, with my weak powers, have advanced.

Full well do I know that in order to attain any definite goal it is imperative that *one* person should do the thinking and commanding and carry most of the responsibility. But those who are led should not be driven, and they should be allowed to choose their leader. It seems to me that the distinctions separating the social classes are false; in the last analysis they rest on force. I am convinced that degeneracy follows every autocratic system of violence, for violence inevitably attracts moral inferiors. Time has proved that illustrious tyrants are succeeded by scoundrels.

For this reason I have always been passionately opposed to such regimes as exist in Russia and Italy today. The thing which has discredited the European forms of democracy is not the basic theory of democracy itself, which some say is at fault, but the instability of our political leadership, as well as the impersonal character of party alignments.

I believe that you in the United States have hit upon the right idea. You choose a President for a reasonable length of time and give him enough power to acquit himself properly of his responsibilities. In the German Government, on the other hand, I like the state's more extensive care of the individual when he is ill or unemployed. What is truly valuable in our bustle of life is not the nation, I should say, but the creative and impressionable individuality, the personality—he who produces the noble and sublime while the common herd remains dull in thought and insensible in feeling.

This subject brings me to that vilest offspring of the herd mind—the odious militia. The man who enjoys marching in line and file to the strains of music falls below my contempt; he received his great brain by mistake—the spinal cord would have been amply sufficient. This heroism at command, this senseless violence, this accursed bombast of patriotism—how intensely I despise them! War is low and despicable, and I had rather be smitten to shreds [*sic*] than participate in such doings.

Such a stain on humanity should be erased without delay. I think well enough of human nature to believe that it would have been wiped out long ago had not the common sense of nations been systematically corrupted through school and press for business and political reasons.

The most beautiful thing we can experience is the mysterious. It is the source of all true art and science. He to whom this emotion is a stranger, who can no longer pause to wonder and stand rapt in awe, is as good as dead: his eyes are closed. This insight into the mystery of life, coupled though it be with fear, has also given rise to religion. To know that what is impenetrable to us really exists, manifesting itself as the highest wisdom and the most radiant beauty which our dull faculties can comprehend only in their most primitive forms—this knowledge, this feeling, is at the center of true religiousness. In this sense, and in this sense only, I belong in the ranks of devoutly religious men.

I cannot imagine a God who rewards and punishes the objects of his creation, whose purposes are modeled after our own—a

God, in short, who is but a reflection of human frailty. Neither can I believe that the individual survives the death of his body, although feeble souls harbor such thoughts through fear or ridiculous egotism. It is enough for me to contemplate the mystery of conscious life perpetuating itself through all eternity, to reflect upon the marvelous structure of the universe which we can dimly perceive, and to try humbly to comprehend even an infinitesimal part of the intelligence manifested in nature.

Shortly before he left Berlin for the United States, Einstein was visited by two foreigners, the writer James Murphy and J.W.N. Sullivan, a mathematician and science journalist. Both had come to Berlin to meet leading German scientists, and their encounter with Einstein led to an intriguing exchange of opinions published under the title "Science and God: A German Dialogue" in *The Forum* (Einstein/Murphy/Sullivan). Among the numerous topics they discussed was the problem of suffering and its relationship to religion. Einstein, a great admirer of *The Brothers Karamazov*, disputed the claim that Dostoevsky was concerned with the problem of suffering. Dostoevsky's aim, he thought, was to present us with the mystery of spiritual existence and not to make this into a problem. Asked by Murphy how far he thought modern science might be able to go toward establishing practical ideals of life on the ruins of religious ideals, Einstein replied: "I do not believe that a moral philosophy can ever be founded on a scientific basis. . . . The valuation of life and all its nobler expressions can only come out of the soul's yearning toward its own destiny. Every attempt to reduce ethics to scientific formulas must fail. Of that I am perfectly convinced. On the other hand, it is undoubtedly true that scientific study of the higher kinds and general interest in scientific theory have great value in leading men toward a worthier valuation of the things of the spirit" (Einstein/Murphy/Sullivan, 374).

Around this same time, a New York publisher issued a small selection of Einstein's writings under the title *Cosmic Religion with Other Opinions and Aphorisms* (Einstein 1931b). The title was adopted from a term introduced in his 1931 essay "Religion and Science," and afterward became a synonym for his nonanthropomorphic views on religion.

RELIGION AND SCIENCE
NEW YORK TIMES MAGAZINE, 9 NOVEMBER 1930; EINSTEIN 1931B, 43–54

Everything that men do or think concerns the satisfaction of the needs they feel or the escape from pain. This must be kept in mind when we seek to understand spiritual or intellectual movements and the way in which they develop. For feeling and longing are the motive forces of all human striving and productivity—however nobly these latter may display themselves to us.

What, then, are the feelings and the needs which have brought mankind to religious thought and to faith in the widest sense? A moment's consideration shows that the most varied emotions stand at the cradle of religious thought and experience.

In primitive peoples it is, first of all, fear that awakens religious ideas—fear of hunger, of wild animals, of illness and of death. Since the understanding of causal connections is usually limited on this level of existence, the human soul forges a being, more or less like itself, on whose will and activities depend the experiences which it fears. One hopes to win the favor of this being by deeds and sacrifices, which, according to the tradition of the race, are supposed to appease the being or to make him well disposed to man. I call this the religion of fear.

This religion is considerably stabilized—though not caused—by the formation of a priestly caste which claims to mediate between the people and the being they fear, and so attains a position of power. Often a leader or despot, or a privileged class whose power is maintained in other ways, will combine the function of the priesthood with its own temporal rule for the sake of greater security; or an alliance may exist between the interests of the political power and the priestly caste.

A second source of religious development is found in the social feelings. Fathers and mothers, as well as leaders of great human communities, are fallible and mortal. The longing for guidance, for love and succor, provides the stimulus for the growth of a social or moral conception of God. This is the God of Providence, who protects, decides, rewards, and punishes. This is the God who, according to man's widening horizon, loves and provides for the life of the race, or of mankind, or who even loves life itself. He is the comforter in unhappiness and in unsatisfied longing, the protector of the souls of the dead. This is the social or moral idea of God.

It is easy to follow in the sacred writings of the Jewish people the development of the religion of fear into the moral religion, which is carried further in the New Testament. The religions of all the civilized peoples, especially those of the Orient, are principally moral religions. An important advance in the life of a people is the transformation of the religion of fear into the moral religion. But one must avoid the prejudice that regards the religions of primitive peoples as pure fear religions and those of the civilized races as pure moral religions. All are mixed forms, though the moral element predominates in the higher levels of social life. Common to all these types is the anthropomorphic character of the idea of God.

Only exceptionally gifted individuals or especially noble communities rise *essentially* above this level; in these there is found a third level of religious experience, even if it is seldom found in a pure form. I will call it the cosmic religious sense. This is hard to make clear to those who do not experience it, since it does not involve an anthropomorphic idea of God: the individual feels the vanity of human desires and aims, and the nobility and marvelous order which are revealed in nature and in the world of thought. He feels the individual destiny as an imprisonment and seeks to experience the totality of existence as a unity full of significance. Indications of this cosmic religious sense can be found even on earlier levels of development—for example, in the Psalms of

David and in the Prophets. The cosmic element is much stronger in Buddhism, as, in particular, Schopenhauer's magnificent essays have shown us.

The religious geniuses of all times have been distinguished by this cosmic religious sense, which recognizes neither dogmas nor God made in man's image. Consequently there cannot be a church whose chief doctrines are based on the cosmic religious experience. It comes about, therefore, that we find precisely among the heretics of all ages men who were inspired by this highest religious experience; often they appeared to their contemporaries as atheists, but sometimes also as saints. Viewed from this angle, men like Democritus, Francis of Assisi, and Spinoza are near to one another.

How can this cosmic religious experience be communicated from man to man, if it cannot lead to a definite conception of God or to a theology? It seems to me that the most important function of art and of science is to arouse and keep alive this feeling in those who are receptive.

Thus we reach an interpretation of the relation of science to religion which is very different from the customary view. From the study of history, one is inclined to regard religion and science as irreconcilable antagonists, and this for a reason that is very easily seen. For any one who is pervaded with the sense of causal law in all that happens, who accepts in real earnest the assumption of causality, the idea of a Being who interferes with the sequence of events in the world is absolutely impossible. Neither the religion of fear nor the social-moral religion can have any hold on him. A God who rewards and punishes is for him unthinkable, because man acts in accordance with an inner and outer necessity, and would, in the eyes of God, be as little responsible as an inanimate object is for the movements which it makes.

Science, in consequence, has been accused of undermining morals—but wrongly. The ethical behavior of man is better based on sympathy, education, and social relationships, and requires no support from religion. Man's plight would, indeed, be sad if he had to be kept in order through fear of punishment and hope of rewards after death.

It is, therefore, quite natural that the churches have always fought against science and have persecuted its supporters. But, on the other hand, I assert that the cosmic religious experience is the strongest and the noblest driving force behind scientific research. No one who does not appreciate the terrific exertions, and, above all, the devotion without which pioneer creations in scientific thought cannot come into being, can judge the strength of the feeling out of which alone such work, turned away as it is from immediate practical life, can grow. What a deep faith in the rationality of the structure of the world and what a longing to understand even a small glimpse of the reason revealed in the world there must have been in Kepler and Newton to enable them to unravel the mechanism of the heavens, in the long years of lonely work!

Anyone who only knows scientific research in its practical applications may easily come to a wrong interpretation of the state of mind of the men who, surrounded by skeptical contemporaries, have shown the way to kindred spirits scattered over all countries in all centuries. Only those who have dedicated their lives to similar ends can have a living conception of the inspiration which gave these men the power to remain loyal to their purpose in spite of countless failures. It is the cosmic religious sense which grants this power.

A contemporary has rightly said that the only deeply religious people of our largely materialistic age are the earnest men of research.

In its commentary on the article, the *New York Times* adopted a fairly neutral stance on this new-fangled cosmic religion, but other authorities could not well afford this luxury. Rabbi Nathan Krass of midtown Manhattan's Temple Emanuel wrote that "the religion of Albert Einstein will not be approved by certain sectarians but it must and will be approved by the Jews" (Clark, 426). Speaking before 1,200 members of the Catholic Teachers Association, Dr. Fulton Sheen countered that the *Times* had degraded itself by publishing "the sheerest kind of stupidity and nonsense" (ibid.).

For Sheen, as for many other American critics, Einstein's deleterious influence had less to do with politics or "cultural relativism" than with the scientist's heretical views on religious matters.

Two Articles of Political Faith: Democracy and Militant Pacifism

If Einstein affirmed his faith in democracy and his adamant rejection of militarism in "What I Believe," it remained unclear whether these commitments carried any concrete implications for political activism. In later years, he held up Gandhi as the supreme embodiment of a political figure guided by a higher moral vision, but in 1930 he was particularly impressed by the case of Rosika Schwimmer, who the previous year had been denied American citizenship because of her refusal to bear arms in defense of the Constitution of the United States. She appealed, but her case was turned down by the Supreme Court. Einstein learned about it the following year and wrote to Frida Perlen, chairperson of the Stuttgart chapter of the oldest international women's peace organization, to commend Schwimmer for her courage.

LETTER TO THE WOMEN'S INTERNATIONAL LEAGUE FOR
PEACE AND FREEDOM, 3 JULY 1930
EINSTEIN ARCHIVES 46-804; EINSTEIN 1933, 24–25

I consider Madame Schwimmer's stand of great value and deserving the support of all true humanitarians.

Governments represent the people, and the people are still under the spell of obsolete traditions of military duty. World peace, a crying necessity, is not to be achieved unless spiritually progressive forces refuse to yield to public authorities, controlled by factions which should be defeated. Those convinced of this necessity should consider it their duty publicly to uphold this conviction; thereby they bring upon themselves a conflict with public authority. A result can be obtained only if a large number of influential personalities have the moral courage for such an attitude.

Such an attitude is revolutionary. But only through acts of rebellion can the fettered individual break chains which, though forged in laws, have grown into unendurable bondage. In this situation, too, such a recourse is unavoidable.

Credit is due Madame Schwimmer for having realized this and having acted courageously in accordance with her convictions.

Schwimmer's activism may well have influenced Einstein to speak out six months later as a forceful advocate of militant pacifism. In fact, she served as interpreter when he delivered his so-called "Two Percent Speech" in New York in December 1930 (see below). Two months earlier, Einstein had joined an illustrious group of intellectuals who signed a pacifist appeal issued by the Joint Peace Council, which brought together such groups as the Quakers, the War Resisters' International, and the Women's International League for Peace and Freedom. The signers of the Manifesto of the Joint Peace Council appealed to the nations that had joined the 1928 Kellogg-Briand Pact repudiating war as an instrument of policy to end once and for all the "slavery" of conscription and the accompanying "education of the mind and body in the technique of killing" (N & N 1960, 113–114).

Technical education was on Einstein's mind when he stepped to the microphone at the opening ceremony of the German Radio Exhibition in August 1930. Listeners throughout Berlin heard him speak on this occasion about the promise of radio for promoting democracy and mutual understanding among nations. Throughout his life Einstein took an avid interest in technological innovations, which he generally saw as a boon to human progress. On this occasion he underscored the importance of the scientific knowledge that made radio technology possible, sprinkling a touch of scorn on those who failed to appreciate this interdependence. Greeted by stirring applause, the celebrated scientist began with an amusing twist by addressing the "Verehrte An- und Abwesende," thereby deftly acknowledging his two audiences—those seated before him and those listening at home.

Radio's Challenge to Promote International Understanding, 22 August 1930
Einstein 2003, CD1, no. 3

Honored Listeners, Present and Absent!

When you listen to the radio, don't forget how humanity came to possess this wonderful means of communication. The source of all technical achievements is the divine curiosity and playfulness of the tinkering and reflective researcher and no less the constructive fantasy of the technical inventor. . . .

Remember thankfully the army of nameless technicians who simplified the instruments of radio communication and adapted them to mass production so that they became accessible to everybody. Anyone who uses them without further reflection about the wonders of science and technology should be ashamed of themselves for having no more conception of these than a cow has about the botany of the plants that it consumes with such relish.

Remember also that it is technical experts who first make true democracy possible. Not only do they ease our daily tasks, they also make accessible to everyone the works of highest art and intellect, enjoyment of which was until recently the privilege of the favored classes. In doing so, they have awakened people from their sluggish torpor.

In particular, radio has a unique function to fulfill in promoting international understanding. Until the present day people only got to know each other through the distorting mirror of their own daily press. Radio reveals them to one another in most vivid form and, for the most part, from an ingratiating side. Thus it contributes to eradicating feelings of mutual alienation, which can so easily turn into mistrust and enmity. Bearing this in mind, view for yourselves the fruits of the creative work put on display at this exhibit for the amazement of its visitors.

Three months later, the Einsteins departed Europe for the United States, arriving in New York aboard the S.S. *Belgenland*. During a fifteen-minute interview, which he granted to the crush of reporters who greeted him and which he compared to a Punch and Judy show (see Plate 12), he made his only overtly political statement when asked what he thought of Hitler: "[He] is living on the empty stomach of Germany. As soon as economic conditions in Germany improve he will cease to be important" (*New York Times*, 12 December 1930). Afterward, Einstein broadcast a greeting and challenge from shipside to the American people that would remain a central theme during the course of his visit.

Defeating the Specter of Militarism, 11 December 1930
Einstein Archives 36-306; N & N 1960, 115–116

As I am about to set foot again on United States soil, after an absence of ten years, the thought uppermost in my mind is that this country, today the most influential on earth because of the peaceful work that is being done here, constitutes a bulwark of the democratic way of life. Here, everyone stands up proudly and jealously for his civil rights. Everyone, irrespective of birth, has the opportunity, not merely on paper but in actual practice, to develop his energies freely for the benefit of the community as a whole.

Your country has demonstrated, by the work of its heads and its hands, that individual freedom provides a better basis for productive labor than any form of tyranny; whenever men are inspired by a healthy pride in their society, they feel a responsibility to make sacrifices for its sake. The sense of solidarity one finds in the American people is the basis for your firm belief in a great international community of all nations and cultures. The belief has led to the creation of cultural institutions whose blessings extend to all countries of the world.

Your country possesses the power to help defeat the specter that menaces our age: militarism. Your political and economic position is so powerful today that, if you are serious in your endeavors, you can break the tradition of war from which Europe has suffered throughout its history and from which the rest of the world has also suffered, even if to a lesser degree. Destiny has placed this historic mission in your hands. By fulfilling that mission you will build an enduring monument to your country and your generation.

Inspired by these hopes, I salute you and the soil of your country. I eagerly look forward to renewing old friendships and to broadening my understanding in the light of what I shall see and learn while I am among you.

Just three days later, at a meeting of the New History Society held at the Ritz-Carlton Hotel, Einstein extemporaneously delivered his famous "Two Percent Speech," in which he advocated a militant brand of pacifism. The New History Society, to whom the address was directed, was an offshoot of the pacifist Baha'i religious movement.

When the woman who had been called upon to translate the speech broke down after the first sentence, Rosika Schwimmer, the well-known Hungarian-born pacifist, volunteered to continue, for which a grateful Einstein dubbed her his "saving angel" (*New York Times*,

15 December 1930). The speech's call to action created a minor sensation at the time as well as certain misgivings among some leading pacifists.

THE TWO PERCENT SPEECH, 14 DECEMBER 1930
EINSTEIN 1933, 34–37

When those who are bound together by pacifist ideals hold a meeting they are always consorting with their own kind only. They are like sheep huddled together while the wolves wait outside. I think pacifist speakers have this difficulty: they usually reach their own crowd, who are pacifists already. The sheep's voice does not get beyond this circle and therefore is ineffective. That is the trouble with the pacifist movement.

Real pacifists, those who are not up in the clouds but who think and count realities, must fearlessly try to do things of practical value to the cause and not merely speak about pacifism. Deeds are needed. Mere words do not get pacifists anywhere. They must initiate action and begin with what can be done at once.

As to what our next step should be, I should like you to realize that under our present system of military duty everyone is compelled to commit a crime—the crime of killing people for his country. The aim of all pacifists must be to convince others of the immorality of war and rid the world of the shameful slavery of military service. I wish to suggest two ways to achieve that aim.

The first has already been put into practice. It is uncompromising war resistance, refusal to do military service under any circumstances. In countries where conscription is established, the real pacifist must refuse military duty. A large number of pacifists in many countries are refusing at great personal sacrifice to serve their military term in peace-time. By doing so, they indicate that they will not fight if there should be war.

In countries where compulsory service does not obtain, real pacifists must in time of peace publicly declare that they will not take up arms under any circumstances. This, too, is an effective way of announcing one's war resistance. I earnestly advise recruiting people with this idea all over the world. For the timid who say, "What is the use? We might be shut up in prison," I add: even if only two percent of those supposed to perform military service should declare themselves war resisters and assert, "We are not going to fight. We need other methods of settling international disputes," the governments would be powerless—they could not put such masses into jail.

As a second line of action for war resisters I suggest something which appears to be less illegal. That is, to try to establish through international legislation the right to refuse military service in peace-time. Those who are unwilling to accept the obligation might advocate legislation which would permit them to do some strenuous or even dangerous work, each for his country or for mankind, in place of military service, to prove that their war resistance is unselfish—a consequence of their belief that international differences can be settled other than by fighting; to prove that they do not oppose war for their personal comfort or because of cowardice or because they do not want to serve their country or humanity. If we take upon ourselves such dangerous occupations we shall be advancing far on the road to the pacification of the world.

I am convinced that such legislation is possible. I suggest that your organization take up this idea at your coming meetings. I think the League of Nations could be induced to attempt the international legalization of refusal of military service in time of peace.

I further suggest that pacifists of every country start collecting funds to support those who want to refuse military service but who cannot carry out their resistance for lack of financial means. I advocate an international organization and an international pacifist fund to support the active war resisters of our day.

My final advice is that those who want to accomplish these aims must have the courage to initiate, to carry on; then the world will have to take notice. They will be heard, and if they are heard they will be effective. If they are less loud their voices will continue to reach only their own kind. They will remain sheep, pacifist sheep.

While the evocative imagery of helpless sheep proved highly effective, some felt Einstein had minimized the threat of slaughter that might come to those in the flock who strayed from the path of legal protest. Romain Rolland wrote Einstein earlier that people needed to be told that the refusal to serve in their nation's armed forces might involve great sacrifice and even martyrdom (12 October 1930, N & N 1960, 112). Rolland later told H. Runham Brown, Honorary Secretary of the War Resisters' International, that "war will by no means be abolished if 'two per cent of the population of the world refuse to fight.' Einstein seems to overlook the fact that the technique of war has changed since 1914, and is still changing. The tendency has been to employ small armies of technicians who know how to run air squadrons armed with gas and bacteriological torpedoes and other weapons of mass destruction" (20 February 1931, N & N 1960, 118–119).

Undoubtedly recalling the problems caused in 1921 when his comments about American life and culture were reported in the press (see chapter 2), Einstein produced a far more flattering account following his return trip in 1930–31. For reasons that are unclear, however, the following text apparently first surfaced in the 1934 edition of *Mein Weltbild*. Ironically, there and in Einstein 1954, it was conflated with the 1921 interview in *Nieuwe Rotterdamsche Courant* that appears in chapter 2, making this into a "thrice-told story."

IMPRESSIONS OF THE U.S.A., CA. 1931
EINSTEIN ARCHIVES 28-168; EINSTEIN 1954, 3–7

I must redeem my promise to say something about my impressions of this country. That is not altogether easy for me. For it is not easy to take up the attitude of impartial observer when one is received with such kindness and undeserved respect as I have been in America. First of all let me say something on this score.

The cult of individuals is always, in my view, unjustified. To be sure, nature distributes her gifts unevenly among her children. But there are plenty of the well-endowed, thank God, and I am firmly convinced that most of them live quiet, unobtrusive lives. It strikes me as unfair, and even in bad taste, to select a few of them for boundless admiration, attributing superhuman powers of mind and character to them. This has been my fate, and the contrast between the popular estimate of my powers and achievements and the reality is simply grotesque. The awareness of this strange state of affairs would be unbearable but for one pleasing consolation: it is a welcome symptom in an age which is commonly denounced as materialistic that it makes heroes of men whose goals lie wholly in the intellectual and moral sphere. This proves that knowledge and justice are ranked above wealth and power by a large section of the human race. My experience teaches me that this idealistic outlook is particularly prevalent in America, which is decried as a singularly materialistic country. After this digression I come to my proper theme, in the hope that no more weight will be attached to my modest remarks than they deserve.

What first strikes the visitor with amazement is the superiority of this country in matters of technology and organization. Objects of everyday use are more solid than in Europe, houses much more practically designed. Everything is designed to save human labor. Labor is expensive, because the country is sparsely inhabited in comparison with its natural resources. The high price of labor was the stimulus which evoked the marvelous development of technical devices and methods of work. The opposite extreme is illustrated by over-populated China or India, where the low price of labor has stood in the way of the development of machinery. Europe is halfway between the two. Once the machine is sufficiently highly developed it becomes cheaper in the end than the cheapest labor. Let the Fascists in Europe, who desire on narrow-minded political grounds to see their own particular countries more densely populated, take heed of this. However, the anxious care with which the United States keep out foreign goods by means of prohibitive tariffs certainly contrasts oddly with the general picture . . . but an innocent visitor must not be expected to rack his brains too much, and when all is said and done, it is not absolutely certain that every question admits of a rational answer.

The second thing that strikes a visitor is the joyous, positive attitude to life. The smile on the faces of the people in photographs is symbolical of one of the greatest assets of the American. He is friendly, self-confident, optimistic—and without envy. The European finds intercourse with Americans easy and agreeable.

Compared with the American the European is more critical, more self-conscious, less kind-hearted and helpful, more isolated, more fastidious in his amusements and his reading, generally more or less of a pessimist.

Great importance attaches to the material comforts of life, and equanimity, unconcern, security are all sacrificed to them. The American lives even more for his goals, for the future, than the European. Life for him is always becoming, never being. In this respect he is even further removed from the Russian and the Asiatic than the European is.

But there is one respect in which he resembles the Asiatic more than the European does: he is less of an individualist than the European—that is, from the psychological, not the economic, point of view.

More emphasis is laid on the "we" than the "I." As a natural corollary of this, custom and convention are extremely strong, and there is much more uniformity both in outlook on life and in moral and esthetic ideas among Americans than among Europeans. This fact is chiefly responsible for America's economic superiority over Europe. Cooperation and the division of labor develop more easily and with less friction than in Europe, whether in the factory or the university or in private charity. This social sense may be partly due to the English tradition.

In apparent contradiction to this stands the fact that the activities of the State are relatively restricted as compared with those in Europe. The European is surprised to find the telegraph, the telephone, the railways, and the schools predominantly in private hands. The more social attitude of the individual, which I mentioned just now, makes this possible here. Another consequence of this attitude is that the extremely unequal distribution of property leads to no intolerable hardships. The social conscience of the well-to-do is much more highly developed than in Europe. He considers himself obliged as a matter of course to place a large portion of his wealth, and often of his own energies, too, at the disposal of the community; public opinion, that all-powerful force, imperiously demands it of him. Hence the most important cultural functions can be left to private enterprise and the part played by the government in this country is, comparatively, a very restricted one.

The prestige of government has undoubtedly been lowered considerably by the Prohibition law. For nothing is more destructive of respect for the government and the law of the land than passing laws which cannot be enforced. It is an open secret that the dangerous increase of crime in this country is closely connected with this.

There is also another way in which Prohibition, in my opinion, undermines the authority of the government. The public house is a place which gives people the opportunity to exchange views and ideas on public affairs. As far as I can see, such an opportunity is lacking in this country, the result being that the Press, which is mostly controlled by vested interests, has an excessive influence on public opinion.

The overestimation of money is still greater in this country than in Europe, but appears to me to be on the decrease. It is at last beginning to be realized that great wealth is not necessary for a happy and satisfactory life.

In regard to artistic matters, I have been genuinely impressed by the good taste displayed in the modern buildings and in articles of common use; on the other hand, the visual arts and music have little place in the life of the nation as compared with Europe.

I have a warm admiration for the achievements of American institutes of scientific research. We are unjust in attempting to ascribe the increasing superiority of American research work exclusively to superior wealth; devotion, patience, a spirit of comradeship, and a talent for cooperation play an important part in its successes.

One more observation, to finish. The United States is the most powerful among the technically advanced countries in the world today. Its influence on the shaping of international relations is absolutely incalculable. But America is a large country and its people have so far not shown much interest in great international problems, among which the problem of disarmament occupies first place today. This must be changed, if only in America's own interest. The last war has shown that there are no longer any barriers between the continents and that the destinies of all countries are closely interwoven. The people of this country must realize that they have a great responsibility in the sphere of international politics. The part of passive spectator is unworthy of this country and is bound in the end to lead to disaster all round.

Although he touched upon the visual arts, Einstein neglected to mention the cinema in his reflections, an omission of some significance. In Pasadena he met Upton Sinclair, who took him to see Sergei Eisenstein's *Que Viva Mexico*. He found the bullfighting scenes revolting, but otherwise enjoyed Sinclair's company, noting in his diary that the latter had a "favorable opinion of Russia because . . . they educate the masses and bring them to life" (N & N 1960, 120). He and Elsa were also invited by Charlie Chaplin to view the premier showing of

Chaplin's *City Lights* in Hollywood, where they had earlier taken in *All Quiet on the Western Front*. The film version of Erich Maria Remarque's best-selling pacifist novel had been censored in Germany, where the German League of Human Rights conducted a campaign for its release. After viewing the film, Einstein issued a strong statement in support of this effort. "The suppression of this film," he wrote, "marks a diplomatic defeat for our government in the eyes of the whole world. Its censorship proves that the government has bowed to the voice of the mob in the street and reveals so great a weakness that a reversal of policy must be emphatically demanded" (ibid.). As a result of this campaign, the German government agreed to let the film be shown later that year.

Debacle in Geneva

Einstein issued a number of public statements regarding the forthcoming General Disarmament Conference in Geneva, an event he characterized as the last chance to prevent another world war. His views regarding the impending disaster were particularly prescient, though relatively few of the world's statesmen took them to heart.

The first three texts below were all written in 1931, several months before the conference opened. The first of these, apparently intended for an American audience, may have been delivered in January 1932 at Whittier College, as suggested in N & N 1960, 658. Einstein refers to a Quaker meeting on the Disarmament Conference in his diary notes of 18 January 1932, adding that he gave a brief speech there, and Elsa Einstein referred to this event in a letter to Rosika Schwimmer (N & N 1960, 158). A particularly noteworthy feature in this address is Einstein's forthright declaration that the abolition of war can only succeed through the surrender of "a portion of [each nation's] sovereignty in favor of international institutions." This would become the touchstone of his political philosophy after World War II when he began making strong appeals for world government.

America and the Disarmament Conference of 1932, 16 June 1931
Einstein Archives 28-152; Einstein 1954, 100–102

The Americans of today are filled with the cares arising out of the economic conditions in their own country. The efforts of their responsible leaders are directed primarily to remedying the serious unemployment at home. The sense of being involved in the destiny of the rest of the world, and in particular of the mother country of Europe, is even less strong than in normal times.

But the free play of economic forces will not by itself automatically overcome these difficulties. Regulative measures by the community are needed to bring about a sound distribution of labor and consumers' goods among mankind; without this even the people of the richest country suffocate. The fact is that since the amount of work needed to supply everybody's needs has been reduced through the improvement of technical methods, the free play of economic forces no longer produces a state of affairs in which all the available labor can find employment. Deliberate regulation and organization are becoming necessary to make the results of technical progress beneficial to all.

If the economic situation cannot be cleared up without systematic regulation, how much more necessary is such regulation for dealing with the international problems of politics! Few of us still cling to the notion that acts of violence in the shape of wars are either advantageous or worthy of humanity as a method of solving international problems. But we are not consistent enough to make vigorous efforts on behalf of the measures which might prevent war, that savage and unworthy relic of the age of barbarism. It requires some power of reflection to see the issue clearly and a certain courage to serve this great cause resolutely and effectively.

Anybody who really wants to abolish war must resolutely declare himself in favor of his own country's resigning a portion of its sovereignty in favor of international institutions: he must be ready to make his own country amenable, in case of a dispute, to the award of an international court. He must, in the most uncom-promising fashion, support disarmament all round, as is actually envisaged in the unfortunate Treaty of Versailles; unless military and aggressively patriotic education is abolished, we can hope for no progress.

No event of the last few years reflects such disgrace on the lead-ing civilized countries of the world as the failure of all disarma-ment conferences so far; for this failure is due not only to the intrigues of ambitious and unscrupulous politicians but also the indifference and slackness of the public in all countries. Unless this is changed we shall destroy all the really valuable achieve-ments of our predecessors.

I believe that the American people are only imperfectly aware of the responsibility which rests with them in this matter.

They no doubt think "Let Europe go to the dogs, if she is de-stroyed by the quarrelsomeness and wickedness of her inhabi-tants. The good seed of our Wilson has produced a mighty poor crop in the stony ground of Europe. We are strong and safe and in no hurry to mix ourselves up in other people's affairs."

Such an attitude is neither noble nor far-sighted. America is partly to blame for the difficulties of Europe. By ruthlessly pressing her claims she is hastening the economic and therewith the moral decline of Europe; she has helped to Balkanize Europe and there-fore shares the responsibility for the breakdown of political moral-ity and the growth of that spirit of revenge which feeds on despair. This spirit will not stop short of the gates of America—I had almost said, has not stopped short. Look around, and beware!

The truth can be briefly stated: the Disarmament Conference comes as a final chance, to you no less than to us, of preserving the best that civilized humanity has produced. And it is on you, as the strongest and comparatively soundest among us, that the eyes and hopes of all are focused.

As an advocate of militant pacifism, Einstein was prepared to fight for a new world order that rejected war as a means for settling disputes among nations. With the world in an economic depression and fascism on the rise, he felt strongly that the forthcoming Geneva Disarmament Conference represented the last chance to prevent another major war. Among his many statements written in anticipation of the Conference, perhaps the most optimistic was the one he wrote for *The Nation*. Encouraged by the ongoing bilateral talks preceding the opening of the Conference, he saw a ray of hope for this undertaking, but only if the statesmen came to the bargaining table with a firm determination to implement universal disarmament. At the same time, Einstein stressed the need for educational reforms aimed at overcoming deeply engrained traditions that glorified military life.

The version presented here follows the text in *The Nation*, which differs significantly in wording from that in N & N 1960.

The 1932 Disarmament Conference, 4 September 1931
The Nation 133 (23 September 1931), no. 3455, 300;
N & N 1960, 146–147

What the inventive genius of mankind has bestowed upon us in the last hundred years could have made human life care free and happy if the development of the organizing power of man had been able to keep step with his technical advances. As it is, the hardly bought achievements of the machine age in the hands of our generation are as dangerous as a razor in the hands of a three-year-old child. The possession of wonderful means of production has not brought freedom—only care and hunger.

Worst of all is the technical development which produces the means for the destruction of human life, and the dearly created products of labor. We older people lived through that shudderingly [*sic*] in the World War. But even more terrible than this destruction seems to me the unworthy servitude into which the individual is swept by war. Is it not terrible to be forced by the community to deeds which every individual feels to be most despicable crimes?

Only a few have had the moral greatness to resist; they are in my eyes the true heroes of the World War.

There is one ray of hope. It seems to me that today the responsible leaders of the several peoples have, in the main, the honest will to abolish war. The opposition to this unquestionably necessary advance lies in the unhappy traditions of the people which are passed on like an inherited disease from generation to generation because of our faulty educational machines. Of course the main supports of this tradition are military training and its glorification, and not less important, the press which is so dependent upon the military and the larger industries. Without disarmament there can be no lasting peace. On the contrary, the continuation of military armaments in their present extent will with certainty lead to new catastrophies.

Hence the Disarmament Conference in Geneva in February, 1932, will be decisive for the fate of the present generation and the one to come. If one thinks back to the pitiful results achieved by the international conferences thus far held, it must be clear that all thoughtful and responsible human beings must exercise all their powers again and again to inform public opinion of the vital importance of the conference of 1932. Only if the statesmen have, to urge them forward, the will to peace of a decisive majority in their respective countries, can they arrive at their important goal. For the creation of this public opinion in favor of disarmament every person living shares the responsibility, through every deed and every word.

The failure of the conference would be assured if the delegates were to arrive in Geneva with fixed instructions and aims, the achievement of which would at once become a matter of national prestige. This seems to be universally recognized, for the meetings of the statesmen of any two states, of which we have seen a number of late, have been utilized for discussions of the problem of disarmament in order to clear the ground for the conference. This procedure seems to me a very happy one, for two persons, or two groups, ordinarily conduct themselves most sensibly, most honorably, and with the greatest freedom from passion if no third person listens in, whom the others believe they must consider or conciliate in their speeches. We can only hope for a favorable outcome in this most vital conference if the meeting is prepared for exhaustively in this way by advance discussions in order that surprises shall be made impossible, and if, through honest good will, an atmosphere of mutual confidence and trust can be effectively created in advance.

Success in such great affairs is not a matter of cleverness, or even shrewdness, but instead a matter of honorable conduct and mutual confidence. You cannot substitute intellect for moral conduct in this matter—I should like to say, thank God that you cannot!

It is not the task of the individual who lives in this critical time merely to await results and to criticize. He must serve this great cause as well as he can. For the fate of all humanity will be that fate which it honestly earns and deserves.

A few months before the Geneva Disarmament Conference opened, Einstein was approached by Imre Révész, who asked him for yet another diagnosis of the problems facing the negotiators. Through his Cooperation Press Service, Révész had the following article placed in several leading newspapers, including Vienna's *Neue Freie Presse* and the *New York Times*. In it Einstein reiterated in even more graphic language how international security can be reached only when all nations agree to limit their national sovereignty and to abide by the decisions of an International Court of Arbitration. At the same time, he pointed to compulsory military service as "the principal cause of the moral decline of the white race," a curse dating back to the *levée en masse* instituted by Lazare Carnot, minister of war during the French Revolution. To oppose this meant—for Einstein—first rejecting the persecution of conscientious objectors. Accordingly, much as he had a year earlier in signing the Manifesto of the Joint Peace Council, he called upon the nations that had subscribed to the antiwar Kellogg-Briand Pact of 1928 to go one step farther by agreeing never to compel their citizens to participate in warfare or to support it in any way.

The version presented here follows the text in the *New York Times*, which differs somewhat from that in Einstein 1954.

THE ROAD TO PEACE
NEW YORK TIMES, 22 NOVEMBER 1931; EINSTEIN 1954, 95–100

Let me begin by stating this political conviction: that the State exists for man, not man for the State. The same may be said of economic institutions. This is an old principle, laid down by those who rated human personality [individuality] as the highest of human values. I should hesitate to restate it if it were not constantly in danger of being forgotten, especially in this era of organization and standardization. I believe that the most important mission of the State is to protect the individual and to make it possible for him to develop into a creative personality.

The State should be our servant; we should not be slaves of the State. The State violates this precept when it forces us to perform military service, especially when this servile employment has for its purpose the destruction of men of other countries or the infringement of their freedom. We should make only such sacrifices for the State as contribute to the free development of the human individual. These statements may be accepted as obvious by every American; but they are not so accepted by every European; therefore our hope is that the fight against war will be strongly supported by Americans.

Now as to the disarmament conference. When we think of it should we laugh, weep or hope? Imagine a city inhabited by irritable and quarrelsome people. The constant danger to life is a complete obstacle to healthy development. The authorities ought to remedy these horrible conditions, but none of the municipal councilors or other citizens will agree to renounce the right to carry daggers. After years of preparation, the authorities finally decide to discuss the subject and the issue is set forth for debate in this form: How long and how sharp shall be the dagger which each citizen may carry in his belt when he walks about the town? As long as the crafty citizens take no action (through the laws, the courts and the police) against stabbing, conditions will continue as they are. And fixing the length and sharpness of the permitted daggers will merely play into the hands of the strongest and most pugnacious, and leave the weaker at their mercy.

The purpose of this comparison will be seen at once. We have, it is true, a League of Nations and a World Court. But the League is not much more than a meeting-place and the World Court has no means of enforcing its decisions. These institutions afford security to no State in case it is attacked. If this is kept in mind there will be less severity than is generally the case at present in judging France's attitude, her refusal to disarm without security.

If we do not agree to limit the sovereignty of individual States, and if, at the same time, all States do not guarantee to take common action against any one of them that openly or surreptitiously violates a decision of the World Court, we cannot escape from a situation of general anarchy and menace. By no artifice can the unlimited sovereignty of the individual States be reconciled with security against attack. Are further catastrophes needed to bring the nations to the point of undertaking the obligation to enforce every decision of the recognized international judicial authority? It can hardly be said that the course of events thus far gives ground for hoping for something better in the near future. But every friend of culture and justice must do all he can to convince his fellows of the necessity of such an international solidarity.

Not without a certain justification will it be objected that this conception places too great stress upon organization and neglects the spiritual—particularly the moral—side. It is asserted that mental disarmament must precede material disarmament. It is also rightly said that the greatest obstacle to international order is the enormously heightened nationalism which receives the attractive but misapplied name of patriotism. In the last century and a half this fetish has everywhere attained a sinister, an uncommonly pernicious, power. In order to give proper weight to this objection we must realize that the organizational and the spiritual aspects mutually condition one another. Not only do organizations depend upon traditional emotional attitudes, to which they owe their origin and their assurance of duration; existing organizations in turn exert powerful influence upon the emotional attitudes of peoples.

The present nationalism, everywhere so highly developed, seems to me to be closely connected with the obligation of universal military service or (as it is more euphemistically called) the people's army. The State which demands military service of its citizens is obliged to cultivate among them a nationalistic state of mind which serves as a psychological preparation for their military usefulness. Before the youth in its schools it must glorify, side by side with religion, its instrument of brute force!

Therefore, in my opinion, the introduction of universal military service is the principal cause of the moral decline of the white race—a decline which raises serious doubts as to the continuance of our culture, indeed of our existence. This curse (together with great social blessings) had its inception in the French Revolution, and in a short time it spread to all other peoples.

Consequently he who would foster the international outlook and counteract national chauvinism must combat universal military service. To refuse on moral grounds to perform military service may expose one to severe persecution; is this persecution any less shameful for society than the persecution to which the religious martyrs were subjected in earlier centuries? Can we (as in the Kellogg pact) outlaw war and at the same time hand over the individual unprotected to the war machinery of the State?

If at the disarmament conference we do not want to limit ourselves to the organizational and technical aspects, but wish to consider also the psychological aspects in a direct way and for educational reasons, we must seek internationally to provide a legal way in which the individual can refuse to perform military service. Such a measure would undoubtedly have a powerful moral effect. . . .

To sum up briefly: Mere agreements for reduction of armaments afford no sort of security. A court of obligatory arbitration must have at its disposal an executive authority, backed by all the participating States, which can step in and take economic and military sanctions against disturbers of the peace. Universal military service, as the chief generator of unhealthy nationalism, must be combated. Especially must the objector to military service be protected by international agreement. . . .

Einstein visited Oxford University in May 1932, by which time the Disarmament Conference had been in session for several weeks. He and other British pacifists were dismayed by the complex negotiations over the quantity and quality of weapons to be allowed, the kind of issues Einstein had dismissed in the article above as rules "fixing the length and sharpness of the permitted daggers." Before departing England, he was approached by Lord Ponsonby, a former MP and leading pacifist, who persuaded Einstein to accompany him on a trip to Geneva aimed at focusing public attention on the languishing negotiations at the Disarmament Conference.

Some sixty reporters attended the press conference announced by Ponsonby, who began with a brief statement proclaiming that "the time has come for the peoples of the world to take the matter into their own hands" by demanding "complete disarmament within five years and . . . the immediate renunciation of war under any circumstances" (N & N 1960, 168). Afterward, Einstein offered these remarks before answering questions.

PRESS CONFERENCE ON THE GENEVA DISARMAMENT
CONFERENCE, 23 MAY 1932
N & N 1975, 184–186

I came here to speak to you because friends convinced me that it
was my duty to do so. I should like to make a few, simple remarks
concerning my impressions of the situation. I need hardly mention
that I agree with my friend here in every detail.

To begin, I should like to say that, if the implications were not
so tragic, the methods used at the Disarmament Conference could
only be called absurd. One does not make wars less likely to occur
by formulating rules of warfare. One must start with the unqualified determination to settle international disputes by way of arbitration. What is involved are questions of morality and good will;
it is not a problem for the so-called technical experts.

I do not intend to deal at length with the methods which should
be employed, since my friend and associate has already described
them briefly. I would rather restrict myself to answering questions.

[What is your opinion of moral disarmament?]

The question of whether material or moral disarmament should
come first reminds me of the question of which came first, the
chicken or the egg. Of course, moral disarmament is essential; but
as long as governments are preparing for war, they will not be
capable of honestly promoting moral disarmament.

Moral disarmament, like the problem of peace as a whole, is
made difficult because men in power never want to surrender any
part of their country's sovereignty, which is exactly what they
must do if war is to be abolished. I am convinced, therefore, that
the solution to the peace problem cannot be left in the hands of
governments; rather, we must see to it that people who are intellectually and morally independent pool their influence and resources in fighting militarism. That is how we must counteract the
doubt and disillusionment which are currently undermining the
all-important influence of intellectual leaders.

Since it has become evident that the negotiations in Geneva

have not taken the sort of course that could lead to a practical disarmament program in time to spare the world from the horror of a new war, we believe that the time has come for the people of the world to take this matter into their own hands by demanding complete disarmament within five years and the immediate renunciation of all methods of warfare under any circumstances. These include the termination of compulsory military service, the immediate cessation of recruiting efforts, and a halt to production of ammunition and weapons of war.

The people of the world must be ready to attain these goals by refusing either to produce or to transport war materiel as well as to serve in the armed forces.

[What do you think of the Geneva Protocol outlawing aggression, which was never fully ratified?]

We must not yield one iota. The greatest crime is to compromise in order to gain a small point. We must continue to fight for the ratification of the Geneva Protocol.

[What about the role of women and the working class in the fight against war?]

Both are important. I have in mind the International Women's Co-operative Guild, the Women's International League for Peace and Freedom and others that could greatly help us to bring about real progress. Similarly, working-class organizations and parties should be keenly interested in the peace movement. They should realize, however, that there is real danger in accepting those compromises which sacrifice the ultimate goal for the sake of temporary sham accomplishments, such as "humanization of war, social patriotism," etc. People must be persuaded to refuse all military service. Only concrete actions such as this will make a real impression.

I am absolutely convinced that we should use every possible means to strengthen the war resistance movement. Its moral significance cannot be overestimated. Unlike anything else, this movement, at one and the same time, inspires individual courage, challenges the conscience of men, and undermines the authority of the military system. I have the greatest admiration for the 78,000 French teachers who have refused to teach along chauvinist lines.

[If you do not believe that compromise is possible, are you not sounding the death knell of the Disarmament Conference?]

That, indeed, is my intention. In my opinion, the conference is working in the direction of a bad compromise. Any agreement that may now be reached as to which type of arms would be permissible in war will not be kept when it comes to the test of war. War cannot be humanized. It can only be abolished.

The General Disarmament Conference went into recess in August 1932, reconvening in January 1933. In the meantime, Edouard Herriot, the new French Premier, offered a plan for settling international disputes. Einstein regarded this as an important step forward, in particular since it represented a willingness on the part of France to renounce a part of its sovereignty. By this time he had begun to articulate many of the key principles underlying the arguments he made in the years ahead to promote world government. The following statement on the Herriot plan, solicited by Imre Révész for his Cooperation Press Service, reveals how Einstein managed to extract important positive consequences from the failure of the Geneva Conference.

STATEMENT ON THE HERRIOT PLAN, 18 NOVEMBER 1932
EINSTEIN ARCHIVES 28-215; N & N 1960, 205–206

I am convinced that Herriot's plan represents an important step forward with regard to how, in the future, international disputes should be settled. I also consider Herriot's plan to be preferable to other proposals that have been made. In striving for comprehension, the way a given problem is defined is always crucial. That means, the question to be asked is not under what conditions are armaments permissible and how wars should be fought. Rather, the point of departure must be the following resolution:

"We are prepared to submit all international disputes to the judgment of an arbitration authority which has been established by the common consent of all of us. To make this possible it is necessary that certain conditions be satisfied which will guarantee our security. The problem is to reach agreement among ourselves about these conditions."

The renunciation of unlimited sovereignty by individual nations is the indispensable prerequisite to a solution of the problem. It is the great achievement of Herriot, or rather of France, that they have announced their willingness, in principle, for such a renunciation.

I also agree with Herriot's proposal that the only military force that should be permitted to have truly effective weapons is a police force which would be subject to the authority of international organs and would be stationed throughout the world.

My main objections to Herriot's plans are these: The police formations should not be composed of national troop units which are dependent on their own governments. Such a force, to function effectively under the jurisdiction of a supranational authority, must be—both men and officers—international in composition.

There is a second important point of disagreement with the French plan—namely, its support of the militia system. A militia system implies that the entire population will be trained in military concepts. It further implies that youth will be educated in a spirit which is at once obsolete and fateful. What would the more advanced nations say if they were confronted with the request that every citizen must serve as a policeman for a certain period of his life? To raise the question is to answer it.

These objections should not appear to detract from my belief that Herriot's proposals must be gratefully welcomed as a courageous and significant step in the right direction.

Winston Churchill, then in "the wilderness years" of his career, thought little of the Herriot plan and criticized the British government sharply for undermining the French armed forces at Geneva (Churchill, 75–77). On the latter score, Einstein saw eye to eye with him. When the new Nazi government came to the Geneva Conference table in 1933, the French were under considerable pressure to scale back and allow the Germans military parity. Only a firm military commitment on the part of the British government might have satisfied the French demands for security, and this was not forthcoming. Another stumbling block proved to be the insistence of Senator Claude Swanson, a member of the American delegation, on expanding his country's navy. Throughout that year and the next, as Einstein foresaw, the numerous paper agreements reached at the conference accomplished nothing toward reconciling German and French interests and mitigating the horrors of modern warfare.

America and Europe in an Hour of Crisis

In December 1932 Einstein was making plans to spend a third winter in Pasadena when he learned that a group called the Woman Patriot Corporation had protested against his return to the United States. Probably he was told relatively little about the charges leveled against him in the lengthy memorandum filed by Mrs. Frothingham with the State Department. According to this document, Einstein was not merely a pernicious influence; he was the ringleader of an anarcho-communist program whose aim was to shatter the military machinery of national governments as a preliminary condition for world revolution. As for Einstein's physics, his theory of relativity was as useless as the old riddle, "How many angels can stand on the point of a needle if the angels do not occupy space?" Its real purpose was subversion: to promote lawless confusion and shatter the Church as well as the State (Sayen, 6).

Thanks to the Freedom of Information Act and internet, millions can now read Mrs. Frothingham's froth-filled brief, one of the earliest efforts to warn the U.S. government of the dangers Einstein posed to all right-thinking Americans. Even if relatively little of this reached Einstein's ears, whatever he did hear was enough to prompt him to issue one of his more amusing replies to such far-right critics in a message for the Associated Press.

REPLY TO THE WOMEN OF AMERICA, 3 DECEMBER 1932
EINSTEIN ARCHIVES 28-213; EINSTEIN 1954, 7–8

Never yet have I experienced from the fair sex such energetic rejection of all advances; or if I have, never from so many at once.

But are they not quite right, these watchful citizenesses? Why should one open one's doors to a person who devours hardboiled capitalists with as much appetite and gusto as the Cretan Minotaur in days gone by devoured luscious Greek maidens, and on top of that is low-down enough to reject every sort of war, except the unavoidable war with one's own wife? Therefore give heed to your clever and patriotic womenfolk and remember that the Capitol of mighty Rome was once saved by the cackling of its faithful geese.

A week later, Einstein left his summer home outside Berlin for Pasadena, never to set foot in Germany again. At Caltech, his host, Robert Millikan, was well aware of the kind of charges that were being leveled against Einstein. This circumstance placed him in a delicate situation owing to the fact that this visit to Pasadena was sponsored by the Oberländer Trust of the Carl Schurz Memorial Foundation for the express purpose of promoting German-American understanding. Anxious to avoid any further bad publicity, Millikan tried to reassure a representative of the Oberländer Trust that Einstein would not cause a scene in Pasadena. After witnessing how his famous guest dealt with the local press on his arrival, Millikan confided to the representative that Einstein had "handled himself with a skill which, I am sure if your trustees had seen, would help to relieve their minds" that he might furnish "additional ammunition to those who have been spreading these grotesquely foolish reports about his connection with influences aimed at the undermining of American institutions and ideals" (Clark, 455).

Einstein's address was broadcast by NBC from the Pasadena Civic Auditorium as part of a "Symposium on America and the World Situation" sponsored by the Southern California College Student Body Presidents' Association. The speech was published in German and in an English translation prepared by Caltech physicist Richard Tolman. The translation that follows is a new one.

On German-American Understanding, 23 January 1933
Bulletin of the California Institute of Technology 42
(February 1933), no. 138, 4–8

. . . Before getting to my actual subject, I must clear away two obstacles.

The first of these is the stumbling block of the black tie. When men gather on ceremonial occasions in black tie, they usually create an atmosphere which shuts out the harsh realities of life. It is the atmosphere of high-sounding phrases, which tend to cluster around the black tie. Away with it!

The second stumbling block lies in words and other symbols which are laden with affect. While words for the most part serve to convey ideas, there are some which cause such turbulent feelings that their part in transmitting ideas becomes secondary. One has only to think of the word *heretic* at the time of the Inquisition—of the word *Communist* for the present-day American—of the word *bourgeois* in Russia—of the word *Jew* for reactionary circles in Germany—of the words *honor, prestige, fatherland* in nearly all the countries of the world. Such words are well suited to drive out reason and replace it with emotion. This I should like to call the obstacle of the taboo.

We should rid ourselves as much as possible of both these stumbling blocks, and without a second thought, quietly and with a cool head, confront naked reality face to face.

No one can deny that we are presently not only passing through an international economic crisis, but also through just as severe a crisis in international confidence, as a matter of fact in international cooperation in all areas. This crisis has produced tensions between governments and peoples, which can only be alleviated by the dispassionate investigation of its causes. . . .

This short speech can only indicate trains of thought, not implement them. It fulfills its purpose if it stimulates dispassionate reflection. There is deep wisdom in language. The word *Verständigung* literally means bringing about mutual understanding. In fact one can best advance understanding by contributing to the comprehension of an existing situation. The will, the readiness, and the ability peacefully to solve such questions, which are of far-reaching importance in the life of nations, arise out of such understanding.

In the part of his speech at the Pasadena symposium that is not reproduced above, Einstein dealt with the causes of the tense relations between the United States and Germany and expressed his hope that he might shed some light on these. In large part he attributed the friction to the inability of the Germans, and the Europeans in general, to meet their obligations on the war debt. This situation he regarded as inevitable because the Americans, fearing a weakening of their domestic economy, were unwilling to accept payment in the form of imported goods. He also countered the German assumption that war debts had precipitated the worldwide depression by noting that causes purely internal to the American economy had led to its sudden collapse and to much of the subsequent international economic distress.

Surprisingly, even in the dying days of the Weimar Republic, Einstein expressed some confidence that the tide of extreme nationalism was ebbing in Germany. In response to a question submitted to him in Pasadena about Nazi chances of gaining control of the German government, Einstein thought "the prospects of this extreme group have recently dimmed significantly" (Einstein Archives 28-216). Though Hitler's party remained the strongest in the Reichstag, modest Nazi reverses at the polls in November 1932 may have buoyed Einstein's hopes. In fact, the swift course of the events that followed soon had him whistling a very different tune.

CHAPTER 6

Hitler's Germany and the Threat to European Jewry, 1933–1938

Einstein was spending the winter at Caltech in Pasadena when he learned that Hitler had been appointed chancellor of Germany on 30 January 1933. Less than a month later the Reichstag was in flames. Recognizing that the struggling Weimar Republic had received a mortal blow, he now prepared for the long fight to save Europe from fascism. His first actions in support of this cause were to sever his ties with Germany by revoking his citizenship and issuing two political statements characterizing his views on political freedom as well as condemning the acts of brutality perpetrated by the Nazi government against its Jewish citizens.

While aimed at capturing public attention, Einstein's blunt language triggered a reaction that went well beyond the conventional political arena. Thus, when he politely tendered his resignation from the Prussian Academy, this noble institution, which long upheld the tradition of neutrality on political matters, issued a public pronouncement to the press regarding its illustrious former member. This official statement began by deploring Einstein's participation in "atrocity-mongering" in America and France and ended by stating that, due to his unacceptable political behavior, the Prussian Academy of Sciences had no reason to regret his decision to resign. On learning of this pronouncement, Einstein was shocked and forthrightly defended his actions, while blaming the German press for presenting a distorted version of his views. Since he still harbored warm feelings toward some of his colleagues in the Academy, this news prompted some painful personal reflections.

Another victim of Hitler's savagery was Einstein's erstwhile militant pacifism. In the face of the German dictator's increasingly hostile behavior, Einstein rejected passive resistance on the individual level and plumped for rapprochement among the powers arrayed against Germany. He continued to believe that, in the middle-term, only renunciation of competitive armaments could solve the question of international security. For the time being, however, Einstein supported the idea of an international police force backed by an international court of arbitration, a theme that he had already touched on during the First World War (letter to Heinrich Zangger, 21 August 1917, chapter 1).

In the United States, one of the most important personal and collegial bonds that Einstein forged was with Rabbi Stephen S. Wise, president of the American Jewish Congress. Wise firmly believed in the principle of "healing and transforming the world" through social activism and Zionism, all the while engaged in a delicate balancing act. Although active in protesting Nazi mistreatment of German Jews, he abstained from publicly calling upon the U.S. government to admit additional refugees from Germany, fearing that such a demand would lead to further restrictions on immigration and an anti-Semitic backlash at home.

Oppression of his fellow Jews caused Einstein to identify not only with their plight but to take pride in their accomplishments. His tributes to the achievements of Jews in the foreword to *The World as I See It*, the first collection of his nonscientific writings, and in a lengthy article for an American magazine provide mournful testimony to his solidarity with their fate. In protesting against Nazi persecution of non-Jews on political grounds, he recognized as well the salami tactics of the new rulers in Germany, in which the opposition was to be lopped off slice by slice, beginning with the left. The resulting flood of refugees lucky enough to escape became a pressing concern for Einstein, who saw in them the bearers of Germany's most precious cultural values.

In attempting to divine reasons for the rise of Hitler, Einstein concluded that many factors beyond the megalomaniacal personality of the Nazi leader were to blame, not least the predisposition on the part of Germans to embrace a herd mentality, inculcated by a long tradition of stubborn obedience. Yet, if militarism was deeply rooted in German culture, he also warned of the "narrow nationalism within our own ranks," excoriating right-wing Zionism for engaging in land speculation and for denying Arab rights in Palestine. Social justice, as Einstein saw it, was the defining characteristic of Judaism, and from this he drew often uncomfortable political consequences. In the spring of 1938 he minced no words in voicing his opposition to the creation of a Jewish state with even the most modest military posture, fearing that the all-important spiritualization of the community would otherwise suffer. Einstein's position was set against a backdrop of hardening feelings among the Arabs, increasing favoritism on the part of the British Mandatory Authority toward the Arabs, and practical Zionist aims focusing on a nation-state or nothing. In this context his emphasis on the moral foundations of Zionism as well as his internationalist and pacifist views were increasingly at odds with the mainstream of the Zionist movement.

By late 1938, when he wrote "Why do They Hate the Jews?," Einstein turned his troubled gaze back to Germany. To his mind the fury of the Nazi regime reflected the fear of all authoritarian regimes toward minorities who demonstrated intellectual independence. Unlike a large proportion of the German population, the Jews could not "be driven into uncritical acceptance of dogma." Political life was, in Einstein's view, a constant struggle between optimists, who favored a form of government that maximized individual freedom, and pessimists, who preferred the security of a strong state that upheld the status quo. If the United States paid lip service to the libertarian tradition of its founders, Einstein was quick to point out that advocates of state authority

"Comrades! As an old-time believer in democracy, one who is not a recent convert . . ."

1. First page of manuscript version of speech to the New Fatherland Association in Berlin, 13 November 1918 (see chapter 1).

2. "A New Giant of World History: Albert Einstein." Title page of *Berliner Illustrirte Zeitung*, 14 December 1919. This is the image that made Einstein a public celebrity in Weimar Germany. Photo by Suse Byk.

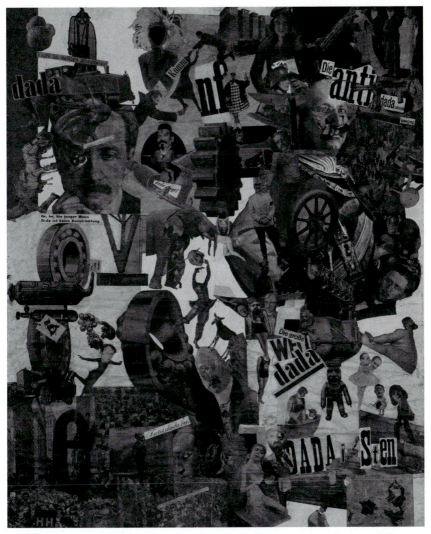

3. "A Slice with the Dada Kitchen Knife through the last Weimar Beer-Belly Cultural Epoch of Germany." Collage by Hannah Höch featured at the First International Dada Fair, Berlin, 1920.

4. Aboard the S.S. *Rotterdam* en route to the United States for a fund-raising trip for the Zionist movement, late March 1921. This was Einstein's first trip to America. With Menachem Ussishkin, Chaim and Vera Weizmann, Elsa Einstein, and Benzion Mossinson.

5. Addressing the Conference of Jewish Students in Germany, Berlin, 27 February 1924.

6. Surveying the wartime destruction of a village near Dormans, France, 9 April 1922. Photo from the popular French periodical *L'Illustration*, 15 April 1922.

7. With Paul Langevin (to Einstein's left, clutching a black hat) at "No More War" demonstration in Berlin, 29 July 1923.

8. On Einstein's first and only trip to Palestine, visiting the British High Commissioner's residence at Augusta Victoria Hospital, Jerusalem, February 1923. Included in the front row are Elsa Einstein, High Commissioner Herbert Samuel, and Beatrice Samuel.

9. With Elsa and members of Rishon Letzion, one of the first two Zionist settlements in Palestine, February 1923.

10. In Geneva with members of the Committee for Intellectual Cooperation including Marie Curie (second from left) and H. A. Lorentz (third from left), July–August 1924.

11. With Martin Hobohm at the lectern and Emil Julius Gumbel (on Einstein's left) at a meeting of the German League of Human Rights in Berlin, 27 April 1931, to support Gumbel's right of free speech.

12. Facing reporters as the S.S. *Belgenland* weighed anchor in New York City Harbor, 11 December 1930.

13. At Royal Albert Hall, London, on 3 October 1933 to give a speech for the benefit of the Refugee Assistance Fund. *Left to right*, Oliver Locker-Lampson, Einstein, Ernest Rutherford, and Austen Chamberlain.

The Insider Is Cast Out of the Temple of Science

14. Receiving the Max Planck Medal from the hands of its namesake, 28 June 1929.

15. "The concierge of the German Embassy in Brussels is authorized to cure an Asiatic [East European Jew] of the delusion that he is a Prussian." Caricature in the *Deutsche Tageszeitung* of 1 April 1933, which was declared "Day of the Jewish Boycott" by the Nazi regime.

Der Hausknecht der Deutschen Gesandtschaft in Brüssel wurde beauftragt, einen dort herumlungernden Asiaten von der Wahnvorstellung, er sei ein Preuße, zu heilen.

16. "Einstein takes up the sword," *The Brooklyn Eagle*, 1933. Einstein took this new stance shortly after Hitler was granted emergency powers in the wake of the Reichstag fire on the night of 27 February 1933. Drawing by Charles Raymond Macauley.

17. Being sworn in as a U.S. citizen with secretary Helen Dukas and daughter Margot Einstein in Trenton, N.J., 1 October 1940.

18. Arm-in-arm with Meyer Weisgal, secretary-general of the U.S. section of the Jewish Agency, Einstein attends the Anglo-American Committee of Inquiry on Palestine, Washington, D.C., 11 January 1946. On the right is Helen Dukas, his secretary for the last twenty-five years of his life.

19. Einstein appearing before the Anglo-American Committee in Washington, D.C., 11 January 1946. His appeal for a binational solution in Palestine disappointed many Zionists.

20. With Leo Szilard at Saranac Lake, September 1945, reenacting the signing of the August 1939 letter to President Roosevelt urging support of nuclear research. The reenactment was for the weekly newsreel series, the March of Time.

21. At home in Princeton, 10 February 1950, being filmed for his appearance on the series premiere of the television program "Today with Mrs. Roosevelt." On Einstein's left, Elliott Roosevelt.

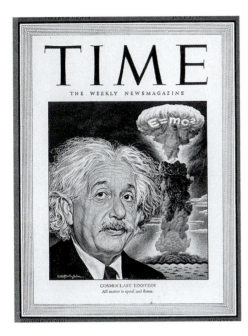

22. "Cosmoclast Einstein," *Time* magazine, 1 July 1946. After Hiroshima, the mainstream media reinforced the image of Einstein as the scientist behind the bomb.

23. Advertisement for *The Anatomy of Peace*, 1945, by Emery Reves.

24. With Rabbi Stephen Wise and the novelist Thomas Mann at the preview of the antiwar film *The Fight for Peace*, New York, 10 May 1938.

25. With future Progressive Party presidential candidate Henry Wallace, Frank Kingdon of the Progressive Citizens of America, and singer/actor/activist Paul Robeson, October 1947.

were numerous and influential. Wary of those in other countries who sought political and spiritual dominion over their own people, he warned that strains of a similar disease were also present in his adopted country.

Einstein's optimism had been sorely tested during the decade of the 1930s. Back in 1930, when he wrote "What I Believe" (chapter 5), he had offered a rare glimpse of the political and ethical values that guided his life. Much had changed when he was asked to write a sequel to this famous text, which appeared in the volume *I Believe* at the end of the decade. Entitled "Ten Fateful Years," it revealed Einstein's feelings about the course of world events in the intervening period. He asked himself whether his reassessment was the result of real changes or of his own deepening pessimism. After dismissing doubts about the subjective nature of his analysis, he went on to conclude that his sense of estrangement reflected the depth of the abyss into which the nations of the world had fallen in this brief period of time.

Responding to a State of Psychic Distemper

Shortly before his departure from California, Einstein issued a statement to the press on his fundamental opposition to the Hitler regime that would serve as his political manifesto for the next twelve years.

POLITICAL MANIFESTO, 11 MARCH 1933
EINSTEIN ARCHIVES 28-235.1; EINSTEIN 1954, 205

As long as I have any choice, I will only stay in a country where political liberty, tolerance, and equality of all citizens before the law prevail. Political liberty implies the freedom to express one's political opinions orally and in writing; tolerance implies respect for any and every individual opinion.

These conditions do not obtain in Germany at the present time. Those who have done most for the cause of international understanding, among them some of the leading artists, are being persecuted there.

Any social organism can become psychically distempered just as any individual can, especially in times of difficulty. Nations usually survive these distempers. I hope that healthy conditions will soon supervene in Germany and that in future her great men like Kant and Goethe will not merely be commemorated from time to time but that the principles which they taught will also prevail in public life and in the general consciousness.

While still aboard the S.S. *Belgenland* on his return trip to Europe, Einstein sent a message to the Prussian Academy in which he communicated his resignation "due to the present prevailing conditions in Germany" (Einstein Archives 36-055). On the same day he also issued a second political statement specifying the nature of his misgivings to the International League for Combating Anti-Semitism. It appeared in several German and French newspapers.

STATEMENT ON CONDITIONS IN GERMANY, 28 MARCH 1933
GRUNDMANN, 368

The actions of brutal force and oppression taken against all free intellects and against the Jews, these actions, which have taken place and still take place in Germany, have fortunately shaken the conscience of all countries that remain faithful to humane thought and to political freedoms. The International League for Combating Anti-Semitism has done a great service in defending justice by establishing the unity of peoples who have not been infected by the poison. We can only hope that the reaction will be strong enough to prevent Europe's relapse into a barbarism of times long past. May all friends of our so imperiled civilization concentrate their efforts to abolish this mental disease from the world. I am with you.

Two days later, readers of the *Kölnische Zeitung* were offered this text under the headline "How Einstein Agitates" and informed that Einstein had contacted the German Embassy in Brussels on his return to Europe to inquire about giving up his citizenship. The editor mockingly characterized this "a reasonable request" in light of his views on "German barbarism and his hope that a kind of European union will come together against us" (Grundmann, 368). A caricature illustrating the intensity of vituperation was published about this time (see Plate 15).

What the right-wing press did not report (and probably did not know) was that Einstein actually exhibited considerable restraint in the matter. When requested to join an official French demonstration against Hitler's regime, he replied that "however much I condemn the terrible aberrations occurring in Germany with the approval of the government, I cannot personally participate in a demonstration which is organized by officials of a foreign government." It would be the same, he wrote, had Zola during the Dreyfus Affair joined a protest sanctioned by the German government (Einstein Archives 28-245).

Not surprisingly though, Einstein's public statements sufficed to arouse considerable agitation within the Prussian Academy, which convened in plenary session on 30 March. Those present decided, however, that in view of Einstein's resignation no further steps were necessary. Ernst Heymann felt differently. Acting in his capacity as one of four permanent secretaries of the Academy, he issued a public statement, written without apparent further consultation, condemning Einstein's behavior. From Le Coq sur Mer on the Belgian coast, where he had taken up temporary residence, Einstein responded somewhat disingenuously by denying all charges.

RESPONSE TO THE DECLARATION OF THE PRUSSIAN
ACADEMY, 5 APRIL 1933
EINSTEIN ARCHIVES 36-062; EINSTEIN 1954, 206–207

I have received information from a thoroughly reliable source that
the Academy of Sciences has spoken in an official statement of
"Albert Einstein's participation in atrocity-mongering in America
and France."

I hereby declare that I have never taken any part in atrocity-
mongering, and I must add that I have seen nothing of any such
mongering anywhere. In general, people have contented themselves with reproducing and commenting on the official statements
and orders of responsible members of the German government,
together with the program for the annihilation of the German Jews
by economic methods.

The statements I have issued to the press were concerned with
my intention to resign my position in the Academy and renounce
my Prussian citizenship; I gave as my reason for these steps that
I did not wish to live in a country where the individual does
not enjoy equality before the law, and freedom of speech and
teaching.

Further, I described the present state of affairs in Germany as a
state of psychic distemper in the masses and made some remarks
about its causes.

In a document which I allowed the International League for
Combating Anti-Semitism to make use of for the purpose of enlisting support and which was not intended for the Press at all, I
also called upon all sensible people, who are still faithful to the
ideals of civilization in peril, to do their utmost to prevent this
mass-psychosis, which manifests itself in such terrible symptoms
in Germany today, from spreading any further.

It would have been an easy matter for the Academy to get hold
of a correct version of my words before issuing the sort of statement about me that it has. The German Press has reproduced a

deliberately distorted version of my words, as indeed was only to be expected with the Press muzzled as it is today.

I am ready to stand by every word I have published. In return, I expect the Academy to communicate this statement of mine to its members and also to the German public before which I have been slandered, especially as it has itself had a hand in slandering me before that public.

While also expressing his outrage in private correspondence, Einstein tempered his tone when addressing or mentioning colleagues for whom he still had respect. The physicist Max von Laue, for example, had been Einstein's staunchest defender during the early Weimar years when opponents of his general theory began their crusade (see "My Response: On the Anti-Relativity Company," chapter 2). Even more important to Einstein had been his friendship with Max Planck, to whom he wrote the following.

LETTER TO MAX PLANCK, 6 APRIL 1933
EINSTEIN ARCHIVES 19-391; N & N 1960, 217–218

. . . I have never taken part in any "atrocity-mongering." I will give the Academy the benefit of assuming it made these slanderous statements only under outside pressure. But even if that should be so, its conduct will hardly be to its credit; some of its more decent members will certainly feel a sense of shame even today.

You have probably learned that these false accusations were used as an excuse for the confiscation of my property in Germany. My Dutch colleagues joined in an effort to help me over the initial financial difficulties. It was fortunately not necessary for me to accept their help since I had been careful to prepare for such an emergency. It will certainly be easy for you to imagine how the public outside of Germany feels about the tactics employed against me. Surely there will come a time when decent Germans will be ashamed of the ignominious way in which I have been treated.

I cannot help but remind you that, in all these years, I have only enhanced Germany's prestige and have never allowed myself to be alienated by the systematic attacks on me in the rightist press, especially those of recent years when no one took the trouble to stand up for me. Now, however, the war of annihilation against my defenseless fellow Jews compels me to employ, in their behalf, whatever influence I may possess in the eyes of the world.

That you may better appreciate my feelings, I ask you to imagine yourself for the moment in this situation: Assume that you were a university professor in Prague and that a government came into power which would deprive Czechs of German origin of their livelihood and at the same time employ crude methods to prevent them from leaving the country. Assume further that guards were posted at the frontiers to shoot all those who, without permission, attempted to leave the country that waged a bloodless war of annihilation against them. Would you then deem it decent to remain a silent witness to such developments without raising your voice in support of those who are being persecuted? And is not the destruction of the German Jews by starvation the official program of the present German Government?

If you were to read what I actually said (not distorted accounts), you would doubtless realize that I expressed myself in a thoughtful and moderate way. I say this not to apologize but to demonstrate vividly the ignoble and ignominious manner in which the German authorities have behaved toward me.

I am happy that you have nevertheless approached me as an old friend and that, in spite of severe pressures from without, the relationship between us has not been affected. It remains as fine and genuine as ever, regardless of what has taken place "on a lower level," so to speak. The same holds true for Laue, for whom I have the very highest respect. . . .

In the meantime, Einstein had received an officious response from the Prussian Academy to his rejoinder of 5 April. Heymann joined forces with his counterpart in the natural sciences, Heinrich von Ficker, to deny that the Academy had been unduly influenced by the German press, noting that their public statement was based principally on reports in French and Belgian newspapers. As for "atrocity-mongering," Einstein had "done nothing to counteract unjust suspicions and slanders . . . [making] statements . . . which were bound to be exploited and abused by the enemies not merely of the present German Government but of the whole German people" (Einstein 1954, 208–209). To this Einstein gave the following scathing response.

Reply to the Prussian Academy, 12 April 1933
Einstein Archives 36-067; Einstein 1954, 209

I have received your communication of the 7th of this month and deeply deplore the mental attitude displayed in it.

As regards the facts, I can only reply as follows: What you say about my behavior is, at bottom, merely another form of the statement you have already published, in which you accuse me of having taken part in atrocity-mongering against the German people. I have already, in my last letter, characterized this accusation as slanderous.

You have also remarked that a "good word" on my part for "the German people" would have produced a great effect abroad. To this I must reply that such a testimony as you suggest would have been equivalent to a repudiation of all those notions of justice and liberty for which I have stood all my life. Such testimony would not be, as you put it, a good word for the German people; on the contrary, it would only have helped the cause of those who are seeking to undermine the ideas and principles which have won for the German people a place of honor in the civilized world. By giving such testimony in the present circumstances I should have been contributing, even if only indirectly, to moral corruption and the destruction of all existing cultural values.

It was for this reason that I felt compelled to resign from the Academy, and your letter only shows me how right I was to do so.

A few days after this final exchange, Einstein wrote to his good friend Paul Ehrenfest, offering a synopsis of what had happened in Germany since the Nazi's seizure of power.

Letter to Paul Ehrenfest, 14 April 1933
Einstein Archives 10-246; N & N 1960, 219

I am sure you know how firmly convinced I am of the causality of all events. You are hence well aware that I never act out of blind passions. I have come to regard developments in Germany in the following way:

A small group of pathological demagogues was able to capture and exploit the support of a population which is completely uneducated politically. This group is now moving in a direction that will become increasingly destructive. There is the danger that even countries outside of Germany will be contaminated, particularly since those belonging to this group in Germany are masters of propaganda. This is why I felt it absolutely necessary to do what I could in order to mobilize some decent individuals outside of Germany, and I have done so with great caution insofar as I did not say anything for which I would not have been willing to assume responsibility. If powerful moral and economic pressure had not been brought to bear so quickly, the Jews would have suffered even more than they actually have. Our friends in Germany need not do anything to protect me; in fact, such action would needlessly endanger them. However, it must be said that, in general, the lack of courage on the part of the educated classes in Germany has been catastrophic. Let me remind you of how pitifully the Academy of Arts conducted itself in the cases of Heinrich Mann and Käthe Kollwitz. Another case in point is the stupid attitude of the Secretary of the Prussian Academy, who permitted the release of the calumnious press notice about myself. . . .

Max von Laue vehemently protested against the statement issued by Secretary Heymann which expressed the Academy's displeasure over Einstein's "activities as an agitator in foreign countries" (1 April 1933, Grundmann, 369). Nevertheless, he felt that academics should keep their distance from the political arena, pointing out that "here they are making nearly all German academics responsible when you do something political" (letter from Laue, 14 May 1933). Einstein defended himself as follows.

LETTER TO MAX VON LAUE, 26 MAY 1933
EINSTEIN ARCHIVES 16-089; N & N 1960, 218

. . . I do not share your view that the scientist should observe silence in political matters, i.e., human affairs in the broader sense. The situation in Germany shows whither such restraint will lead: to the surrender of leadership, without any resistance, to those who are blind or irresponsible. Does not such restraint signify a lack of responsibility? Where would we be had men like Giordano Bruno, Spinoza, Voltaire and Humboldt thought and behaved in such a fashion? I do not regret one word of what I have said and am of the belief that my actions have served mankind. Do you think that I regret not being able to stay in your country under present circumstances? It would have been impossible for me, even if they had wrapped me in cotton. . . .

———————————————— ◈ ————————————————

As he emphasized to Ehrenfest, Einstein tried to use cautious language in his public statements regarding events and circumstances inside Germany. A case in point was his speech six months later before a gathering of more than 10,000 at the Royal Albert Hall in London. Organized by the Refugee Assistance Fund and supported by various relief agencies, including the Quakers, this meeting raised money for scholars who were being forced out of Germany (see Plate 13). Mainly remembered for its imagery of the scientific thinker as a lighthouse keeper lost in undisturbed reverie while pursuing demanding problems, the speech was published in Einstein 1956 under the title "Civilization and Science."

SPEECH IN ROYAL ALBERT HALL, 3 OCTOBER 1933
EINSTEIN ARCHIVES 28-253; N & N 1960, 237–239

I am glad you have given me an opportunity of expressing to you my deep sense of gratitude as a man, a good European and a Jew. Through your well-organized program of relief you have rendered great service not only to those scholars who have been the innocent victims of persecution, but to all of humanity and to science. You have shown that you, and the British people as a whole, have remained faithful to the tradition of tolerance and justice which your country has proudly upheld for centuries.

It is precisely in times of economic distress, such as we experience everywhere today, that we may recognize the effectiveness of the vital moral force of a people. Let us hope that, at some future time, when Europe is politically and economically united, the historian rendering judgment will be able to say that, in our own days, the liberty and honor of this continent were saved by the nations of Western Europe; that they stood fast in bitter times against the forces of hatred and oppression; that they successfully defended that which has brought us every advance in knowledge and invention: the freedom of the individual without which no self-respecting individual finds life worth living.

It cannot be my task to sit in judgment over the conduct of a nation which for many years counted me among its citizens; it is perhaps futile even to try to evaluate its policies at a time when it is so necessary to act. The crucial questions today are: How can we save mankind and its cultural heritage? How can we guard Europe from further disaster?

There can be no doubt that the present world crisis, and the suffering and privation which it has engendered, are in large measure responsible for the dangerous upheavals we witness today. In such times discontent breeds hatred, and hatred leads to acts of violence, revolution and even war. Thus, we see how distress and evil beget new distress and evil.

Once again, as was the case twenty years ago, leading statesmen are faced with a tremendous responsibility. One can only hope that, before it is too late, they will devise for Europe the kind of international treaties and commitments whose meaning is so completely clear that all countries will come to view any attempt at warlike adventures as utterly futile. However, the work of the statesmen can succeed only if they are backed by the sincere and determined will of the people.

We are concerned not only with the technical problems of securing and maintaining peace, but also with the important task of enlightenment and education. If we are to resist the powers that threaten intellectual and individual freedom, we must be very conscious of the fact that freedom itself is at stake; we must realize how much we owe to that freedom which our forefathers won through bitter struggle.

Without this freedom there would be no Shakespeare, Goethe, Newton, Faraday, Pasteur or Lister. There would be no decent homes for the mass of people, no railways or radios, no protection against epidemics, no low-priced books, no culture, no general enjoyment of the arts. There would be no machines to relieve people of the drudgery required to produce the necessities of life. Were it not for these freedoms, the majority of people would lead lives of oppression and slavery, as they did under the great ancient despotisms of Asia. Only in a free society is man able to create the inventions and cultural values which make life worth while to modern man.

Without doubt the present economic difficulties will bring forth some legislation to the effect that an adjustment between supply and demand of labor as well as between production and consumption will always be brought about through government control. But these problems too must be solved by free men. In the search for a solution we must be careful not to be driven into a kind of slavery which would impede any healthy development.

I should like to give expression to an idea which occurred to me recently. When I was living in solitude in the country, I noticed how the monotony of a quiet life stimulates the creative mind. There are certain occupations, even in modern society, which entail living in isolation and do not require great physical or intellectual effort. Such occupations as the service of lighthouses and lightships come to mind. Would it not be possible to place young people who wish to think about scientific problems, especially of a mathematical or philosophical nature, in such occupations? Very few young people with such ambitions have, even during the most productive period of their lives, the opportunity to devote themselves undisturbed for any length of time to problems of a scientific nature. Even if a young person is fortunate enough to obtain a scholarship for a limited period, he is pressured to arrive as soon as possible at definite conclusions. Such pressure can only be harmful to the student of pure science. In fact, the young scientist who enters a practical profession which earns him a livelihood is in a much better position, assuming, of course, that his profession affords him sufficient time and energy for his scientific work.

Should we merely lament the fact that we live in a time of tension, danger and want? I think not. Man, like every other animal, is passive by nature. Unless goaded by circumstance, he scarcely takes the trouble to reflect upon his condition and tends to behave as mechanically as an automaton. I think I am old enough to be able to say that, as a child and a young man, I passed through such a phase. One thought only of the trivialities of one's personal existence, slicked back one's hair, and strove to talk and act like one's fellows. Only with difficulty did one perceive what lay behind the conventional mask of behavior and speech. It protected the real person as though he were wrapped in cotton wool.

How different it is today! In the stark lightning flashes of these tempestuous times, one is able to see human beings and human values in all their nakedness. Every nation and every human being now clearly exposes his virtues and weaknesses, aims and passions. In the rush of contemporary events, ordinary behavior becomes meaningless; conventions fall away like dry husks.

Men in distress become aware of the inadequacies of economic institutions and the need for supranational political commitments. Only when subjected to peril and social upheaval do nations feel induced to adopt progressive measures; one can only hope that the present crisis will lead to a better world.

But over and above this rather abstract approach, we must not lose sight of those supreme and everlasting values which alone lend meaning to life and which we should strive to pass on to our children as a heritage purer and richer than that which we received from our own parents. Noble endeavors such as yours will serve this end.

The original draft (Einstein Archives 28-253) of this speech contained two paragraphs which Einstein subsequently deleted, persuaded perhaps to omit them from the address. They are not reproduced above. In one passage Einstein pointed out that "there is still time to avoid a disastrous fire. If I may express a hope, I should like to suggest that a fireman rather than a lawyer be consulted in this matter, and that we recognize above all the point of view of those neighbors whose homes are most directly threatened."

In private, Einstein expressed his fears more concretely. Writing in spring 1933 to a colleague in England, he said that he had "heard from very reliable sources that war materiel, especially airplanes, was being produced in great haste. If one grants [the Nazis] one or two years' time the world can once again be in a fine state of affairs because of the Germans" (letter to Frederick Lindemann, 7 May 1933).

Pacifism Reconsidered

In the course of the earlier Solvay conferences, Einstein had the opportunity to meet and befriend the Belgian royal couple, Queen Elisabeth and King Albert, whom he referred to as "the Kings." They, in turn, were pleased to offer the Einsteins refuge when they returned to Europe in late March of 1933. The controversy sparked by Einstein's forceful public remarks regarding the Hitler regime made him more vulnerable than ever. Guards were posted outside the seaside retreat at Le Coq sur Mer, as rumors of a Nazi assassination plot against Germany's most distinguished ex-citizen swirled about the small country. An entirely different political concern, however, brought Einstein and the royal family together again.

In June 1933 Alfred Nahon, a French pacifist living in Belgium, appealed to Einstein to testify at the trial of two Belgian conscientious objectors. News of this potentially embarrassing circumstance reached the royal court, and Einstein was informed that "the husband of the second fiddler" (N & N 1960, 227) wanted to see him urgently. (Queen Elisabeth and her physicist friend enjoyed making music together.) After conferring with King Albert, Einstein decided that the time was ripe to announce that his days as a militant pacifist were over (see Plate 16).

LETTER TO ALFRED NAHON, 20 JULY 1933
EINSTEIN ARCHIVES 51-231; N & N 1960, 229

What I shall tell you will greatly surprise you. Until quite recently we in Europe could assume that personal war resistance constituted an effective attack on militarism. Today we face an altogether different situation. In the heart of Europe lies a power, Germany, that is obviously pushing toward war with all available means. This has created such a serious danger to the Latin countries, especially Belgium and France, that they have come to depend completely on their armed forces. As for Belgium, surely so small a country cannot possibly misuse its armed forces; rather it needs them desperately to protect its very existence. Imagine Belgium occupied by present-day Germany! Things would be far worse than in 1914 and they were bad enough even then. Hence I must tell you candidly: were I a Belgian, I should not, in the present circumstances, refuse military service; rather, I should enter such service cheerfully in the belief that I would thereby be helping to save European civilization.

This does not mean that I am surrendering the principle for which I have stood heretofore. I have no greater hope than that the time may not be far off when refusal of military service will once again be an effective method of serving the cause of human progress. . . .

Einstein's letter ended with a request that its contents be conveyed to fellow pacifists. One month later it was published in *La Patrie*

humaine, provoking general dismay and anger. Many found it difficult to believe that the man they had so idolized had now abandoned them and their cause. Romain Rolland wrote to Stefan Zweig proclaiming Einstein a "genius in his science and a fool in every other field," and someone whose friendship is such that one has no need of enemies (Grundmann, 361).

Most of these attacks took place privately, but by no means all of them. Brent Dow Allinson, an outspoken pacifist, published an article entitled "Speak, Einstein, for the Peace of Europe" in the November 1934 issue of *Polity*. Allinson's challenge was taken up directly.

A Re-examination of Pacifism
Polity 3 (January 1935), no. 1, 4–5; N & N 1960, 254–256

Herr Allinson has placed me on the defendant's bench in a friendly manner. I am glad of this, since it gives me a favorable opportunity to say publicly certain things that ought to be said.

Mr. Allinson's accusation, put briefly and plainly, is something like this: "A few years ago you publicly urged the refusal to do military service. Now—although the international situation has become worse and more acute in an undreamed of manner—you envelop yourself in silence, perhaps even retracting your former utterances. Is this because your understanding or your courage or possibly both have suffered from the onslaught of events? If not, then show us unhesitatingly that you still belong to the Brotherhood of the Valiant."

Now my answer. I stand firmly by the principle that a real solution of the problem of pacifism can be achieved only by the organization of a super-national court of arbitration, which, differing from the present League of Nations in Geneva, would have at its disposal the means of enforcing its decisions. In short, an international court of justice with a permanent military establishment— or better, police force. An excellent expression of this conviction of mine is contained in Lord Davies' book "Force" (London,

Ernst Benn, Ltd., 1934), the reading of which I strongly recommend to everyone who is seriously concerned with this fundamental problem of mankind.

Taking as starting point this fundamental conviction, I stand for every measure which appears to me capable of bringing mankind nearer to this goal. Up to a few years ago, the refusal to bear arms by courageous and self-sacrificing persons *was* such a measure; it is no longer—especially in Europe—a means to be recommended. When the great Powers had nearly equally democratic governments, and when none of these Powers founded its future plans on military aggression, the refusal to do military service on the part of a fairly large number of citizens might have induced the governments of these Powers to look favorably on international legal arbitration. Moreover, such refusals were apt to educate public opinion to real pacifism. The public came to consider as oppression any pressure brought by the State upon its citizens to force them to fulfil their military obligations, besides considering such pressure unethical from the moral standpoint. Under these circumstances, such refusals worked for the highest good.

Today, however, we are brought face to face with the fact that powerful States make independent opinions in politics impossible for their citizens, and lead their own people into error through the systematic diffusion of false information. At the same time, these States become a menace to the rest of the world by creating military organizations which encompass their entire populations. This false information is spread by a muzzled press, a centralized radio service, and school education ruled by an aggressive foreign policy. In States of that description, refusal to perform military service means martyrdom and death for those courageous enough to object. In those States in which citizens still cling to some of their political rights, refusal to do military service means weakening the power of resistance of the remaining sane portions of the civilized world.

Because of this, no reasonable human being would today favor the refusal to do military service, at least not in Europe, which is at present particularly beset with dangers.

I do not believe that under present circumstances passive resistance is an effective method, even if carried out in the most heroic manner. Other times, other means, even if the final aim remains the same.

The confirmed pacifist must therefore at present seek a plan of action different from that of former, more peaceful times. He must try to work for this aim: That those States which favor peaceful progress may come as close together as possible in order to diminish the likelihood that the warlike programs of political adventurers whose States are founded on violence and brigandage will be realized. I have in mind, in the first place, well-considered and permanent concerted action on the part of the United States and the British Empire, together with France and Russia when possible.

Perhaps the present danger will facilitate this *rapprochement* and thus bring about a pacifistic solution of international problems. This would be the hopeful side to the present dark situation; here consistent action can contribute much toward influencing public opinion in the right direction.

Lord David Davies was the founder of the New Commonwealth Society, which advocated the creation of an international police force to enforce security agreements. Einstein met Davies in Glasgow in the spring of 1933 and soon thereafter joined forces with him. In a letter to Davies of 20 July 1933, he confessed "I could not have expressed my own position as well or as completely as you have: No disarmament without security; no security without a mandatory international court of arbitration and an international standing army" (N & N 1960, 226). One month later, as a "foundation member" of the New Commonwealth Society, Einstein issued a statement declaring that the goal of creating an international police force cannot be attained gradually. "Renunciation of competitive armaments by individual countries will be possible only when the problem of security is completely solved. Such a program requires that the participating countries voluntarily renounce a part of their sovereignty. At present, such renunciation meets resistance due to the vain nationalism especially of the great powers" (N & N 1960, 226–227).

The Bitter Fruit of Cultural Disintegration

Two months after Hitler's rise to power, Rabbi Stephen Wise organized a mass protest rally at Madison Square Garden in the face of opposition by those within the Jewish community who argued that such protests would provoke far stronger Nazi counteraction. Simultaneous rallies were held in seventy other metropolitan areas in the United States and in Europe.

Detached from the infighting of rival Jewish groups, Einstein viewed the situation in Germany in broader terms: the Nazi regime not only imperiled the Jewish population—its rapid rearmament posed a general threat to European peace.

LETTER TO STEPHEN S. WISE, 6 JUNE 1933
EINSTEIN ARCHIVES 35-133

. . . On the one hand, what is done *visibly* by Jews in the rest of the world has a certain effect on the German authorities, on the other hand it serves them as fodder for anti-Semitic baiting and as a pretext for acts of violence against German Jews. It could, however, be that demonstrations by American Jews might be important for mobilizing non-Jewish America against the German rulers. For this reason alone I can understand the value of organizing such public meetings by Jewish Americans (demonstrations by non-Jews, if feasible, would be more advantageous).

At any rate, far more important is a quiet *organized* effort directed toward American authorities and the American press. Once again the simple truth must be emphasized (in the press as well) that the Germans are secretly rearming on a large scale. Factories are running day and night (airplanes, light bombs, tanks, and heavy ordnance, produced in part on Swedish territory). A million troops are being covertly trained. Reporting such breaches of the Versailles Treaty is considered high treason and individuals are silenced through terror of all sorts. The revenge mentality displayed by intellectuals explains why they have not opposed the dishonorable and criminal regime, alternatively did not oppose it as long as this might have met with some measure of success. One will have to pay a dreadful sacrifice in men and materiel if one accommodates these people for just one year. Then they will speak a different language than Hitler recently (more like Japan's!) and "convert" even cautious individuals in Germany to their adventure.

When high officials in America realize that in these circumstances "magnanimous" tolerance would be a crime that will soon avenge itself bitterly, a crime above all against France, they will certainly not flinch from a detailed surveillance of the entire German industry, and find the means to tie the hands of these people. An accurate estimate of the German danger would also predispose the American authorities to introduce something effective via Geneva for the protection of the Jews of Germany.

The press must beat the danger of the German military into the heads of the public. The consequences of a new catastrophic war in Europe must be shown quite plainly—combined with the Japanese threat, the result of such a war would, due to the paralysis of Europe, become much more threatening than it already is.

When the American Jewish Congress under Wise's leadership organized a general boycott of German goods in August, Einstein urged more straightforward measures. In the face of the continued indifference of the democracies, he called for direct support of a resettlement program whose scope extended beyond the Zionist undertaking in Palestine.

LETTER TO STEPHEN S. WISE, 18 NOVEMBER 1933
EINSTEIN ARCHIVES 35-134

... Have you read Schwarzschild's article in the *Tagebuch* of 16 September, in which it is plain to see that the boycott against German goods on which we had placed such high hopes has remained completely ineffective? If, as seems likely, the great-power democracies act with vigorous neutrality against Hitler Germany, this breeding ground of disease will soon pose a grave moral and political danger for the rest of the world, not to speak of the unspeakable misery meted out to German Jews. I believe, however, that for the moment we Jews are the least able to arouse the somnolent conscience of the non-Jewish public through *direct* action. In my opinion, the best thing we can do is provide generous support for Jewish emigration from Germany, combined with occupational retraining. I'd like to call your attention to a fine German Jew, Consul Heinrich Goldberg in Paris (Blvd. des Italien 5, c/o Agricultural Association of German Émigré Jews) who is working on this problem with diligence and skill. I believe that it would be very much in the interest of the Jewish cause if you were to get in touch with this man. One must face the fact that Palestine does not even remotely have the capacity [to absorb Jewish emigration] and it would be disastrous should all available funds drift into the pockets of the Zionist Organization.

Leopold Schwarzschild was editor of *Das Tage-Buch*, one of Germany's most influential left-liberal periodicals.

Wise was greatly admired for his sense of moral rectitude and independence. A Reform Jew, he was atypical in his support for Zionism. He turned down an offer to serve as rabbi of the prestigious Temple Emanu-El in New York, fearing that he would be constrained in speaking his mind. Soon after, he founded the Free Synagogue in New York, which had free pews for all without fixed dues and maintained an extensive program of social welfare. His oratorical skills and commanding presence, apparent in a photograph of him with Einstein and the novelist Thomas Mann in 1938, made him a favorite with the American Reform Jewish community (see Plate 24). Einstein expressed his own admiration for Wise on the occasion of the latter's birthday.

Tribute to Stephen S. Wise on his 60th Birthday, March 1934
Einstein Archives 28-268

It is not an easy task to be a good teacher of children. It is difficult to recognize and cultivate good qualities and contain the bad ones. Yet it is easier to lead children since they yearn for direction.

To be a teacher of adults is far more difficult because they themselves wish to lead and dominate, though they are often most desperately in need of direction. Only true intellectual and personal superiority enables a teacher of adults to exercise a beneficial influence. By clothing his utterances in tasteful and attractive form he becomes a sought-after figure, one who is not a preacher in the wilderness, a preacher who stands before empty pews. It is not a small task to offer truth in such a way that it does not appear stern and forbidding but instead desirable. The task is particularly difficult for the Jewish religious leader, since the tradition of his people does not endow him with divine powers that elevate his utterances above the criticism of the congregation.

Stephen Wise is a man on whom nature has conferred those abilities of a leader and at the same time that devotion to recognizably worthy goals without which a meaningful social life is not possible. In rare fashion he combines love of the spiritual and moral ideal with an unusual sense for the practical and necessary.

Above all, what I admire in him is his bold activity toward building the self-respect of the Jewish people, combined with profound tolerance and penetrating understanding for everything human. Thus it is that, though a fighter, he comes across in the end as a balancing and conciliatory force.

His devotion to the Zionist task has attracted to this goal many friends among Jews and non-Jews in America. For this we thank him.

Today we wish him and us that his unusual good health and vitality bear up against all exertions and tasks. All of us are convinced that he will take care of the rest.

Another Zionist close to Einstein was his son-in-law, Rudolf Kayser, who published a biography of the physicist under a pseudonym (Reiser 1930). A leading essayist and critic during the Weimar period, Kayser also compiled one of the earliest collections of Einstein's nonscientific writings in *Mein Weltbild*, first published in 1934. That same year an English edition appeared under the title *The World as I See It*, which contains a brief preface by Kayser—signed merely with the mysterious initials J. H.—worth citing for the sense it conveys of the atmosphere during those years. "This man," he writes, "is being drawn, contrary to his own intention, into the whirlpool of political passions and contemporary history. As a result, Einstein is experiencing the fate that so many of the great men of history experienced: his character and opinions are being exhibited to the world in an utterly distorted form. To forestall this fate is the real object of this book."

Although still in print in a revised form, the later editions of *The World as I See It* do not present the foreword Einstein wrote for the original edition of 1934. What readers have therefore missed is his stirring tribute to the intellectual and cultural achievements of the German Jews written at the very time when many were being driven out of their country.

FOREWORD TO *THE WORLD AS I SEE IT*
EINSTEIN 1934

The following pages are dedicated to an appreciation of the achievements of the German Jews. It must be remembered that we are concerned here with a body of people amounting, in numbers, to no more than the population of a moderate-sized town, who have held their own against a hundred times as many Germans, in spite of handicaps and prejudices, through the superiority of their ancient cultural traditions. Whatever attitude people may take up towards this little people, nobody who retains a shred of sound judgment in these times of confusion can deny them respect. In these days of the persecution of the German Jews especially it is time to remind the western world that it owes to the Jewish people (a) its religion and therewith its most valuable moral ideals, and (b), to a large extent, the resurrection of the world of Greek thought. Nor should it be forgotten that it was a translation of the Bible, that is to say, a translation from Hebrew, which brought about the refinement and perfection of the German language. Today the Jews of Germany find their fairest consolation in the thought of all they have produced and achieved for humanity by their efforts in modern times as well; and no oppression however brutal, no campaign of calumny however subtle will blind those who have eyes to see the intellectual and moral qualities inherent in this people.

Hitler initially sought to consolidate power by eliminating political threats from the left. The first concentration camp, in Dachau near Munich, was set up in March 1933 to break the back of Communists, Social Democrats, and the labor movement. In September 1935 the Nazis completed the disenfranchisement of the Jews by decreeing the Nuremberg Laws. This legislation was based on a theory of racial inferiority, which served among other things as a pretext for expropriating Jewish assets. The ensuing flood of Jewish and non-Jewish refugees from Germany became one of Einstein's major preoccupations in the mid-thirties.

The following CBS radio broadcast was sponsored by the American Christian Committee for German Refugees and the Emergency Committee in Aid of Polish Refugees from Nazism. The former governor of New York, Alfred E. Smith, shared the microphone.

Victims of Fascism, 22 October 1935
Einstein Archives 28-317; N & N 1960, 262–263

The process of cultural disintegration, which has assumed such dangerous proportions in Central Europe during the past few years, must alarm anyone who is sincerely interested in the welfare of humanity. Because neither the establishment of an international organization nor the sense of responsibility among nations has progressed sufficiently to allow joint action against the disease, efforts in two different directions are being made to protect our cultural values from the danger that threatens them.

The first and most important of these efforts must be an attempt to bring about a consolidation, within the framework of the League of Nations, of those nations which have not been directly affected by recent developments in Europe. Such a consolidation should have as its aim the common defense of peace and the establishment of military security. The second effort is to help those individuals who have been compelled to emigrate from Germany, either because their lives were endangered or their livelihood was taken away from them. The situation of these people is particularly precarious because of the economic crisis throughout the world and the high level of unemployment in almost all countries, which has frequently led to regulations prohibiting the employments of aliens.

It is well known that German Fascism has been particularly violent in its attack upon my Jewish brothers. We have here the spectacle of the persecution of a group which constitutes a religious community. The alleged reason for this persecution is the desire to purify the "Aryan" race in Germany. As a matter of fact, no such "Aryan" race exists; this fiction has been invented solely to justify the persecution and expropriation of the Jews.

The Jews of all countries have come to the assistance of their impoverished brothers as best they could and have also helped the non-Jewish victims of Fascism. But the combined forces of the Jewish community have not nearly sufficed to help all these victims of Nazi terror. Hence, the emergency among the non-Jewish emigrants—that is to say, people of partly Jewish origin, liberals, socialists and pacifists, who are endangered because of their previous political activities or their refusal to comply with Nazi rules— is often even more serious than that of the Jewish refugees. . . .

To help these victims of Fascism constitutes an act of humanity, an attempt to save important cultural values and, not the least, a gesture of considerable political significance. . . . To allow the condition of these victims to deteriorate further would not only be a heavy blow to all who believe in human solidarity but would encourage those who believe in force and oppression. . . .

Einstein attributed Hitler's megalomania to feelings of inadequacy. Yet he also stressed the complicity of the German people, weakened by economic uncertainty and indoctrinated in mindless acceptance of empty authority, in enabling his rise. In keeping with his cultural Zionist views, he also stressed the incompatibility of Jewish and German cultures. The following manuscript, written after Hitler declared himself Führer on the death of President Hindenburg in August 1934, was apparently never published.

ON HITLER, 1935
EINSTEIN ARCHIVES 28-322; N & N 1960, 263–264

To the everlasting shame of Germany, the spectacle unfolding in the heart of Europe is tragic and grotesque; and it reflects no credit on the community of nations which calls itself civilized!

For centuries the German people have been subject to indoctrination by an unending succession of schoolmasters and drill sergeants. The Germans have been trained in hard work and made to learn many things, but they have also been drilled in slavish submission, military routine and brutality. The postwar democratic Constitution of the Weimar Republic fitted the German people about as well as the giant's clothes fitted Tom Thumb. Then came inflation and depression, with everyone living under fear and tension.

Hitler appeared, a man with limited intellectual abilities and unfit for any useful work, bursting with envy and bitterness against all whom circumstance and nature had favored over him. Springing from the lower middle class, he had just enough class conceit to hate even the working class which was struggling for greater equality in living standards. But it was the culture and education which had been denied him forever that he hated most of all. In his desperate ambition for power he discovered that his speeches, confused and pervaded with hate as they were, received wild acclaim by those whose situation and orientation resembled his own. He picked up this human flotsam on the streets and in the taverns and organized them around himself. This is the way he launched his political career.

But what really qualified him for leadership was his bitter hatred of everything foreign and, in particular, his loathing of a defenseless minority, the German Jews. Their intellectual sensitivity left him uneasy and he considered it, with some justification, as un-German.

Incessant tirades against these two "enemies" won him the support of the masses to whom he promised glorious triumphs and a golden age. He shrewdly exploited for his own purposes the centuries-old German taste for drill, command, blind obedience and cruelty. Thus he became the *Fuehrer*.

Money flowed plentifully into his coffers, not least from the propertied classes who saw in him a tool for preventing the social and economic liberation of the people which had its beginning under the Weimar Republic. He played up to the people with the kind of romantic, pseudo-patriotic phrasemongering to which they had become accustomed in the period before the World War, and with the fraud about the alleged superiority of the "Aryan" or "Nordic" race, a myth invented by the anti-Semites to further their sinister purposes. His disjointed personality makes it impossible to know to what degree he might actually have believed in the nonsense which he kept on dispensing. Those, however, who rallied around him or who came to the surface through the Nazi wave were for the most part hardened cynics fully aware of the falsehood of their unscrupulous methods.

Resisting Temptation While Preserving the Essence

Many of Einstein's political pronouncements during the 1930s dealt with cultural Zionism and the problems facing Jewish settlers in Palestine. He saw the labor movement as the key to peaceful relations in the Middle East: "I consider the formation of labor unions . . . necessary for Palestine. For the working class is not only the very soul of construction work, it is the only truly effective bridge between Jews and Arabs" (letter to Irma Lindheim, 2 February 1933). A statement on behalf of the Zionist organization "Laboring Palestine" underscored this conviction: "It is the workers alone who are in a position to establish healthy relations with the Arabs, which is the most important task of Zionism. Bureaucracies come and go, but human ties are decisive in the life of nations" (Einstein Archives 28-224). In the worldview of the Revisionists, on the other hand, he found a free-enterprise ideology antithetical to the socialist goals of Zionism, as well as an intransigent unwillingness to accommodate Arab interests.

At the third annual "Third Seder" celebration held 20 April 1935 at the Manhattan Opera House under the auspices of the National Labor Committee for Palestine, Einstein confided that he feared inner enemies more than those from without since "only the former can seriously threaten our well-being and thus our future." He expressed confidence, however, because "the secret of our apparently inexhaustible vitality lies in our strong tradition of social justice and of modest service both to our immediate community and society as a whole" ("Jewish-Arab Amity Urged by Einstein," *New York Times*, 21 April 1935). The following text is drawn from this *New York Times* article. Other speakers included his friend Rabbi Stephen S. Wise and the former president of Hadassah, Irma Lindheim.

THE GOAL OF JEWISH-ARAB AMITY, 20 APRIL 1935
EINSTEIN ARCHIVES 28-305; *NEW YORK TIMES*, 21 APRIL 1935

If Palestine is to become a Jewish national centre, then the Palestinian settlement must develop into a model way of life for all Jewry through the cultivation of spiritual values. I am convinced that the Histadrut is the embodiment of the best energies working in this direction. It is the strongest bulwark against all tendencies to poison the life of the community. It forms the most effective check on Revisionism, a movement which seeks to lead our youth astray with phrases borrowed from our worst enemies, and hinders the labor of most devoted pioneers.

Under the guise of nationalist propaganda Revisionism seeks to support the destructive speculation in land; it seeks to exploit the people and deprive them of their rights. Revisionism is the modern embodiment of those harmful forces which Moses with foresight sought to banish when he formulated his model code of social law. Furthermore, the state of mind fed by Revisionism is the most serious obstacle in the way of our peaceable and friendly cooperation with the Arab people, who are racially our kin. . . .

Histadrut, the Jewish Federation of Labor, was praised by Einstein a year later as an organization "which has sought a reasonable policy of cooperation with the Arab masses" (Einstein Archives 28-364).

At a fund-raising dinner sponsored by the United Jewish Appeal
(UJA) five-hundred guests heard Einstein warn that "moral disintegration and intensified national egoism" threatened not only the future of
Jewish settlements in Palestine but also the fate of the Jewish people as a
whole. To counter these trends he called for a renewal of the Jewish
ideals of "social justice, voluntary self-denial and joyful service to one's
fellow-human beings as well as to mankind as a whole" ("Einstein
Finds Judaism in Peril," *New York Times*, 27 June 1935). Fellow speakers struck equally somber notes about what faced European Jews. One
of the national co-chairmen of the UJA warned that the spread of Nazi
ideology to West European countries had made thousands of unab-
sorbed refugees in those countries the focus of increasing anti-Semitism,
while the other stated that nearly 300,000 Jews would soon be forced
to leave Germany "as an alternative to sheer starvation" (ibid.).

JUDAISM IN PERIL
NEW YORK TIMES, 27 JUNE 1935

The basic nature of Judaism automatically imposes upon us as a
community two kinds of obligations. First, we must be alert and
prepared for sacrifice on the side of all those aspirations which
have a moral goal similar to that described above; secondly, we
must keep our community healthy and creative so that it may con-
serve and increase its value for the achievement of the great goal
of humanity.

In such times as these consideration for the preservation of
Jewry comes foremost. Today it is our duty to put all available
Jewish forces, if possible exclusively, at the disposal of the current
relief activities. Particularly essential is it to provide help for the
upbuilding of Palestine, for this work showed itself in this time of
persecution as a refuge and as a moral support unparalleled among
all other organizations, no matter how meritorious they may be.

Without Palestine, the horrors of recent years would have been
even more terrible and the demoralization of the Jewish people far

greater. Today, therefore, we should with particular gratitude remember Theodore Herzl and the prophetic visionary insight with
which he foresaw the threatening danger in all its magnitude. He
died early, a sacrifice to his superhuman efforts in the battle
against misunderstanding and indifference. May our present generation at least prove worthy of him.

In spite of the hardships visited on the Jews of Europe, Einstein did
not regard Palestine "exclusively or even primarily as a place of refuge
for the oppressed." In remarks to the National Labor Committee for
Palestine in early April 1936, he argued that "the work of construction
in Palestine should serve above all as the embodiment of the social ideal
that constitutes the most important element in Jewish tradition"
(Einstein Archives 28-348).

When nine Jews were murdered in Jaffa, Palestine, in mid-April
1936, a wave of violence grew into a full-scale Arab rebellion. To give
the conflict the appearance of an organized nationwide protest, the
Arab Higher Committee declared a general strike, which persisted for
half a year; the rebellion continued until 1939.

While hostilities were still raging, the Peel Commission issued a report in 1937 concluding that there was no common ground between
the parties, that the British Mandate was in shambles, and that binationalism was an illusory goal. The commission recommended that
Palestine be partitioned, a position that was accepted with major qualifications by the international Zionist Organization and rejected outright by the Arabs. The following statement, which most clearly
presents Einstein's cultural objectives in Palestine and his rejection of
the trappings of a Jewish nation-state, was delivered at the sixth annual
"Third Seder" celebration of the National Labor Committee for
Palestine at the Hotel Commodore in New York and published in the
weekly organ of the Zionist Organization of America.

Our Debt to Zionism, 17 April 1938
New Palestine 28 (29 April 1938), no. 16, 2; Einstein 1954, 188–190

Rarely since the conquest of Jerusalem by Titus has the Jewish community experienced a period of greater oppression than prevails at the present time. In some respects, indeed, our own time is even more troubled, for man's possibilities of emigration are more limited today than they were then.

Yet we shall survive this period, too, no matter how much sorrow, no matter how heavy a loss in life it may bring. A community like ours, which is a community purely by reason of tradition, can only be strengthened by pressure from without. For today every Jew feels that to be a Jew means to bear a serious responsibility not only to his own community, but also toward humanity. To be a Jew, after all, means first of all, to acknowledge and follow in practice those fundamentals in humaneness laid down in the Bible—fundamentals without which no sound and happy community of men can exist.

We meet today because of our concern for the development of Palestine. In this hour one thing, above all, must be emphasized: Judaism owes a great debt of gratitude to Zionism. The Zionist movement has revived among Jews the sense of community. It has performed productive work surpassing all the expectations any one could entertain. This productive work in Palestine, to which self-sacrificing Jews throughout the world have contributed, has saved a large number of our brethren from direst need. In particular, it has been possible to lead a not inconsiderable part of our youth toward a life of joyous and creative work.

Now the fateful disease of our time—exaggerated nationalism, borne up by blind hatred—has brought our work in Palestine to a most difficult stage. Fields cultivated by day must have armed protection at night against fanatical Arab outlaws. All economic life suffers from insecurity. The spirit of enterprise languishes and a certain measure of unemployment (modest when measured by American standards) has made its appearance.

The solidarity and confidence with which our brethren in Palestine face these difficulties deserve our admiration. Voluntary contributions by those still employed keep the unemployed above water. Spirits remain high, in the conviction that reason and calm will ultimately reassert themselves. Everyone knows that the riots are artificially fomented by those directly interested in embarrassing not only ourselves but especially England. Everyone knows that banditry would cease if foreign subsidies were withdrawn.

Our brethren in other countries, however, are in no way behind those in Palestine. They, too, will not lose heart but will resolutely and firmly stand behind the common work. This goes without saying.

Just one more personal word on the question of partition. I should much rather see reasonable agreement with the Arabs on the basis of living together in peace than the creation of a Jewish state. Apart from practical consideration, my awareness of the essential nature of Judaism resists the idea of a Jewish state with borders, an army, and a measure of temporal power no matter how modest. I am afraid of the inner damage Judaism will sustain— especially from the development of a narrow nationalism within our own ranks, against which we have already had to fight strongly, even without a Jewish state. We are no longer the Jews of the Maccabee period. A return to a nation in the political sense of the word would be equivalent to turning away from the spiritualization of our community which we owe to the genius of our prophets. If external necessity should after all compel us to assume this burden, let us bear it with tact and patience.

One more word on the present psychological attitude of the world at large, upon which our Jewish destiny also depends. Anti-Semitism has always been the cheapest means employed by selfish minorities for deceiving the people. A tyranny based on such deception and maintained by terror must inevitably perish from the poison it generates within itself. For the pressure of accumulated injustice strengthens those moral forces in man which lead to a liberation and purification of public life. May our community through its suffering and its work contribute toward the release of those liberating forces.

This speech met with a very mixed reception. A number of American Zionists, including Louis Brandeis and Rabbi Wise, who opposed partition, were gratified, as was the president of the Arab National League in New York, who subsequently wrote Einstein to arrange a meeting with him. Chaim Weizmann, on the other hand, was furious at Einstein's outspokenness. Even before the speech was published, he dashed off a long letter, accusing Einstein of a breach of faith. Nevertheless, Einstein held firm to his convictions for the next decade (see chapter 7).

By far the best-known among Einstein's writings on Jewish affairs is the following article first published in *Collier's* just two weeks after the devastating attacks carried out during the "Night of Broken Glass." This was the Nazi pogrom directed against Jews throughout Germany during the night of 10 November 1938 when Jewish shops were vandalized, synagogues sacked, Jews beaten on the street, and men rounded up for registration and internment in concentration camps.

This tableau of horror went unmentioned only because Einstein wrote the original German text for this article some months earlier, during August 1938, giving it the simple title "Antisemitismus" (Einstein Archives 120-936). Indeed, a particularly striking feature of this essay is its matter-of-fact analysis of anti-Semitism. Einstein found that this social disease shares elements of envy and hatred with other forms of prejudice. The temptation for human beings to succumb to such feelings is at times great, but he emphasized the need to resist such weaknesses if freedom and independence of mind are to be gained. These virtues are the pride of the Jewish people, whose ideal is the striving for social justice, a view he expressed with unabashed, almost chauvinistic delight. The article ends with a stern warning against authoritarian tendencies in the United States, encouraged, as Einstein saw it, by those who profit from a concentration of economic power.

WHY DO THEY HATE THE JEWS?

COLLIER'S 102 (26 NOVEMBER 1938), NO. 22, 9–10 AND 38; EINSTEIN 1954, 191–198

I should like to begin by telling you an ancient fable, with a few minor changes—a fable that will serve to throw into bold relief the mainsprings of political anti-Semitism:

The shepherd boy said to the horse: "You are the noblest beast that treads the earth. You deserve to live in untroubled bliss; and indeed your happiness would be complete were it not for the treacherous stag. But he practiced from youth to excel you in fleetness of foot. His faster pace allows him to reach the water holes before you do. He and his tribe drink up the water far and wide, while you and your foal are left to thirst. Stay with me! My wisdom and guidance shall deliver you and your kind from a dismal and ignominious state."

Blinded by envy and hatred of the stag, the horse agreed. He yielded to the shepherd lad's bridle. He lost his freedom and became the shepherd's slave.

The horse in this fable represents a people, and the shepherd lad a class or clique aspiring to absolute rule over the people; the stag, on the other hand, represents the Jews.

I can hear you say: "A most unlikely tale! No creature would be as foolish as the horse in your fable." But let us give it a little more thought. The horse had been suffering the pangs of thirst, and his vanity was often pricked when he saw the nimble stag outrunning him. You, who have known no such pain and vexation, may find it difficult to understand that hatred and blindness should have driven the horse to act with such ill-advised, gullible haste. The horse, however, fell an easy victim to temptation because his earlier tribulations had prepared him for such a blunder. For there is much truth in the saying that it is easy to give just and wise counsel—to others!—but hard to act justly and wisely for oneself. I say to you with full conviction: We all have often played the tragic role of the horse and we are in constant danger of yielding to temptation again.

The situation illustrated in this fable happens again and again in the life of individuals and nations. In brief, we may call it the process by which dislike and hatred of a given person or group are diverted to another person or group incapable of effective defense. But why did the role of the stag in the fable so often fall to the Jews? Why did the Jews so often happen to draw the hatred of the masses? Primarily because there are Jews among almost all nations and because they are everywhere too thinly scattered to defend themselves against violent attack.

A few examples from the recent past will prove the point: Toward the end of the nineteenth century the Russian people were chafing under the tyranny of their government. Stupid blunders in foreign policy further strained their temper until it reached the breaking point. In this extremity the rulers of Russia sought to divert unrest by inciting the masses to hatred and violence toward the Jews. These tactics were repeated after the Russian government had drowned the dangerous revolution of 1905 in blood—and this maneuver may well have helped to keep the hated regime in power until near the end of the World War.

When the Germans had lost the World War hatched by their ruling class, immediate attempts were made to blame the Jews, first for instigating the war and then for losing it. In the course of time, success attended these efforts. The hatred engendered against the Jews not only protected the privileged classes, but enabled a small, unscrupulous and insolent group to place the German people in a state of complete bondage.

The crimes with which the Jews have been charged in the course of history—crimes which were to justify the atrocities perpetrated against them—have changed in rapid succession. They were supposed to have poisoned wells. They were said to have murdered children for ritual purposes. They were falsely charged with a systematic attempt at the economic domination and exploitation of all mankind. Pseudo-scientific books were written to brand them an inferior, dangerous race. They were reputed to foment wars and revolutions for their own selfish purposes. They were presented at once as dangerous innovators and as enemies of true progress. They were charged with falsifying the culture of nations by penetrating the national life under the guise of becoming assimilated. In the same breath they were accused of being so stubbornly inflexible that it was impossible for them to fit into any society.

Almost beyond imagination were the charges brought against them, charges known to their instigators to be untrue all the while, but which time and again influenced the masses. In times of unrest and turmoil the masses are inclined to hatred and cruelty, whereas in times of peace these traits of human nature emerge but stealthily.

Up to this point I have spoken only of violence and oppression against the Jews—not of anti-Semitism itself as a psychological and social phenomenon existing even in times and circumstances when no special action against the Jews is under way. In this sense, one may speak of latent anti-Semitism. What is its basis? I believe that in a certain sense one may actually regard it as a normal manifestation in the life of a people.

The members of any group existing in a nation are more closely bound to one another than they are to the remaining population. Hence a nation will never be free of friction while such groups continue to be distinguishable. In my belief, uniformity in a population would not be desirable, even if it were attainable. Common convictions and aims, similar interests, will in every society produce groups that, in a certain sense, act as units. There will always be friction between such groups—the same sort of aversion and rivalry that exists between individuals.

The need for such groupings is perhaps most easily seen in the field of politics, in the formation of political parties. Without parties the political interests of the citizens of any state are bound to languish. There would be no forum for the free exchange of opinions. The individual would be isolated and unable to assert his convictions. Political convictions, moreover, ripen and grow only through mutual stimulation and criticism offered by individuals of similar disposition and purpose; and politics is no different from any other field of our cultural existence. Thus it is recognized, for example, that in times of intense religious fervor different sects are likely to spring up whose rivalry stimulates religious life in general. It is well known, on the other hand, that centralization—that is, elimination of independent groups—leads to one-sidedness and barrenness in science and art because such centralization checks and even suppresses any rivalry of opinions and research trends.

Just What Is a Jew?

The formation of groups has an invigorating effect in all spheres of human striving, perhaps mostly due to the struggle between the convictions and aims represented by the different groups. The Jews too form such a group with a definite character of its own, and anti-Semitism is nothing but the antagonistic attitude produced in the non-Jews by the Jewish group. This is a normal social reaction. But for the political abuse resulting from it, it might never have been designated by a special name.

What are the characteristics of the Jewish group? What, in the first place, is a Jew? There are no quick answers to this question. The most obvious answer would be the following: A Jew is a person professing the Jewish faith. The superficial character of this answer is easily recognized by means of a simple parallel. Let us ask the question: What is a snail? An answer similar in kind to the one given above might be: A snail is an animal inhabiting a snail shell. This answer is not altogether incorrect; nor, to be sure, is it exhaustive; for the snail shell happens to be but one of the material products of the snail. Similarly, the Jewish faith is but one of the characteristic products of the Jewish community. It is, furthermore, known that a snail can shed its shell without thereby ceasing to be a snail. The Jew who abandons his faith (in the formal sense of the word) is in a similar position. He remains a Jew.

Difficulties of this kind appear whenever one seeks to explain the essential character of a group.

The bond that has united the Jews for thousands of years and that unites them today is, above all, the democratic ideal of social justice, coupled with the ideal of mutual aid and tolerance among all men. Even the most ancient religious scriptures of the Jews are steeped in these social ideals, which have powerfully affected Christianity and Mohammedanism and have had a benign influence upon the social structure of a great part of mankind. The introduction of a weekly day of rest should be remembered here—a profound blessing to all mankind. Personalities such as Moses, Spinoza and Karl Marx, dissimilar as they may be, all lived and sacrificed themselves for the ideal of social justice; and it was the tradition of their forefathers that led them on this thorny path. The unique accomplishments of the Jews in the field of philanthropy spring from the same source.

The second characteristic trait of Jewish tradition is the high regard in which it holds every form of intellectual aspiration and spiritual effort. I am convinced that this great respect for intellectual striving is solely responsible for the contributions that the Jews have made toward the progress of knowledge in the broadest sense of the term. In view of their relatively small number and the considerable external obstacles constantly placed in their way on all sides, the extent of those contributions deserves the admiration of all sincere men. I am convinced that this is not due to any special wealth of endowment, but to the fact that the esteem in which intellectual accomplishment is held among the Jews creates an atmosphere particularly favorable to the development of any talents that may exist. At the same time a strong critical spirit prevents blind obeisance to any mortal authority.

I have confined myself here to these two traditional traits, which seem to me the most basic. These standards and ideals find expression in small things as in large. They are transmitted from parents to children; they color conversation and judgment among friends; they fill the religious scriptures; and they give to the community life of the group its characteristic stamp. It is in these distinctive ideals that I see the essence of Jewish nature. That these ideals are but imperfectly realized in the group—in its actual everyday life—is only natural. However, if one seeks to give brief expression to the essential character of a group, the approach must always be by the way of the ideal.

Where Oppression Is a Stimulus

In the foregoing I have conceived of Judaism as a community of tradition. Both friend and foe, on the other hand, have often asserted that the Jews represent a race; that their characteristic behavior is the result of innate qualities transmitted by *heredity* from one generation to the next. This opinion gains weight from the fact that the Jews for thousands of years have predominantly married within their own group. Such a custom may indeed *preserve* a homogeneous race—if it existed originally; it cannot *produce* uniformity of the race—if there was originally a racial intermixture. The Jews, however, are beyond doubt a mixed race, just as are all other groups of our civilization. Sincere anthropologists are agreed on this point; assertions to the contrary all belong to the field of political propaganda and must be rated accordingly.

Perhaps even more than on its own tradition, the Jewish group has thrived on oppression and on the antagonism it has forever met in the world. Here undoubtedly lies one of the main reasons for its continued existence through so many thousands of years.

The Jewish group, which we have briefly characterized in the foregoing, embraces about sixteen million people—less than one per cent of mankind, or about half as many as the population of present-day Poland. Their significance as a political factor is negligible. They are scattered over almost the entire earth and are in no way organized as a whole—which means that they are incapable of concerted action of any kind.

Were anyone to form a picture of the Jews solely from the utterances of their enemies, he would have to reach the conclusion that they represent a world power. At first sight that seems downright absurd; and yet, in my view, there is a certain meaning behind it. The Jews as a group may be powerless, but the sum of the achievements of their individual members is everywhere considerable and telling, even though these achievements were made in the face of obstacles. The forces dormant in the individual are mobilized, and the individual himself is stimulated to self-sacrificing effort, by the spirit that is alive in the group.

Hence the hatred of the Jews by those who have reason to shun popular enlightenment. More than anything else in the world, they fear the influence of men of intellectual independence. I see in this the essential cause for the savage hatred of Jews raging in present-day Germany. To the Nazi group the Jews are not merely a means for turning the resentment of the people away from themselves, the oppressors; they see the Jews as a nonassimilable element that cannot be driven into uncritical acceptance of dogma, and that, therefore—as long as it exists at all—threatens their authority because of its insistence on popular enlightenment of the masses.

Proof that this conception goes to the heart of the matter is convincingly furnished by the solemn ceremony of the burning of the books staged by the Nazi regime shortly after its seizure of power. This act, senseless from a political point of view, can only be understood as a spontaneous emotional outburst. For that reason it seems to me more revealing than many acts of greater purpose and practical importance.

In the field of politics and social science there has grown up a justified distrust of generalizations pushed too far. When thought is too greatly dominated by such generalizations, misinterpretations of specific sequences of cause and effect readily occur, doing injustice to the actual multiplicity of events. Abandonment of generalization, on the other hand, means to relinquish understanding altogether. For that reason I believe one may and must risk generalization, as long as one remains aware of its uncertainty. It is in this spirit that I wish to present in all modesty my conception of anti-Semitism, considered from a general point of view.

In political life I see two opposed tendencies at work, locked in constant struggle with each other. The first, optimistic trend proceeds from the belief that the free unfolding of the productive forces of individuals and groups essentially leads to a satisfactory state of society. It recognizes the need for a central power, placed above groups and individuals, but concedes to such power only organizational and regulatory functions. The second, pessimistic trend assumes that free interplay of individuals and groups leads to the destruction of society; it thus seeks to base society exclusively upon authority, blind obedience, and coercion. Actually this trend is pessimistic only to a limited extent; for it is optimistic in regard to those who are, and desire to be, the bearers of power and authority. The adherents of this second trend are the enemies of the free groups and of education for independent thought. They are, moreover, the carriers of political anti-Semitism.

Here in America all pay lip service to the first, optimistic, tendency. Nevertheless, the second group is strongly represented. It appears on the scene everywhere, though for the most part it hides its true nature. Its aim is political and spiritual dominion over the people by a minority, by the circuitous route of control over the means of production. Its proponents have already tried to utilize the weapon of anti-Semitism as well as of hostility to various other groups. They will repeat the attempt in times to come. So far all such tendencies have failed because of the people's sound political instinct.

And so it will remain in the future, if we cling to the rule: Beware of flatterers, especially when they come preaching hatred.

A Painful Retrospective

Einstein summed up his assessment of the political situation at the end
of the 1930s with a harsh denunciation of Prime Minister Neville
Chamberlain and the Western powers, chiding his closest friend,
Michele Besso, for his groundless optimism.

LETTER TO MICHELE BESSO, 10 OCTOBER 1938
EINSTEIN ARCHIVES 7-376.1; EINSTEIN/BESSO, DOC. 129

You have faith in the English and even in Chamberlain? O sancta
simplicitas. He sacrifices Eastern Europe in the hope that Hitler
will spend his fury on Russia. But we will see here too that cunning
has short legs. He has forced the Left into a corner in France and
there lifted those into the saddle whose motto is: "Better Hitler
than the Reds." This was already evident in the politics of annihilation against Spain. In the last moment he saved Hitler by placing
the wreath of peace-lover on his brow and encouraged France to
betray the Czechs. . . . The only anxiety which has fueled his humiliating flights was that Hitler could lose the ground under his
feet. I don't give a farthing more for Europe's future. America
joined boldly in the strangulation maneuver against Spain. Here
too in fact money rules, as well as the fear of the Bolshevik or
more generally the fear of the propertied for their privileges. . . .

Private thoughts were mirrored in public statements as well. When
Clifton Fadiman, book editor of the *New Yorker* from 1934 to 1943,
approached him for a sequel to the article "What I Believe," written
eight years earlier (chapter 5), Einstein produced the following
retrospective.

Ten Fateful Years, 1938
Fadiman 1939, 367–369; Einstein 1956, 6–8

Reading once again the lines I wrote almost ten years ago to the volume *Living Philosophies*, I receive two strangely contrasting impressions. What I wrote then still seems essentially as true as ever; yet, it all seems curiously remote and strange. How can that be? Has the world changed so profoundly in ten years, or is it merely that I have grown ten years older, and my eyes see everything in a changed, dimmer light? What are ten years in the history of humanity? Must not all those forces that determine the life of man be regarded as constant compared with such a trifling interval? Is my critical reason so susceptible that the physiological change in my body during those ten years has been able to influence my concept of life so deeply? It seems clear to me that such considerations cannot throw light upon a change in the emotional approach to the general problems of life. Nor may the reasons for this curious change be sought in my own external circumstances; for I know that these have always played a subordinate part in my thoughts and emotions.

No, something quite different is involved. In these ten years confidence in the stability, yes, even the very basis for existence, of human society has largely vanished. One senses not only a threat to man's cultural heritage, but also that a lower value is placed upon all that one would like to see defended at all costs.

Conscious man, to be sure, has at all times been keenly aware that life is an adventure, that life must, forever, be wrested from death. In part the dangers were external: one might fall downstairs and break one's neck, lose one's livelihood without fault, be condemned though innocent, or ruined by calumny. Life in human society meant dangers of all sorts; but these dangers were chaotic in nature, subject to chance. Human society, as a whole, seemed stable. Measured by the ideals of tastes and morals it was decidedly imperfect. But, all in all, one felt at home with it and, apart from the many kinds of accidents, comparatively safe in it.

One accepted its intrinsic qualities as a matter of course, as the air one breathed. Even standards of virtue, aspiration, and practical truth were taken for granted as an inviolable heritage, common to all civilized humanity.

To be sure, the [First] World War had already shaken this feeling of security. The sanctity of life vanished and the individual was no longer able to do as he pleased and to go where he liked. The lie was raised to the dignity of a political instrument. The War was, however, widely regarded as an external event, hardly or not at all as the result of man's conscious, planful action. It was thought of as an interruption of man's normal life from the outside, universally considered unfortunate and evil. The feeling of security in regard to human aims and values remained, for the main part, unshaken.

The subsequent development is sharply marked by political events that are not as far-reaching as the less easily grasped socio-psychological background. First a brief, promising step forward characterized by the creation of the League of Nations through the grandiose initiative of Wilson, and the establishment of a system of collective security among the nations. Then the formation of Fascist states, attended by a series of broken pacts and undisguised acts of violence against humanity and against weaker nations. The system of collective security collapsed like a house of cards—a collapse the consequences of which cannot be measured even today. It was a manifestation of weakness of character and lack of responsibility on the part of the leaders in the affected countries, and of shortsighted selfishness in the democracies—those that still remain outwardly intact—which prevented any vigorous counterattack.

Things grew even worse than a pessimist of the deepest dye would have dared prophesy. In Europe to the east of the Rhine free exercise of the intellect exists no longer, the population is terrorized by gangsters who have seized power, and youth is poisoned by systematic lies. The pseudo-success of political adventurers has dazzled the rest of the world; it becomes apparent everywhere that this generation lacks the strength and force which enabled previous generations to win, in painful struggle and at great sacrifice, the political and individual freedom of man.

Awareness of this state of affairs overshadows every hour of my present existence, while ten years ago it did not yet occupy my thoughts. It is this that I feel so strongly in rereading the words written in the past.

And yet I know that, all in all, man changes but little, even though prevailing notions make him appear in a very different light at different times, and even though current trends like the present bring him unimaginable sorrow. Nothing of all that will remain but a few pitiful pages in the history books, briefly picturing to the youth of future generations the follies of its ancestors.

This gloomy picture, written as Europe stood on the edge of the abyss, suggests that Einstein was already bracing himself for what was to come. If so, he did not have long to wait. His reaction to the events of World War II and the Holocaust constitute the principal themes in chapter 7.

The Fate of the Jews, 1939–1949

After the Nazi seizure of power, Einstein had celebrated the Jewish cultural tradition and the fact that it had held its own in an increasingly hostile host society. As German policy turned from persecution to slaughter, Einstein's despair turned into a fierce resolve. As he saw it, the madness that gripped his native country not only uprooted vast numbers of human beings but sought the extermination of the "fundamental structure of modern civilization." Invoking an interpretation of anti-Semitism complementary to that in his essay a half year earlier ("Why do They Hate the Jews?," chapter 6), he argued that Nazism's hatred for its "irreconcilable opponents," the Jews, was rooted in contempt for their tradition of independent judgment and communal service. Liberal experimentation in Weimar Germany, he claimed, had been doomed to failure by the perpetuation of its authoritarian ruling caste.

The magnitude of the refugee problem and uncharitable restrictions on admission to Western countries greatly increased the importance of Palestine as a potential place of refuge. Under the terms of the British White Paper of 1939, however, Jewish immigration there was limited to 75,000 over the next five years, a quota "calculated to ensure that the Jews would constitute no more than a third of Palestine's population" (Segev, 440). Shortly before the deadline in spring 1944, congressional hearings were opened into the question of American support for Jewish aspirations in Palestine. Philip Hitti, a prominent expert on the Arab world, presented his perspective to a House committee. Resolutions in support of continued Jewish immigration were introduced in both houses of Congress, but withheld from consideration because of fears in the American military high command of offending Arab rulers. After Hitti's testimony was published in a Princeton newspaper, Einstein and Erich Kahler, a colleague and close friend, countered with a case for lifting the restrictions. Kahler was not only a fervent advocate of the Zionist cause; his support for a world constitution, advocacy of civil liberties, and membership in antiwar groups closely mirrored Einstein's political positions. Kahler did not, however, apparently share Einstein's ability of rising above acrimonious academic discourse, as evidenced by the sharply polemical tone that pervades their arguments against Hitti.

In the meantime, Einstein's moderate position on the nature of the Jewish homeland was increasingly being overwhelmed by events. At the Biltmore Conference in May 1942, the chairman of the Jewish Agency's Palestine Executive and later first prime minister of Israel, David Ben-Gurion, had called openly for that which had long been features of the political Zionist agenda: creation of a Jewish army and transformation of postwar Palestine into an independent Jewish commonwealth. Though non-Zionist organizations for the first time joined with their Zionist counterparts in advocating the establishment of such a state, Einstein resisted the calls for a purely political solution. As the full horror of the systematic genocide of European Jews was revealed, however, the dynamic in favor of a Jewish state became increasingly irresistible.

In other respects, however, the impact of the Holocaust was decisive for Einstein. He never again budged from his conviction of the German people's collective complicity in the crimes of their government. His instinctive distrust of national sovereignty now also lent itself readily to accepting the principle of international intervention where a state's aggressive tendencies could not be curbed domestically.

While he had perhaps too exalted a view of Roosevelt's commitment to the principle of Jewish immigration to Palestine, he had no illusions about what he called "the game in Washington." This, he pointed out, could be "summed up with the maxim— never let the right hand know what the left is doing. One thumps the table with the right hand, while with the left one helps England . . . in its insidious attack" (letter to Max Born, 1 June 1948). He was equally pointed in condemning the claim of an American anti-Zionist organization that support for Jewish settlement in Palestine implied disloyalty to the United States. Such charges, he argued, were but a "pitiable attempt" to curry favor with the enemy.

Special blame was reserved, however, for the party that Einstein saw as the chief villain of the piece. In his testimony before the Anglo-American Committee in 1946, he made no bones about his contempt for the British policy of divide and conquer in which the Mandatory Authority pitted Jew against Arab. At the same time, he caused consternation among many Zionists by reiterating his resistance to the idea of a Jewish state and calling for a binational solution to the Palestine problem. Yet two years later he welcomed the creation of Israel in the expectation that its supreme moral test might be met: that the rescue of "our endangered brethren, dispersed in many lands" not be won at the price of abandoning the ethical ideal of "peace, based on understanding and self-restraint, and not on violence."

Persecution and the Rage against Reason

As in the foreword to *The World as I See It* (chapter 6), Einstein returned in the following essay to the theme of the debt owed by European civilization to the ancient cultural tradition of the Jews. This tradition is imbedded in the Jewish covenant of adherence to "truth, justice, and freedom" which leads to the ennoblement of all mankind. By dispossessing and exiling German Jews, a process that found its culmination in the brutality of the "Night of Broken Glass" (November 1938), the Nazis destroyed the equilibrium of all human societies. As if to drive home the point, Hitler subjected the Jews of annexed Austria and the Sudetenland to the same fate in 1938. On 15 March 1939, a week before the following CBS radio address for the United Jewish Appeal was delivered, German troops occupied the rest of Czechoslovakia.

A more extensive version of this talk, delivered in Einstein's capacity as honorary chairman of the United Jewish Appeal for Refugees and Overseas Needs, was printed in the *New York Times*, 22 March 1939, under the headline "Einstein Asks Aid for Persecuted."

THE DISPERSAL OF EUROPEAN JEWRY, 21 MARCH 1939
EINSTEIN ARCHIVES 28-476; EINSTEIN 1956, 254–256

The history of the persecutions which the Jewish people have had to suffer is almost inconceivably long. Yet the war that is being waged against us in Central Europe today falls into a special category of its own. In the past we were persecuted *despite* the fact that we were the people of the Bible; today, however, it is just *because* we are the people of the Book that we are persecuted. The aim is to exterminate not only ourselves but to destroy, together with us, that spirit expressed in the Bible and in Christianity which made possible the rise of civilization in Central and Northern Europe. If this aim is achieved Europe will become a barren waste. For human community life cannot long endure on a basis of crude force, brutality, terror, and hate.

Only understanding for our neighbors, justice in our dealings, and willingness to help our fellow men can give human society permanence and assure security for the individual. Neither intelligence nor inventions nor institutions can serve as substitutes for these most vital parts of education.

Many Jewish communities have been uprooted in the wake of the present upheaval in Europe. Hundreds of thousands of men, women, and children have been driven from their homes and made to wander in despair over the highways of the world. The tragedy of the Jewish people today is a tragedy which reflects a challenge to the fundamental structure of modern civilization.

One of the most tragic aspects of the oppression of Jews and other groups has been the creation of a refugee class. Many distinguished men in science, art, and literature have been driven from the lands which they enriched with their talents. In a period of economic decline these exiles have within them the possibilities for reviving economic and cultural effort; many of these refugees are highly skilled experts in industry and science. They have a valuable contribution to make to the progress of the world. They are in a position to repay hospitality with new economic development and the opening up of new opportunities of employment for native populations. I am told that in England the admission of refugees was directly responsible for giving jobs to 15,000 unemployed.

As one of the former citizens of Germany who have been fortunate enough to leave that country, I know I can speak for my fellow refugees, both here and in other countries, when I give thanks to the democracies of the world for the splendid manner in which they have received us. We, all of us, owe a debt of gratitude to our new countries, and each and every one of us is doing the utmost to show our gratitude by the quality of our contributions to the economic, social, and cultural work of the countries in which we reside.

It is, however, a source of gravest concern that the ranks of the refugees are being constantly increased. The developments of the past week have added several hundred thousand potential refugees from Czechoslovakia. Again we are confronted with a major tragedy for a Jewish community which had a noble tradition of democracy and communal service.

The power of resistance which has enabled the Jewish people to survive for thousands of years is a direct outgrowth of Jewish adherence to the Biblical doctrines on the relationship among men. In these years of affliction our readiness to help one another is being put to an especially severe test. Each of us must personally face this test, that we may stand it as well as our fathers did before us. We have no other means of self-defense than our solidarity and our knowledge that the cause for which we are suffering is a momentous and sacred cause.

The obverse to Einstein's pride in Jewish righteousness and intellectual achievement was a fierce antipathy to the moral degradation of a Germany that had so richly benefited from them.

Credo as a Jew
Universal Jewish Encyclopedia, 4 (1941), 32–33

The striving after knowledge for its own sake, the love of justice verging on fanaticism, and the quest for personal independence— these are the motivating traditions of the Jewish people which cause me to regard my adherence thereto as a gift of destiny. Those who rage today against the ideals of reason and of individual freedom, and seek to impose an insensate state of slavery by means of brutal force, rightly see in us their irreconcilable opponents. History has imposed upon us a severe struggle. But as long as we remain devoted servants of truth, justice, and freedom, we shall not only continue to exist as the oldest of all living peoples, but we shall also, as hitherto, create, through productive effort, values which shall contribute to the ennobling of mankind.

Shortly after the end of World War I, Einstein had expressed the conviction that no country had a monopoly on moral behavior (CPAE 9,

Doc. 80). Initially, too, he was optimistic that Germany as a republic might make a clean break with the bellicose posturing of its imperial past. In the following statement, however, he ascribes the persistence of this belligerence to a ruling caste in Germany which successfully defended its material and intellectual hegemony during the 1920s.

The statement commemorates the third anniversary of the Nazi closure of all Czech universities, the execution of nine student leaders in Prague, and the deportation of 1,200 students to the concentration camp of Sachsenhausen.

On the Mentality of the German Ruling Caste, 17 November 1942
Einstein Archives 28-577

It is important that we be vividly aware of the mass murders which the Germans have committed against the civilian population of the occupied countries. The Germans are intent everywhere on exterminating those educated strata of society which represent a nation's independent spirit. They seek not only to exterminate freedom itself but the vital will to be free and independent as well. The motive for shooting student demonstrators in Czechoslovakia is the same as that for persecuting Jews.

The mentality which underlies all these crimes is not limited to the Nazi Party. It is entrenched in the German ruling caste. Earlier the Kaiser appeared to be responsible for all evil, now it seems to be Hitler. To believe this is a disastrous mistake, which has as a consequence the conclusion of a second feeble peace. One will have to disabuse oneself of the misjudgment that this mentality which endangers all of humanity may be abolished in the foreseeable future by the liberal development of Germany from within. Let us hope that political leaders clearly recognize this when the imminent peril has temporarily been surmounted.

After the policy of humiliation, expropriation, and exile visited earlier upon the Jews, Hitler turned to the madness of mass murder. News of wholesale killing of Jews in areas conquered by German troops trickled into the West in the course of 1942. By November of that year, Einstein's friend Stephen Wise broke the news to the public of Hitler's detailed plan to annihilate European Jewry.

Grasping for an explanation of these unspeakable crimes, Einstein attributed them to the hatred of those who bitterly resented the "harmonious partnership" of intellect and creativity in the Jewish people. The following is a transcript of an NBC radio address for the United Jewish Appeal.

The Goal of Human Existence, 11 April 1943
Einstein Archives 28-587; Einstein 1956, 260–261

Our age is proud of the progress it has made in man's intellectual development. The search and striving for truth and knowledge is one of the highest of man's qualities—though often the pride is most loudly voiced by those who strive the least. And certainly we should take care not to make the intellect our god; it has, of course, powerful muscles, but no personality. It cannot lead, it can only serve; and it is not fastidious in its choice of a leader. This characteristic is reflected in the qualities of its priests, the intellectuals. The intellect has a sharp eye for methods and tools, but is blind to ends and values. So it is no wonder that this fatal blindness is handed on from old to young and today involves a whole generation.

Our Jewish forbears [sic], the prophets and the old Chinese sages understood and proclaimed that the most important factor in giving shape to our human existence is the setting up and establishment of a goal; the goal being a community of free and happy human beings who by constant inward endeavor strive to liberate themselves from the inheritance of anti-social and destructive instincts. In this effort the intellect can be the most powerful aid. The fruits of intellectual effort, together with the striving itself, in cooperation with the creative activity of the artist, lend content and meaning to life.

But today the rude passions of man reign in our world, more unrestrained than ever before. Our Jewish people, a small minority everywhere, with no means of defending themselves by force, are exposed to the cruelest suffering, even to complete annihilation, to a far greater degree than any other people in the world. The hatred raging against us is grounded in the fact that we have upheld the ideal of harmonious partnership and given it expression in word and deed among the best of our people.

Deeply troubled by the brutality of events in wartime Europe, Einstein went one step farther than his suggestion in spring 1938 to abandon the principle of nonintervention in the affairs of other nations. Then—even before the Nazi pogroms were systematically unleashed—he had argued that "the necessity of adhering to the principle of noninterference in the internal affairs of other countries" had to be abandoned whenever any government arrogated to itself "the right to conduct a systematic campaign of physical destruction of any segment of the population which resides within its borders" (N & N 1960, 278). Now—a month after the preceding speech (May 1943)—he wrote that "one would truly be doing the peoples of Europe a favor if without any sentimentality one ended the existence of sovereign states" (Einstein Archives 28-593).

The Palestine Problem

Philip K. Hitti, Professor of Semitic Literature and head of the Department of Near Eastern Studies at Princeton University, was author (in 1937) of the first comprehensive study of the Arabs in the English language. In early 1944 he gave testimony before the Committee on International Relations of the U. S. House of Representatives on the question of Jewish immigration to Palestine. His statement was published in the *Princeton Herald* on 7 April under the header "Dr. Hitti Opposes Opening Palestine for Jewish State." Einstein and a colleague at the Institute for Advanced Study, Erich (Eric) Kahler, felt compelled to respond a week later. The debate continued in print with a further exchange in the following two weeks, Hitti arguing that Jews had no historical priority in Palestine, Einstein and Kahler denying this while emphasizing their moral right to live under self-rule in the Holy Land. The following reply was "jointly conceived" by Einstein and Kahler, "and written by Erich Kahler" (Kahler, 123).

The same year, the Christian Council on Palestine and American Palestine Committee published Einstein and Kahler's two articles in a booklet entitled *The Arabs and Palestine.* Two decades later Kahler republished all four essays in *The Jews among the Nations* (1967), 123–149.

PALESTINE, SETTING OF SACRED HISTORY OF THE JEWISH RACE (WITH ERICH KAHLER)
PRINCETON HERALD, 14 APRIL 1944

The presentation of the Palestine problem by Professor Hitti in the last issue of *The Princeton Herald* is so one-sided that it cannot go unanswered. Before considering Professor Hitti's views we want, however, to state that we do not speak in the name of the Zionist movement but as non-partisan Jews and plain human beings.

Professor Hitti defends the Arab stand on ethnical, religious and political grounds. The Arabs, he says, are descendants of the ancient Canaanites who held the land before the Jews. Jerusalem is to the Arabs the third holy city; it is the direction in which the early Arabs prayed, and the land was given to them by Allah as the result of a *jihad*, a holy war.

We do not believe that in our epoch these are the real issues that influence the turn of events, but we have to deal with them as stressed by Professor Hitti.

Both Jews and Arabs are said to stem from a common ancestor, from Abraham who immigrated into Canaan, i.e. Palestine, and so neither of them seem to have been earlier in the land than the others. Recent views assume that only part of the Israelites migrated to Egypt—as reflected in the Joseph story—and part of them remained in Palestine. So part of the Canaanite population encountered by the Jews when they entered the promised land under Joshua were Israelites too. Therefore, the Arabs have no priority on the land.

To the Arabs Jerusalem is only the third holy city, to the Jews it is the first and only holy city and Palestine is the place where their original history, their sacred history took place. Besides, to the Arabs Jerusalem is a holy city only insofar as they trace their tradition back to Jewish origins, insofar as after the Arab conquest of Jerusalem in 637, the "Omar Mosque," the "Dome of the Rock" was erected by the Omayyad Caliph Abd el Malek on the very place where the Jewish Ark of the Covenant and the Temple of Solomon had stood, on a rock *"even hetijah"* (world foundation stone) which was considered by the Jews as reaching down to the bottom of the cosmic ocean, the navel of the world. And Jerusalem was a *gibah*, a direction of prayer, under Mohammed only as long as he counted on the Jews as the main supporters of his new creed; he changed it, when his hopes failed, together with other institutions established out of pure consideration for his Jewish adherents, as for instance the fasting on the Jewish Day of Atonement. The first *gibah* has, therefore, as much validity for the Arabs as the Jewish Day of Atonement—both are today abolished in their religious significance. It seems a little far-fetched to use this abrogated rite as an evidence on which to base the Arab claim to Palestine.

If, finally, the Arab conquest of Palestine is considered holy it would be only fair to admit the corresponding holiness of the peaceful claim and the peaceful reclamation of the country by the Jews. To refer to the legitimacy of a "holy war" sounds rather queer for a people which denounces peaceful immigration as a violation of their rights. No wonder, Professor Hitti, on the one hand, uses as a threat the overwhelming Arab power and, on the other hand, plays on the Nazi insinuations to which the Arab world is said to be highly susceptible: that a tiny Jewish community in Palestine of two or three million at the most would become a danger to four mighty Arab states and fifty million Arabs.

But the Jews do not resort to arguments of power or of priority. One does not get very far with historical rights. Very few peoples of the world would be entitled to their present countries if such a criterion were applied. Professor Hitti says the Arabs cannot understand why the Jewish problem which is not of their making should be solved at their expense. But by their holy war and their conquest of Palestine the Arabs contributed their share to depriving the Jews of their homeland and so to the making of the Jewish problem, even though one must concede that their share is comparatively smaller than that of other peoples. The stand the Arabs take, however, with regard to the Jews is exactly the one which all peoples of the world are taking. No people, unfortunately, understands why it should contribute anything to the solution of the Jewish problem. The surface of the globe is everywhere occupied, and wherever the Jews could be given a piece of land under fair climatic conditions they would encroach on some property rights and sovereignties and would face friction with a population already firmly established on the spot. No country has been found where the Jews could possibly form an autonomous community, however small.

There is still one difference between the other peoples and the Arabs. Every people has one country of its own which it developed with all the care of its generations and none of these countries has any connection with a specifically Jewish tradition or concern. The Arabs possess four major countries—Saudi Arabia, which harbors their holy places, Yemen, Iraq, and Transjordan—if we leave aside

Egypt, which is only partly Arab, Syria and all the North African colonies and provinces as yet not enfranchised from European rule. And the least and obviously most neglected of their settlements was the part they occupied in the tiny Palestinian country; only nine hundred thousand of fifty million Arabs live there. This tiny Palestinian country, on the other hand, is the only place in the world legitimately and most deeply connected with the Jewish people, its religious foundation and its historic tradition as an independent people.

In order to clarify the Palestinian problem let us compare the situation of the Jews with that of the Arabs. The Jews are and have always been numerically a small people. They have never exceeded fifteen and a half million. Deprived of their homeland through the ancient and medieval conquests of Palestine, they lived dispersed all over the world and what they have suffered since by persecutions, expulsions and tortures of all kinds is far beyond anything that other peoples had to endure. Of the fifteen and a half million computed in 1938 at least two millions have been slaughtered or starved to death by the Nazis in the various European countries during the past few years. So the Zionist movement, or better the striving for a haven in the place of Jewish origin, is by no means an "exotic, artificially stimulated movement" as Professor Hitti calls it, but a movement urged forward by utter need and distress. The promise held out to the Jews in the Balfour Declaration after the First World War has been whittled down bit by bit in the course of the British appeasement policy yielding to interests partly British, partly Arabian—a policy bitterly denounced by Churchill himself before he became Prime Minister. Palestine is a link in the lifeline of the British Empire between the Near East and India; and the Jewish people, by necessity a dependable ally of the British, have been sacrificed to the Arabs who, by their numerical and political strength and the trump of the Islam portion of the Indian population, were in a position to sell dearly even their neutrality in the present conflict. The final result has been the complete prohibition of Jewish immigration into Palestine at the very moment when some more hundreds of thousands of Jews are threatened with annihilation by the Hitler armies occupying Hungary and Rumania. . . .

This is the Jewish situation; and there is no guarantee whatever against the persistence or recurrence of antisemitic outbreaks everywhere after this war. Even if we put aside the spiritual, religious and cultural ties making Palestine the only place in the world which persecuted Jews could consider their home and develop with all the devotion a homeland inspires—there is not even any other country acceptable to human beings which the numerous refugee conferences were able to offer to this hounded people. The Jews are prepared for extreme sacrifices and hardest work to convert this narrow strip which is Palestine into a prosperous country and model civilization. What Jewish youth has already achieved in the few decades of Zionist settlement may be gathered from Mr. Lowdermilk's book. They took over from the period of Arabian predominance deserts and rocks and barren soil and turned them into flowering farms and plantations, into forests and modern cities. They created new forms of cooperative settlements and raised the living standard of the Arabian and the Jewish population alike. The Jews are willing and ready to give any guarantee of protection for the holy places and the civil rights, indeed the autonomy, of Arabs and Christians, a guarantee safeguarded by the overwhelming power of their neighbors on whose cooperation they depend. They offer their assistance and their experience for the economic and scientific advancement of the Arab countries, for the lifting of the population to a modern standard of living.

But this, unfortunately, is just what the Arab leaders do not want. For the true source of Arab resistance and hostility toward a Jewish Palestine is neither religious nor political, but social and economic. The Arabian population of Palestine is negligible in comparison with the vast number of Arab elements in the European provinces of North Africa and Asia. The Arabian chieftains did not arouse the Moslem world against Mussolini's regime in

Libya; most of them were on splendid terms with him. The Mufti of Jerusalem and other Arab leaders were greatly honored guests in Rome. The rich Arabian landowners did nothing to improve the nature, the civilizations, or the living standards of their countries. The large Arabian states are underpopulated, the masses of the people are held in a backward and inferior condition. "Life in the Damascus of the eighth century was not greatly different from what it is today," says Professor Hitti in his book about the Arabs. But the big Effendis fear the example and the impulse which the Jewish colonization of Palestine presents to the peoples of the Near East, they resent the social and economic uplift of the Arabian workers in Palestine. They act as all fascist forces have acted: they screen their fear of social reform behind nationalistic slogans and demagoguery. If it were not for these leaders and instigators a perfect agreement and cooperation could be achieved between the Arab and the Jewish people.

The purpose of this statement is not a nationalistic one. We do not, and the vast majority of Jews does not, advocate the establishment of a state for the sake of national greed and self-glorification, which would run counter to all the traditional values of Judaism and which we consider obsolete everywhere. In speaking up for a Jewish Palestine, we want to promote the establishment of a place of refuge where persecuted human beings may find security and peace and the undisputed right to live under a law and order of their making. The experience of many centuries has taught us that this can be provided only by home rule and not by a foreign administration. This is why we stand for a Jewish controlled Palestine, be it ever so modest and small. We do not refer to historic rights, although if there exists something like a historic right on a country, the Jews, at least as well as the Arabs, could claim it on Palestine. We do not resort to threats of power, for the Jews have no power; they are, in fact, the most powerless group on earth. If they had had any power they should have been able to prevent the annihilation of millions of their people and the closing of the last door to the helpless victims of the Nazi. What we appeal to is an elementary sense of justice and humanity. We know how weak such a position is, but we also know that if the arguments of threats of power, of sacred egoisms and holy wars continue to prevail in the future world order, not only the Jews but the whole of humanity will be doomed.

Walter Clay Lowdermilk, author of *Palestine, Land of Promise* (1944), was for a time chief of the Soil Conservation Service of the United States Department of Agriculture.

Hitti's rejoinder was published the following week in the *Princeton Herald* under the headline "Palestinian Arabs Descended from Abraham, says Dr. Hitti." After somewhat dismissively challenging Zionist arguments for settlement that were based on historical priority, humanitarian concerns, and "successful cultivation of the soil," the Arab scholar called for "a *Palestinian* state—neither Jewish nor Moslem—in which all citizens, regardless of faith or origin become equal and free citizens." Einstein did not take up the gauntlet when he and Kahler responded one week later.

Arabs Fare Better in Palestine than in Arab Countries (with Erich Kahler)
Princeton Herald, 28 April 1944

Professor Hitti found some minor "bubbles" of ours to prick while leaving the major ones undisturbed. As we shall presently see, however, even those he pricked still float in the sun. . . .

Professor Hitti terms Jewish immigration into Palestine an "attenuated invasion" and a "creeping conquest." The difference between a regular conquest and this "creeping conquest" is that the one results in the ruin, the other in the rise in the "conquered" population. The improvement of the living conditions of the Arabs through the Zionist enterprise is an established fact confirmed by British official reports. The British Royal Commission that investigated Palestine in the winter 1936–37 made the following statements: "1) The large import of Jewish capital into Palestine had a fructifying effect on the economy of the country. 2) The expansion of Arab industry and citriculture has been largely financed by the capital thus obtained. 3) Jewish example has done much to improve Arab cultivation, especially citrus. 4) Owing to Jewish development and enterprise the employment of Arab labor has increased in urban areas, particularly in the ports. 5) The reclamation and anti-malaria work undertaken in Jewish villages have benefitted all Arabs in the neighborhood. 6) Institutions founded with Jewish funds primarily to serve the National Home, have also served the Arab population. Hadassah, for example, notably at the Tuberculosis Institute at Jerusalem, admits Arab country folk to the clinics of the Rural Sick Benefit Fund and does much infant welfare work for Arab mothers. 7) The general beneficent effect of Jewish immigration on Arab welfare is illustrated by the fact that the increase in the Arab population is most marked in urban areas affected by Jewish development. 8) The whole range of public services has steadily developed to the benefit of the fellaheen (the Arab peasants) . . . the revenue available for those services having been largely provided by the Jews." The Jewish Agency, being intended for the promotion of Jewish enterprise, is, of course, bound to patronize Jewish labor. This is far from signifying a boycott against the Arabs. Arab workers are employed in great numbers in privately owned Jewish plantations and industries. Wages in Palestine are more than double those in Syria and three times as high as those in Iraq.

Let us compare these conditions in Palestine with those in Arab ruled countries. "The situation of the fellaheen in Iraq is very poor," says W. C. Lowdermilk, "in fact, even in overpopulated China I never saw conditions so bad as those I found in the underpopulated but potentially rich lands of the Tigris-Euphrates Valley." Another expert on the country, Ernest Main, reports: "The fellaheen and coolie classes were living on less than a penny a day per head.... There are probably about two million people in the country living on such standards, and it can be imagined what purchasing power they possess, and what revenue they can offer." As to the conditions of the peasants in Trans-Jordan which is included in the British Mandate for Palestine, the High Commissioner, Sir Arthur Wauchope, pointed out: ... "Owing to the tax-payers' poverty (the Government) could only be carried on by means of grants-in-aid"—by charity, Professor Hitti would say. If the Arab peasants and workers did not find better living conditions in Palestine it would be hardly understandable that between 1933 and 1936, for instance, more than 30,000 Arab workers from Iraq, Syria, Trans-Jordan and even the Arabian desert migrated to Palestine. On the other hand, there was twice as much Arab emigration from the Arab countries as from Palestine.

To reproach Palestinian economy with being not self-supporting, as Professor Hitti does, is equivalent to blaming a child for being dependent on its family. Jewish economy had to be built up from scratch, land had to be purchased at prices far higher than the land was actually worth, three or four times as high as a similar type of land would sell in Syria or in Southern California. Machinery, fertilizer, raw materials were lacking. And still, even Professor Hitti has to admit that imports decreased by fifty percent from 1927 to 1937. On the prospects of the country we may refer to the testimony of Sir Charles Warren, one of the British scholars of the Palestine Exploration Fund, who wrote as early as 1875: "Give Palestine a good government and increase the commercial life of the people and they may increase tenfold and yet there is room." And no suspicion of bias can certainly arise as to the statement of T. E. Lawrence, "Lawrence of Arabia," one of the most ardent friends the Arabs ever had: "Palestine was a decent country (in ancient times), and could so easily be made so again. The sooner the Jews farm it all the better: their colonies are bright spots in a desert."

There is one point in which we may agree with Professor Hitti: The Jews too have their diehards and their terrorists—although proportionately far less than other peoples. We do not shield nor excuse these extremists. They are a product of the bitter experience that in our present world only threat and violence are rewarded and that fairness, sincerity and consideration get the worst of it. As far as Dr. Weizmann is concerned, however, we have to correct Professor Hitti's quotation. He never threatened the Arabs with expulsion. The passage to which Professor Hitti refers reads: "There will be complete civil and political equality of rights for all citizens without distinction of race or religion, and, in addition, the Arabs will enjoy full autonomy in their own internal affairs. But if any Arabs do not wish to remain in a Jewish state, every facility will be given to them to transfer to one of the many and vast Arab countries."

There was a time, in 1919, when a perfect Arab-Jewish-British agreement was worked out by the late King Feisal, a nobler brand of leader than the present chiefs, Dr. Weizmann and T. E. Lawrence. Feisal declared that "the Arabs, especially the educated among us, look with deepest sympathy on the Zionist movement. . . . Interested parties have been enabled to make capital out of what they call our differences. . . . I wish to give you my firm conviction that these differences . . . are easily dispelled by mutual good-will."

Let us close our discussion with the fervent hope that this spirit of the great Arab leader will dominate the postwar arrangements and that matters will be decided not on the narrow scope of vested interests and local prevalences but from the broad point of view of human welfare at large.

Einstein's precise contribution to the two preceding essays is unknown. It remains a curious fact, though, that in their joint response to Hitti,

Einstein at least did not address that very element in the Arab professor's presentation to which he, the cultural Zionist, had always assigned pride of place. Hitti summarized it as follows in his second essay: "The great contribution of Israel throughout the ages has been in the spiritual and intellectual rather than the political realm."

Bearing the Brunt

In 1944 Einstein was interviewed about the social obligations of scientists and the possibility of reeducating German citizens after the end of the war. He responded to the first issue by stating that it was the duty of every citizen to take part in political affairs: "If this duty is neglected by intelligent citizens capable of judgment, there can be no sound democracy." Regarding the second issue, he merely asserted that "the Germans can be killed or constrained, but they cannot be reeducated to a democratic way of thinking and acting within a foreseeable period of time" ("Our Goal Unity, but Germans Are Unfit," *Free World* 8 (October 1944), no. 4, 370–371). In commemorating the martyrs who fought to their deaths in the Warsaw ghetto, Einstein pronounced the German people collectively guilty for the mass murder.

To the Heroes of the Battle of the Warsaw Ghetto
Bulletin of the Society of Polish Jews, 1944; Einstein 1956, 265

They fought and died as members of the Jewish nation, in the struggle against organized bands of German murderers. To us these sacrifices are a strengthening of the bond between us, the Jews of all the countries. We strive to be one in suffering and in the effort to achieve a better human society, that society which our prophets have so clearly and forcibly set before us as a goal.

The Germans as an entire people are responsible for these mass murders and must be punished as a people if there is justice in the world and if the consciousness of collective responsibility in the nations is not to perish from the earth entirely. Behind the Nazi party stands the German people, who elected Hitler after he had in his book and in his speeches made his shameful intentions clear beyond the possibility of misunderstanding. The Germans are the only people who have not made any serious attempt of counteraction leading to the protection of the innocently persecuted. When they are entirely defeated and begin to lament over their fate, we must not let ourselves be deceived again, but keep in mind that they deliberately used the humanity of others to make preparation for their last and most grievous crime against humanity.

Einstein was never an admirer of the nation-state. The annihilation of European Jews stiffened his resolve that the principle of nonintervention in a nation's internal affairs was not an absolute and had to be overturned (see his comment of spring 1938 which follows "The Goal of Human Existence" above). A more immediate consequence of the Nazi terror campaign was Einstein's increased determination to insist on unrestricted Jewish immigration to Palestine.

His resolve on both issues is apparent in his unpublished preface to the Black Book, a publication venture that was intended as a compilation of Nazi crimes against the Jewish people. Sponsoring organizations were the World Jewish Congress, the Jewish Anti-Fascist Committee of the U.S.S.R., the American Committee of Jewish Writers, Artists, and Scientists, of which Einstein served as honorary president, and the National Council of Palestine. After Einstein submitted the draft for a preface, the editor requested changes which Einstein refused to make, arguing that it would "limit the effectiveness of what was written." He suggested that another foreword supplement his own or that his contribution be replaced (letter to Ursula Wassermann, 26 August 1945). The volume was published in 1946 (*The Black Book: The Nazi Crime against the Jewish People*) without a preface.

PREFACE TO A JEWISH BLACK BOOK, BEFORE 24 AUGUST 1945
EINSTEIN ARCHIVES 28-654; EINSTEIN 1956, 258–259

This book is a collection of documentary material on the systematic work of destruction by which the German Government murdered a great proportion of the Jewish people. Responsibility for the truth of the facts set forth is borne by the Jewish organizations that have joined to create the present work and present it to the public.

The purpose of this publication is manifest. It is to convince the reader that an international organization for safeguarding the sanctity of life can effectively fulfill its purpose only if it does not limit itself to protecting countries against military attack but also extends its protection to national minorities within the individual countries. For in the last reckoning it is the individual who must be protected against annihilation and inhuman treatment.

It is true that this goal can be attained only if the principle of non-intervention, which has played such a fateful role in the last decades, is cast overboard. Yet today no one can doubt the need for this far-reaching step any longer. For even those who envision only the attainment of protection against military attack from the outside must today realize that the disasters of war are preceded by certain internal developments in the various countries, and not merely by military and armaments preparations.

Not until the creation and maintenance of decent conditions of life for all men are recognized and accepted as a common obligation of all men and all countries—not until then shall we, with a certain degree of justification, be able to speak of mankind as civilized.

Percentagewise the Jewish people have lost more than any other people affected by the disasters of recent years. If a truly just settlement is to be striven for, the Jewish people must be given special consideration in the organization of the peace. The fact that the Jews, in the formal political sense, cannot be regarded as a nation, insofar as they possess no country and no government, ought to be no impediment. For the Jews have been treated as a uniform group, as though they were a nation. Their status as a uniform political group is proved to be a fact by the behavior of their enemies. Hence in striving toward a stabilization of the international situation they should be considered as though they were a nation in the customary sense of the word.

Another factor must be emphasized in this connection. In parts of Europe Jewish life will probably be impossible for years to come. In decades of hard work and voluntary financial aid the Jews have restored the soil of Palestine to fertility. All these sacrifices were made because of trust in the officially sanctioned promise given by the governments in question after the last war, namely that the Jewish people were to be given a secure home in their ancient Palestinian country. To put it mildly, the fulfillment of this promise has been but hesitant and partial. Now that the Jews— especially the Jews in Palestine—have in this war too rendered a valuable contribution, the promise must be forcibly called to mind. The demand must be put forward that Palestine, within the limits of its economic capacity, be thrown open to Jewish immigration. If supranational institutions are to win that confidence that must form the most important buttress for their endurance, then it must be shown above all that those who, trusting to these institutions, have made the heaviest sacrifices are not defrauded.

Official American and British Attitudes

The sympathy that Einstein felt for Franklin Roosevelt is suggested in a quote ascribed to him by his friend Frieda Bucky: "I'm so sorry that Roosevelt is president—otherwise I would visit him more often" (Calaprice, 96). When he died unexpectedly on 12 April 1945, a memorial meeting was organized in his honor, and Einstein was asked to pen some lines to be read at the event, which he did not attend. The occasion was hosted by the German-language newspaper *Aufbau*; its publisher, the New World Club; and the Society for Ethical Culture.

Commemorative Words for FDR
Aufbau 11 (27 April 1945), no. 17, 7

We Jews and immigrants are particularly bound and indebted to the dear departed one. It seldom happens that a person who has his heart in the right place also has the political genius and the will power needed in order to have a decisive and lasting influence on the course of history. President Roosevelt recognized the inevitable early on and took care that America would emerge victorious against the menacing danger represented by Germany. Within the limits of political possibilities, he worked successfully for the security of the weak and for a revitalization of the economy. The burden he carried was heavy, but his sense of humor gave him an inner freedom seldom found among those who are constantly faced with the most critical decisions. He was unbelievably bound and determined to attain his final goals, yet amazingly flexible in overcoming the strong resistance any farsighted statesman faces in a democratic country, where even those in the highest office have a limited amount of authority. No matter when this man might have been taken from us, we would have felt we had suffered an irreplaceable loss. It is tragic that he did not live to lend his unique abilities to solving the problem of international security. It is also tragic especially for us Jews that he, with his lively sense for justice, was unable to take part in the decisive negotiations which will determine whether our unspeakably hard-tested people will find a refuge, whether the gates of Palestine will be opened for the refugees and persecuted among us. For all people of good will Roosevelt's death will be felt like that of an old and dear friend. May he have a lasting influence on our thoughts and convictions.

In autumn 1945, Lessing Rosenwald, the president of the American Council for Judaism, an anti-Zionist organization, submitted a seven-point memorandum for a "just and peaceful resolution" of the Palestinian problem to President Truman. The memorandum was published as a full-page advertisement in the 20 November edition of the *New York Post*. Rosenwald's most stinging accusation was that Zionism demanded dual national allegiance of its adherents. The editors of *Aufbau* asked Einstein to respond, and his statement appeared there under the headline "Einstein Condemns Lessing Rosenwald." The same text was conveyed in a letter to the Committee on Unity for Palestine in New York (Einstein 1956, 279).

THE AMERICAN COUNCIL FOR JUDAISM
AUFBAU 11 (14 DECEMBER 1945), NO. 50, 11; EINSTEIN 1956, 273

I am very happy indeed to hear that the platform for which the American Council for Judaism stands is meeting with strong opposition. This organization appears to me to be nothing more than a pitiable attempt to obtain favor and toleration from our enemies by betraying true Jewish ideals and by mimicking those who claim to stand for 100 per cent Americanism. I believe this method to be both undignified and ineffective. Our opponents are bound to view it with disdain and even with contempt, and in my opinion justly. He who is untrue to his own cause cannot command the respect of others. Apart from these considerations, the movement in question is a fairly exact copy of the "Zentralverein Deutscher Staatsbürger Jüdischen Glaubens ("Central Association of German Citizens of Jewish Faith") of unhappy memory, which in the days of our crucial need showed itself utterly impotent and corroded the Jewish group by undermining that inner certitude by which alone our Jewish people could have overcome the trials of this difficult age.

Frank Aydelotte was director of Princeton's Institute for Advanced Study and a member of the Anglo-American Committee of Inquiry on

Jewish Problems in Palestine and Europe. At his invitation, Einstein traveled from Princeton to Washington to testify before the committee on 11 January 1946. Charged with inquiring into the political, economic, and social conditions in Palestine as they related to Jewish immigration and settlement, the committee was also empowered to assess the well-being of Arabs and Jews there, as well as conditions among the surviving Jews of Europe.

Einstein began his testimony by excusing himself for his "faulty English," and was asked to speak more loudly. Photographs taken of him at the session can be seen at Plates 18 and 19 of this volume. The *New York Times*, 12 January 1946, summarized the highlights of the previous day as follows:

1) Britain's colonial policy makes her unfit for further administration of her mandate over Palestine;

2) a trusteeship should be set up by the United Nations to administer Palestine, but it should not be confined to a single power, including the United States;

3) the great majority of Jewish refugees in Europe should be settled in Palestine;

4) there was no need to establish a Jewish commonwealth [state] in Palestine as advocated by the Zionist organizations.

TESTIMONY AT A HEARING OF THE ANGLO-AMERICAN COMMITTEE OF INQUIRY, 11 JANUARY 1946
EINSTEIN 1946, 118–135

EINSTEIN: . . . I wish to explain why I believe that the difficulties between the Jews and the Arabs are artificially created, and are created by the English. I believe, if there would be a really honest government for the people there, and get the Arabs and the Jews together, there would be nothing to fear. I cannot convince you gentlemen, but I can only say what convinces me.

I may first state what I think about British colonial rule. I find that the British colonial rule is based on a native. Do you know what that means? The native was exploited already before the English came into the land. Of course, the English had two interests. The first was to have raw materials for their industry. Also the oil in those countries. I find that everywhere there are big land owners who are exploiters of that race of people. These big land owners, of course, are in a precarious situation because they are always afraid that they will be gotten rid of. The British are always in a passive alliance with those land-possessing owners which suppress the work of the people in the different trades.

It is my impression that Palestine is a kind of small model of India. There is an attempt to dominate, with the help of a few officials, the people of Palestine, and it seems to me that the English rule in Palestine is absolutely of this kind. It is difficult to imagine how it could be otherwise. . . .

Now how can I explain otherwise than that national trouble-making is a British enterprise? It is not so easy to get information about all that is going on that cannot be directly proved. For instance, if there were pogroms against the Jews in Palestine, there was a taking away of arms so that the Jews could not defend themselves. It is hard to prove all this, so I will not insist too much on such a thing. But there are certain things which are for me very strong arguments. . . .

The most important thing for international relations is confidence in international rule. I believe that complete honesty in the procedure is the most important thing to create confidence. So I must add that I believe that the frame of mind of the colonial people of the British is so rigid that I am absolutely convinced that any councils will not have any effect.

I think commissions like this are like a smoke-screen to show good will. I believe that the Palestine people, under severe influence of the United Nations, will be able to create a better state of affairs. But, with the British rule as it is, I believe it is impossible to find a real remedy. I may be wrong, but that is my conviction. . . .

AYDELOTTE: One of the things which we must do is to figure out some kind of a report or some kind of advice with reference to what the authority shall be, or who is to have authority over Palestine. Now, do you think that the United Nations or some other outside authority should force the Arabs to allow unlimited immigration into Palestine, or do you think they should take the Arab point of view into consideration and close off immigration? Just what do you think should be done with Palestine by whatever power has the trusteeship over it?

EINSTEIN: Of course, it is very difficult to answer such a question in a general way. I believe that such a government should be composed for the people concerned. It should be handled from the human standpoint of the matter. For instance, there is great difficulty with the refugees. Of course, there should be done something about them. I believe it is natural to bring the bulk of them to Palestine. In Palestine the Jews who are already there will take care of the ones that are brought in. It is not true that they will be in trouble with the Arabs. I believe that such kind of action should not be taken from a political standpoint but from a human standpoint. It would be best for the population of Palestine to feed those people and take care of them. I believe it is quite natural that they can take into their homes people who have no place to stay.

AYDELOTTE: What would you do if the Arabs refused to consent to bringing these refugees to Palestine? Suppose the Arab population were prepared to resist it by force; would you compel them by force to receive the refugees?

EINSTEIN: That will never be the case if there is not politics. But there are not only Arab politicians, but Jewish politicians, as well.

JAMES MCDONALD: Would you eliminate the Jewish and Arab politicians both?

EINSTEIN: No, you cannot eliminate them. If you eliminate them, ten others grow up in their place. (Laughter.) . . .

AYDELOTTE: An Arab was talking this afternoon before you came in, and he is a man who has lived in Palestine the greater part of his life. He contends that the Arabs are afraid, that they need no instigation to resist Jewish immigration. They are afraid that the Zionists are trying to develop a majority in Palestine so that they will have political control. Of course, the Arabs are in the majority now, but they fear that the Jews may attain a majority and then they would be in the minority.

EINSTEIN: But who has created that mentality? If the people work together and stay in peace together, they will not care anything about the idea of who has the biggest number. The number doesn't count if it is not politically activized. Nobody is interested in how many people in the United States speak French, German, English, or Italian. It is all in the minds of the people. . . .

RICHARD CROSSMAN: Since the British are, according to your point of view, completely incompetent to rule in the various parts of the world where they have ruled—

EINSTEIN: No, oh, no.

CROSSMAN: Well, at least in Palestine—you say they should not rule Palestine. Would you be prepared to advocate publicly that the American people should take over the mandate and assume full military responsibility for unlimited Jewish immigration, and thereby prove—

EINSTEIN: No, I would not do that. I would be King of Palestine if I did that. God forbid! (Laughter.) . . .

CROSSMAN: Your point of view is that you wish to blame the British, and you are not prepared to suggest that the other great democracies, since we have failed, should take responsibility for carrying out the job which we have failed to do. We have failed, according to you. Why shouldn't you take the responsibility and show how wrong we are?

McDONALD: When he says "you," he means the United States.

CROSSMAN: Yes; the United States.

EINSTEIN: It should be done under an international regime, . . .

AYDELOTTE: Dr. Einstein, what is your attitude toward the idea of a political Zionism, a political Jewish state, as versus a cultural center? There are two conceptions of Palestine. You understand them.

EINSTEIN: Yes. I was never in favor of a state. . . .

JUDGE HUTCHESON: It has been told to our committee by the Zionists that the passionate heart of Jewry will never be satisfied until they have a Jewish state in Palestine. It is contended, I suppose, that they must have a majority over the Arabs. It has been told to us by the Arab representatives that the Arabs are not going to permit any such condition as that, that they will not permit having themselves converted from a majority into a minority.

EINSTEIN: Yes.

JOSEPH HUTCHESON: I have asked these various persons if it is essential to the right or the privilege of Jews to go to Palestine, if it is essential to real Zionism, leaving out of the picture the political side of the question, that a setup be fixed so that the Jews may have a Jewish state and a Jewish majority without regard to the Arab view. Do you share that point of view, or do you think the matter can be handled on any other basis?

EINSTEIN: Yes, absolutely. The state idea is not according to my heart. I cannot understand why it is needed. It is connected with many difficulties and a narrow-mindedness. I believe it is bad. . . .

Reaction in the Zionist community to this last comment was one of dismay. Einstein's friend Rabbi Stephen Wise pleaded with him to renounce the concept of binationalism, but Einstein clung to his faith in "a solution on the basis of an honestly bi-national character" (letter of 14 January 1946). One month later he approved a compromise plan by the recently formed Progressive Palestine Association in which the Arab side would allow a maximum of 200,000 European Jewish refugees in exchange for a renunciation of the demand for a Jewish state by the Zionist groups. Einstein wrote the chairman of the association's Washington branch: "I agree whole-heartedly with your program and I am convinced that your work will be of real value for the solution of the hard problems the Palestine situation is presenting" (*New York Times*, 15 February 1946).

In reflecting on his testimony in Washington, Einstein wrote to the British committee member most sympathetic to Zionism: "You will now agree that my remark was not quite unjustified; I mean the remark that in the eyes of the British government the Commission was looked upon only as a smokescreen" (letter to Richard Crossman, 3 June 1946, cited in Sayen, 237). To his closest friend he reiterated the convictions expressed in his testimony and evoked an idea newly dear to his heart—the need for world government (see chapter 8 for more on this topic).

Letter to Michele Besso, 21 April 1946
Einstein Archives 7-381; Einstein/Besso, Doc. 146

The article which you sent me gives a fairly faithful account of my testimony before the Palestine Commission in Washington. If you had an inkling with what treachery the English implement their time-tested principle of "divide et impera," you would not have been so amazed as you apparently were by the harshness of my accusation. My testimony was based on very reliable sources and was far more precise than the article claims.

If you occasionally hear my name mentioned in connection with political excursions, don't think that I spend much time on such things, as it would be sad to waste much energy on the meager soil of politics. From time to time, however, the moment arrives when I cannot help myself, for instance, when one can draw the public's attention to the necessity of a world government, without which all our human grandeur will go to the dogs. . . .

Events were rapidly marginalizing Einstein's hopes. Though the Anglo-American Committee declared its opposition to the White Paper of 1939 and proposed that the immigration of 100,000 European Jews be authorized immediately, the British Mandatory Authority and His Majesty's government rejected the proposal, stating that such immigration was impossible while armed organizations in Palestine—both Arab and Jewish—were disputing its authority and disrupting public order.

Striving for Right and Justice

After the genocide in Europe, developments within the Zionist movement swept aside Einstein's cautionary voice. In late November 1947 the United Nations General Assembly approved a plan to partition Palestine. Heavy fighting broke out with excesses committed on both sides. The virtual state of war that now existed between Jews and Arabs, coupled with the psychological shock provided by the mass murder of the Jews of Europe, magnified the calls for a Jewish state in Palestine. They also provided a veneer of justification for those who, like the Revisionists, advocated terror in its name. In commenting on the increasing acts of brutality, Einstein had written a confidant in early 1946 that while men of reason might say what they think, "the facts . . . will be determined mainly by the cerebellum—that is, by the 'men of action'." To the same correspondent at the beginning of the following year: "With respect to Palestine we have advocated unreasonable and unjust demands under the influence of demagogues and other loud-mouths. Our impotence is bad. If we had power it might be worse still. We imitate the stupid nationalism and racial nonsense of the *goyim* even after having gone through a school of suffering without equal" (letters to Hans Mühsam, 3 April 1946 and 22 January 1947).

The social psychologist Erich Fromm wrote two drafts of the following appeal before it was published as a letter to the editor in the *New York Times* with the prominent rabbi Leo Baeck and Einstein as the only signatories.

APPEAL TO JEWS TO WORK FOR GOAL OF COMMON WELFARE
(WITH LEO BAECK), 12 APRIL 1948
NEW YORK TIMES, 18 APRIL 1948

Both Arab and Jewish extremists are today recklessly pushing Palestine into a futile war. While believing in the defense of legitimate claims, these extremists on each side play into each other's hands. In this reign of terror the needs and desires of the common man in Palestine are being ignored.

We believe that in such a situation of national conflict it is vitally important that each group and particularly its leaders uphold standards of morality and reason in their own ranks rather than confine themselves to accuse their opponents of the violation of these standards. Hence we feel it to be our duty to declare emphatically that we do not condone methods of terrorism and of fanatical nationalism any more if practiced by Jews than if practiced by Arabs. We hope that responsible Arabs will appeal to their people as we do to the Jews.

Were war to occur, the peace would still leave the necessity of the two peoples working together, unless one or the other were exterminated or enslaved. Short of such a calamity, a decisive victory by either would yield a corroding bitterness. Common sense dictates joint efforts to prevent war and to foster cooperation now.

Jewish-Arab cooperation has been for many years the aim of far-sighted Jewish groups opposed to any form of terror. Recently a declaration of such a group was published in the American press under the dateline Jerusalem, March 28, 1948, to which we want to draw attention. We quote here some of the key sentences:

"An understanding between the two peoples is possible, despite the constant refrain that Jewish and Arab aspirations are irreconcilable. The claims of their extremists are indeed irreconcilable, but the common Jew and the common Arab are not extremists. They yearn for the opportunity of building up their common country, the Holy Land, through labor and cooperation."

The signers of the statement represent various groups in Palestine Jewry. Besides Dr. Magnes, the chairman, those who signed were Dr. Martin Buber, Professor of Jewish Philosophy at Hebrew University; . . .

Those who signed this declaration represent at the moment only a minority. However, besides the fact that they speak for a much wider circle of inarticulate people, they speak in the name of principles which have been the most significant contribution of the Jewish people to humanity.

We appeal to the Jews in this country and in Palestine not to permit themselves to be driven into a mood of despair or false heroism which eventually results in suicidal measures. While such a mood is undoubtedly understandable as a reaction to the wanton destruction of six million Jewish lives in the last decade, it is nevertheless destructive morally as well as practically.

We believe that any constructive solution is possible only if it is based on the concern for the welfare and cooperation of both Jews and Arabs in Palestine. We believe that it is the unquestionable right of the Jewish community in Palestine to protect its life and work, and that Jewish immigration into Palestine must be permitted to the optimal degree.

The undersigned plead with all Jews to focus on the one important goal: the survival and permanent development of the Jewish settlement in Palestine on a peaceful and democratic basis, the single one which secures its future in accordance with the fundamental moral and spiritual principles inherent in the Jewish tradition and essential for Jewish hope.

The first draft of the appeal contained a far more damning excerpt from the Jerusalem declaration of 28 March: "We appeal, more particularly, to our Jewish brethren: do not desecrate your name and honour. If we also follow the rabble and the incited mob, not only shall we achieve nothing positive, but we shall only be contributing to a worsening of the situation, to an increase of hatred, and to reprisal after reprisal, without distinction and without mercy. We appeal to public opinion and to the Jewish leadership to take every possible step to prevent these vicious mob attacks. . . . Let these recent regrettable incidents serve as a warning, not to let the mob rule us, not to destroy with our own hands the moral foundations of our life and of our future" (Fromm, 229).

Stubbornly optimistic, Einstein returned to the theme of the harmonious society touched on in "The Goal of Human Existence," a text presented earlier in this chapter.

Before the Monument to the Martyred Jews of the Warsaw Ghetto, 19 April 1948
Einstein Archives 28-815; Einstein 1956, 266–267

The monument before which you have gathered today was built to stand as a concrete symbol of our grief over the irreparable loss our martyred Jewish nation has suffered. It shall also serve as a reminder for us who have survived to remain loyal to our people and to the moral principles cherished by our fathers. Only through such loyalty may we hope to survive this age of moral decay.

The more cruel the wrong that men commit against an individual or a people, the deeper their hatred and contempt for their victim. Conceit and false pride on the part of a nation prevent the rise of remorse for its crime. Those who have had no part in the crime, however, have no sympathy for the sufferings of the innocent victims of persecution and no awareness of human solidarity. That is why the remnants of European Jewry are languishing in concentration camps and the sparsely populated lands of this earth close their gates against them. Even our right, so solemnly pledged, to a national homeland in Palestine is being betrayed. In this era of moral degradation in which we live the voice of justice no longer has any power over men.

Let us clearly recognize and never forget this: That mutual co-operation and the furtherance of living ties between the Jews of all lands is our sole physical and moral protection in the present situation. But for the future our hope lies in overcoming the general moral abasement which today gravely menaces the very existence of mankind. Let us labor with all our powers, however feeble, to the end that mankind recover from its present moral degradation and gain a new vitality and a new strength in its striving for right and justice as well as for a harmonious society.

Only hours before the British Mandate was to come to an end and before the beginning of the Jewish Sabbath, on Friday, 14 May 1948, David Ben-Gurion read the Declaration of the Establishment of the State of Israel to a meeting of representatives of the Jewish community in Palestine (the *Yishuv*) and the Zionist movement. Almost immediately Arab armies began a combined military attack on the new state.

Half a year later, together with a number of other prominent Jewish intellectuals, including Hannah Arendt and Sidney Hook, Einstein signed an open letter calling attention to what the group considered among "the most disturbing political phenomena of our times . . . the emergence in the newly created state of Israel of . . . a political party [Freedom Party] closely akin in its organization, methods, political philosophy and social appeal to the Nazi and Fascist parties." Wishing to dispel the impression of American support for the party in the upcoming Israeli elections, the signatories protested the visit to the United States of the party's leader, Menachem Begin, head of "the former Irgun Zvai Leumi, a terrorist, right-wing, chauvinist organization in Palestine." Among other actions, the Irgun had perpetrated an atrocity at the defenseless village of Deir Yassin, in which some 240 people—men, women, and children—had been killed in April 1948 (letter to the editor of the *New York Times*, 2 December 1948). For Einstein, the Irgun and its successor Freedom Party were only the most egregious example of a tendency that he had always resisted—the assertion of rights through force of arms, though, as in this letter, he recognized the pressing need for self-defense in the face of the combined military attack of five Arab armies and numerous irregulars.

Despite his fear of growing violence on both sides, he retained his optimism. On receiving an honorary degree from the Hebrew University in March 1949, Einstein wrote of his sense of fulfillment with the flourishing of that institution and stressed his hope once again that Jew and Arab might still learn to coexist.

To the University of Jerusalem, 15 March 1949
EINSTEIN ARCHIVES 28-85; EINSTEIN 1956, 272

The little that I could do, in a long life favored by external circumstances to deepen our physical knowledge, has brought me so much praise that for a long time I have felt rather more embarrassed than elated. But from you there comes a token of esteem that fills me with pure joy—joy about the great deeds that our Jewish people have accomplished within a few generations, under exceptionally difficult conditions, by itself alone, through boundless courage and immeasurable sacrifices. The University which twenty-seven years ago was nothing but a dream and a faint hope, this University is today a living thing, a home of free learning and teaching and happy brotherly work. There it is, on the soil that our people have liberated under great hardships; there it is, a spiritual center of a flourishing and buoyant community whose accomplishments have finally met with the universal recognition they deserved.

In this last period of the fulfillment of our dreams there was but one thing that weighed heavily upon me: the fact that we were compelled by the adversities of our situation to assert our rights through force of arms; it was the only way to avert complete annihilation. The wisdom and moderation the leaders of the new state have shown gives me confidence, however, that gradually relations will be established with the Arab people which are based on fruitful cooperation and mutual respect and trust. For this is the only means through which both peoples can attain true independence from the outside world.

Arab-Jewish relations continued to weigh heavily on Einstein's mind, though he thought the prospects much improved now that the British had left the scene. In an NBC radio address delivered in November 1949 for a conference of the United Jewish Appeal in Atlantic City, New Jersey, he once again pointed his finger at the British as the main culprit in stirring up hatred between Jews and Arabs.

THE JEWS OF ISRAEL, 27 NOVEMBER 1949
EINSTEIN ARCHIVES 28-862; EINSTEIN 1956, 274–276

There is no problem of such overwhelming importance to us Jews as consolidating that which has been accomplished in Israel with amazing energy and an unequalled willingness for sacrifice. May the joy and admiration that fill us when we think of all that this small group of energetic and thoughtful people has achieved give us the strength to accept the great responsibility which the present situation has placed upon us.

When appraising the achievement, however, let us not lose sight of the cause to be served by this achievement: rescue of our endangered brethren, dispersed in many lands, by uniting them in Israel; creation of a community which conforms as closely as possible to the ethical ideals of our people as they have been formed in the course of a long history.

One of these ideals is peace, based on understanding and self-restraint, and not on violence. If we are imbued with this ideal, our joy becomes somewhat mingled with sadness, because our relations with the Arabs are far from this ideal at the present time. It may well be that we would have reached this ideal, had we been permitted to work out, undisturbed by others, our relations with our neighbors, for we *want* peace and we realize that our future development depends on peace.

It was much less our own fault or that of our neighbors than of the Mandatory Power that we did not achieve an undivided Palestine in which Jews and Arabs would live as equals, free, in peace. If one nation dominates other nations, as was the case in the British Mandate over Palestine, she can hardly avoid following the notorious device of *Divide et Impera*. In plain language this means: create discord among the governed people so they will not unite in order to shake off the yoke imposed upon them. Well, the yoke has been removed, but the seed of dissension has borne fruit and may still do harm for some time to come—let us hope not for too long.

The Jews of Palestine did not fight for political independence for its own sake, but they fought to achieve free immigration for the Jews of many countries where their very existence was in danger; free immigration also for all those who were longing for a life among their own. It is no exaggeration to say that they fought to make possible a sacrifice perhaps unique in history.

I do not speak of the loss in lives and property [when] fighting an opponent who was numerically far superior, nor do I mean the exhausting toil which is the pioneer's lot in a neglected arid country. I am thinking of the additional sacrifice that a population living under such conditions has to make in order to receive, in the course of eighteen months, an influx of immigrants which comprise more than one third of the total Jewish population of the country. In order to realize what this means you have only to visualize a comparable feat of the American Jews. Let us assume there were no laws limiting the immigration into the United States;

imagine that the Jews of this country volunteered to receive more than one million Jews from other countries in the course of one year and a half, to take care of them, and to integrate them into the economy of this country. This would be a tremendous achievement, but still very far from the achievement of our brethren in Israel. For the United States is a big, fertile country, sparsely populated with a high living standard and a highly developed productive capacity, not to compare with small Jewish Palestine whose inhabitants, even without the additional burden of mass immigration, lead a hard and frugal life, still threatened by enemy attacks. Think of the privations and personal sacrifices which this voluntary act of brotherly love means for the Jews of Israel.

The economic means of the Jewish Community in Israel do not suffice to bring this tremendous enterprise to a successful end. For a hundred thousand out of more than three hundred thousand persons who immigrated to Israel since May 1948 no homes or work could be made available. They had to be concentrated in improvised camps under conditions which are a disgrace to all of us.

It must not happen that this magnificent work breaks down because the Jews of this country do not help sufficiently or quickly enough. Here, to my mind, is a precious gift with which all Jews have been presented: the opportunity to take an active part in this wonderful task.

The hopeful note sounded for the future of Israel was tempered as always by Einstein's concern that the new state meet its supreme moral test of accommodating Israel's security needs with Palestinian Arab interests.

Einstein's mix of hard-headed realism and moral sensibility is succinctly captured in a comment he made when offered the presidency of Israel on the death of Chaim Weizmann three years later, in November

1952. Though the position was a purely ceremonial one, Einstein declined, pleading advancing age and a special affinity for dealing with objective matters. Another more practical consideration intruded. Recognizing that his views on Jewish-Arab relations were not shared by mainstream opinion in the country he viewed as his spiritual home, he explained that "I also gave thought to the difficult situation that could arise if the government or the parliament made decisions which might create a conflict with my conscience; for the fact that one has no actual influence on the course of events does not relieve one of moral responsibility" (letter to the editor of *Ma'ariv*, 21 November 1952). Prime Minister David Ben-Gurion acknowledged this fearless sense of independence when he told his personal secretary: "Tell me what to do if he says yes! . . . If he accepts, we are in for trouble" (Yitzhak Navon, cited in Sayen, 247).

Another bone of contention with Israeli hawks was Einstein's advocacy of nonaligned status for the young state. In the face of a looming conflict with Egypt—which only broke out a year after his death—he feared that Israeli ties to the Western alliance would lead to increased militarization of its society and relegate it to the status of a pawn in the Cold War (see the last two texts in chapter 10).

CHAPTER 8

The Second World War, Nuclear Weapons, and World Peace, 1939–1950

Throughout the war, Leo Szilard—the most politically active, though hardly the best connected, nuclear scientist in the United States—served as Einstein's principal confidant regarding the politics of nuclear energy. Using Einstein's fame and prestige as a springboard, the Hungarian physicist managed to bring the issue of atomic weapons to the attention of high-ranking U.S. officials at an early stage. In March 1945, several months before the detonation of the first atomic bomb, Szilard again sought Einstein's help in an effort to arrange a meeting with President Roosevelt, an initiative that came to naught after FDR's unexpected death in April 1945.

Just after the bombing of Hiroshima and Nagasaki brought the war in the Pacific to a close, Einstein received a letter from another Hungarian émigré, Emery Reves, whom he had known in Berlin (as Imre Révész; see chapter 5). Reves enclosed a copy of his newly released book, *The Anatomy of Peace*, which presents a tightly constructed argument for world government and a vision for a new political order very similar to the one Einstein had described around the time of the 1932 Geneva Disarmament Conference. Reading it quickly, he was stirred to action. For while he was deeply impressed by the logical scope of Reves's argument, Einstein was even more encouraged to learn that the book had already attracted the attention of a number of leading intellectuals who planned to use it as the centerpiece for a broadly based educational campaign to promote the cause of world government. After signing an open letter in support of this effort, he threw himself whole-heartedly behind this movement during the months ahead.

Soon afterward, Einstein made his first public statement on the subject of nuclear warfare. He had already been in contact with the radio talk show host, Raymond Gram Swing, who helped him prepare an article that appeared in the November 1945 issue of the *Atlantic Monthly*. Although prompted by the ghastly new threat of atomic weapons, Einstein emphasized from the outset that the deeper problem was warfare itself rather than this latest advance in military technology. Echoing Reves, he noted: "The release of atomic energy has not created a new problem. It has merely made more urgent the necessity of solving an existing one. . . . As long as there are sovereign nations possessing great power, war is inevitable."

Many found Einstein's views far too idealistic, if not outright dangerous. In a polite rebuttal in the January 1946 issue of the *Atlantic Monthly*, former undersecretary of state Sumner Welles made it plain that, in his view, the physicist was meddling in matters about which he had little competence to form sound judgments. Einstein's response, which went unpublished, underscores why he had so little confidence in politicians and career diplomats. In the meantime, he joined Szilard and other leading American physicists in pushing for a new Atomic Energy Commission under civilian control, a proposal that gained congressional approval despite strong resistance from the U.S. Army.

Einstein remained optimistic in an interview of 23 June 1946, "The Real Problem Is in the Hearts of Men," but U.S.-Soviet relations turned noticeably sour during the months that followed. The primary irritant was a plan for international control of nuclear energy that had been floated by Bernard Baruch, the U.S. representative to the United Nations Atomic Energy Commission, in mid-June. Baruch argued that the United Nations should not allow members to use the veto to protect themselves from penalties for atomic energy violations while granting the United States several years before dismantling its own weapons.

After the Russians rejected the Baruch Plan in March 1947, arguing that the abolition of atomic weapons should precede the establishment of an international authority, Einstein addressed an "Open Letter to the General Assembly of the United Nations," calling for reforms that would strengthen the organization with an eye toward a future world government. This initiative was sharply criticized, however, by four leading Soviet academicians speaking on behalf of their government. Einstein realized that the Soviets had not only rebuffed his proposals to strengthen the U.N. Charter; they had now thrown a roadblock in front of all efforts to reach any meaningful agreement on the control of nuclear technology under U.N. auspices. Nevertheless, he cautioned against proceeding with a massive program of weapons research, fearing that this would escalate into a new arms race between the former wartime allies. At the same time, however, he recognized that the window of opportunity for negotiations was fast closing.

The Soviet Union successfully detonated an atomic bomb in September 1949, and political pressure soon mounted in the U.S. for a crash program to stockpile nuclear weapons. Approached by the Reverend Abraham J. Muste, a leading pacifist, Einstein firmly rejected the position he had once advocated in the early 1930s when he thought that unilateral disarmament was a plausible means toward achieving world peace. Although he shared Muste's goals and realized that lasting peace required eventual disarmament, Einstein argued that the latter could only be achieved through mutual trust and normalized economic relations. Above all, he continued to believe that the only way forward was the one set forth in his "Open Letter to the General Assembly of the United Nations," which concluded: "The UN now and world government eventually must serve one single goal—the guarantee of the security, tranquillity, and the welfare of all mankind."

Fighting Fascism

In mid-July 1939, while vacationing on Long Island, Einstein received a surprise visit from two prominent émigré Hungarian physicists, Leo Szilard and Eugene Wigner. During this visit Einstein learned for the first time that a sustainable nuclear chain reaction was not only physically feasible but also probably imminent. As refugees from fascism, both men were deeply concerned that physicists in Nazi Germany might already have set their sights on building a nuclear bomb. After due deliberations over the next few weeks, Szilard convinced Einstein to help him write a letter to President Franklin D. Roosevelt urging the United States government to promote ongoing research on nuclear fission. (See Plate 20 for a postwar reenactment of their fateful meeting.) This letter was to be conveyed to the president by Alexander Sachs, an economist and close adviser to FDR. Einstein's original draft was dictated in German to another Hungarian refugee, the physicist Edward Teller. Szilard and Sachs then prepared the more elaborate English version of the famous text that carried Einstein's signature.

LETTER TO FRANKLIN DELANO ROOSEVELT, 2 AUGUST 1939
EINSTEIN ARCHIVES 33-088; N & N 1960, 294–296

Some recent work by E. Fermi and L. Szilard, which has been communicated to me in manuscript, leads me to expect that the element uranium may be turned into a new and important source of energy in the immediate future. Certain aspects of the situation seem to call for watchfulness and, if necessary, quick action on the part of the Administration. I believe, therefore, that it is my duty to bring to your attention the following facts and recommendations.

In the course of the last four months it has been made probable—through the work of Joliot in France as well as Fermi and Szilard in America—that it may become possible to set up nuclear chain reactions in a large mass of uranium, by which vast amounts of power and large quantities of new radium-like elements would be generated. Now it appears almost certain that this could be achieved in the immediate future.

The new phenomenon would also lead to the construction of bombs, and it is conceivable—though much less certain—that extremely powerful bombs of a new type may thus be constructed. A single bomb of this type, carried by boat or exploded in a port, might very well destroy the whole port together with some of the surrounding territory. However, such bombs might very well prove to be too heavy for transportation by air.

The United States has only very poor ores of uranium in moderate quantities. There is some good ore in Canada and the former Czechoslovakia, while the most important source of uranium is the Belgian Congo.

In view of this situation you may think it desirable to have some permanent contact maintained between the Administration and the group of physicists working on chain reactions in America. One possible way of achieving this might be for you to entrust with this task a person who has your confidence and who could perhaps serve in an unofficial capacity. His task might comprise the following:

a) To approach Government Departments, keep them informed of the further developments, and put forward recommendations for Government action, giving particular attention to the problem of securing a supply of uranium ore for the United States.

b) To speed up the experimental work which is at present being carried on within the limits of the budgets of University laboratories, by providing funds, if such funds be required, through his contacts with private persons who are willing to make contributions for this cause, and perhaps also by obtaining the cooperation of industrial laboratories which have the necessary equipment.

I understand that Germany has actually stopped the sale of uranium from the Czechoslovakian mines which she has taken over. That she should have taken such early action might perhaps be understood on the ground that the son of the German Under-Secretary of State, von Weizsäcker, is attached to the Kaiser Wilhelm Institute in Berlin, where some of the American work on uranium is now being repeated.

FDR was a busy man, and even more so after Hitler's forces invaded Poland on 1 September. Thus, Sachs did not get to see the president until 11 October. An Advisory Committee on Uranium was immediately formed, chaired by the chief of the Bureau of Standards, Lyman J. Briggs. Ten days later the Briggs committee met with the Hungarian trio of physicists, Szilard, Wigner, and Teller. A report to the president was then filed on 1 November, urging further investigation of nuclear potential and coordination of ongoing research at universities.

After hearing nothing more from the Briggs committee, Szilard decided to contact Einstein once again. Both were persuaded that Sachs should try to approach FDR a second time in view of the urgency of the situation. In the meantime, Szilard had learned more about top secret uranium research in Germany from Peter Debye, who left his post as director of the Kaiser Wilhelm Institute for Physics in Dahlem after refusing to give up his Dutch citizenship (Rhodes 1988, 331–332). These and other matters were alluded to in a letter nominally written to Sachs, but intended as a communiqué for the White House. With Sachs acting both as ghostwriter and intermediary, Einstein signed a second letter urging swift action.

Letter to Alexander Sachs, 7 March 1940
Einstein Archives 39-475; N & N 1960, 299–300

In view of our common concern in the bearings of certain experimental work on problems connected with the national defense,

I wish to draw your attention to the development which has taken place since the conference that was arranged through your good offices in October last year between scientists engaged in this work and governmental representatives.

Last year, when I realized that results of national importance might arise out of the research on uranium, I thought it my duty to inform the Administration of this possibility. You will perhaps remember that in the letter which I addressed to the President I also mentioned the fact that C. F. von Weizsäcker, son of the German Secretary of State, was collaborating with a group of chemists working upon uranium at one of the Kaiser Wilhelm Institutes—namely, the Institute of Chemistry.

Since the outbreak of the war, interest in uranium has intensified in Germany. I have now learned that research there is carried out in great secrecy and that it has been extended to another of the Kaiser Wilhelm Institutes, the Institute of Physics. The latter has been taken over by the government and a group of physicists, under the leadership of C. F. von Weizsäcker, who is now working there on uranium in collaboration with the Institute of Chemistry. The former director was sent away on a leave of absence, apparently for the duration of the war.

Should you think it advisable to relay this information to the President, please consider yourself free to do so. Will you be kind enough to let me know if you are taking any action in this direction?

Dr. Szilard has shown me the manuscript which he is sending to the *Physics Review* in which he describes in detail a method for setting up a chain reaction in uranium. The papers will appear in print unless they are held up, and the question arises whether something ought to be done to withhold publication.

I have discussed with Professor Wigner of Princeton University the situation in the light of the information available. Dr. Szilard will let you have a memorandum informing you of the progress made since October last year so that you will be able to take such action as you think in the circumstances advisable. You will see that the line he has pursued is different and apparently more promising than the line pursued by M. Joliot in France, about whose work you may have seen reports in the papers.

In requesting this second letter, Szilard clearly intended to give the White House a wake-up call after months of inaction. Einstein must have enjoyed the irony in the passage warning the president that Szilard was about to publish details concerning a method for setting off a nuclear chain reaction unless the government interceded, in effect a request to be censored rather than exercising self-censorship, a strange sort of "political blackmail" (Lanouette, 215).

One week later, Sachs conveyed this letter to the president's staff. On 5 April FDR responded by suggesting that a new meeting of the Briggs committee be called, to which Einstein was invited. Although unable to attend, Einstein wrote in support of Sachs's proposal to establish a fund-raising committee so that the researches of Fermi and Szilard could move ahead without delay. The Advisory Committee declined to pursue this initiative, however, and in June 1940 it was left without a raison d'être after Roosevelt created the National Defense Research Committee, chaired by Vannevar Bush. Einstein, who was not privy to any of these developments, remained in the dark regarding nuclear research throughout the war.

By late 1944, however, he had learned through Otto Stern that work on atomic weapons was advancing at a rapid pace. After a discussion with Stern on 11 December, Einstein became so alarmed by these developments that he wrote the following day to Niels Bohr, then residing at the Danish Embassy in Washington, D.C., to convey the conclusions that he and Stern had reached.

LETTER TO NIELS BOHR, 12 DECEMBER 1944
EINSTEIN ARCHIVES 8-095; CLARK, 575—576

... [W]hen the war is over, then there will be in all countries a pursuit of secret war preparations with technological means which will lead inevitably to preventive wars and to destruction even more terrible than the present destruction of life. The politicians do not appreciate the possibilities and consequently do not know the extent of the menace. Every effort must be made to avert such a development. I share your view of the situation but I see no way of doing anything promising. . . .

It seemed to us that there is one possibility, however slight it may be. There are in the principal countries scientists who are really influential and who know how to get a hearing with political leaders. There is you yourself with your international connections, Compton here in the U.S.A., Lindemann in England, Kapitza and Joffe in Russia, etc. The idea is that these men should bring combined pressure on the political leaders in their countries in order to bring about an internationalization of military power—a method that has been rejected for too long as being too adventurous. But this radical step with all its far-reaching political assumptions regarding extranational government seems the only alternative to a secret technical arms race.

We agreed that I should lay this before you. Don't say, at first sight, "Impossible," but wait a day or two until you have got used to the idea. . . .

Throughout the year, Bohr had been involved in high-level negotiations with the American and British governments regarding the future control of atomic energy. These efforts, however, had foundered on Churchill's resistance to Bohr's ideas. Unbeknownst to Bohr, the British prime minister had characterized him in private as "a great advocate of publicity," who had made unauthorized disclosures to Supreme Court Justice Felix Frankfurter. Churchill was also annoyed by Bohr's "close correspondence with a Russian Professor [Peter Kapitza]" and counseled that the Dane "ought to be confined or at any rate made to see that he is very near the edge of mortal crimes" (Rhodes 1988, 537–538).

Bohr visited Einstein on 22 December in order to discuss with him the political problems posed by atomic energy. The gist of their conversation was recorded by Bohr in a memorandum. From this document it is clear that Bohr had been enjoined to convince Einstein not to undertake discussions with other parties that might conceivably "complicate the delicate task of the statesmen." Einstein "assured B[ohr] that he quite realized the situation" (Clark, 577).

The following March, Leo Szilard again contacted Einstein, just months before the bomb would be tested. During the last three years, Szilard had been the chief physicist at the Metallurgical Laboratory in Chicago, where he was privy to classified information. Although he could not share any of this with Einstein, he was able to impress upon him the urgency of the situation and the need for a face-to-face meeting with government leaders. Einstein obliged by composing a letter of introduction for Szilard to present to the president of the United States.

LETTER TO FRANKLIN DELANO ROOSEVELT, 25 MARCH 1945
EINSTEIN ARCHIVES 33-109; N & N 1960, 304–305

I am writing you to introduce Dr. L. Szilard, who proposes to submit to you certain considerations and recommendations. Unusual circumstances which I shall describe further below induce me to take this action in spite of the fact that I do not know the substance of the considerations and recommendations which Dr. Szilard proposes to submit to you.

In the summer of 1939 Dr. Szilard put before me his views concerning the potential importance of uranium for national defense. He was greatly disturbed by the potentialities involved and anxious that the United States Government be advised of them as soon as possible. Dr. Szilard, who is one of the discoverers of the neutron emission of uranium on which all present work on uranium is based, described to me a specific system which he devised and which he thought would make it possible to set up a chain reaction in unseparated uranium in the immediate future. Having known him for over twenty years both from his scientific work and personally, I have much confidence in his judgment, and it was on the basis of his judgment as well as my own that I took the liberty to approach you in connection with this subject. You responded to my letter dated August 2, 1939, by the appointment of a committee under the chairmanship of Dr. Briggs and thus started the government's activity in this field.

The terms of secrecy under which Dr. Szilard is working at present do not permit him to give me information about his work; however, I understand that he now is greatly concerned about the lack of adequate contact between scientists who are doing this work and those members of your Cabinet who are responsible for formulating policy. In the circumstances, I consider it my duty to give Dr. Szilard this introduction and I wish to express the hope that you will be able to give his presentation of the case your personal attention.

This letter was conveyed to Eleanor Roosevelt, who made an appointment for Szilard to meet the president on 8 May. Szilard had prepared a secret memorandum, which had won the approval of A. H. Compton (Clark, 583), but Roosevelt's sudden death on 12 April dashed his plans to present it. Aides later found Einstein's letter in the Oval Office and gave it to President Truman, who arranged an appointment with James Byrnes, Truman's nominee for Secretary of State. Their meeting put an end to Szilard's initiative.

A few days later the Interim Committee on the Use of Atomic Weapons voted unanimously to act against Japan without warning. On 16 July 1945 the first atomic bomb was tested successfully in Alamogordo, New Mexico. Truman got word of this while attending the Potsdam Peace Conference, and hinted to Stalin that the United States now had a powerful new weapon in its arsenal. On 6 August Einstein was resting at Saranac Lake. When he came down to take tea, Helen Dukas told him what she had heard on the radio: a new kind of bomb had been dropped on Japan.

Vision for a New World Order

Just a few weeks later, Einstein received a letter from Emery Reves, the owner of Cooperation Publishing Company, now located in New York. Along with his letter Reves sent a copy of his newly released book, *The Anatomy of Peace*. Although written before the atomic bomb was dropped over Hiroshima, Reves's book provided a probing analysis of the international political arena now that the war had ended. Einstein devoured its contents while on vacation at Saranac Lake, from where he wrote the author.

LETTER TO EMERY REVES, 28 AUGUST 1945
EINSTEIN ARCHIVES 57-292

I have read your book "The Anatomy of Peace" carefully and finished it in 24 hours. I agree with you wholeheartedly in every essential point and I admire sincerely the clarity of your exposition of the most important problem of our time. I appreciated it very much that you criticized all the wrong steps already taken (secrecy in armament under purely national viewpoints, especially about the production of the atomic bomb, occupation of strategic parts of the Pacific under exclusive U.S.A. control).

I believe that Justice Roberts' action is of the greatest value in the matter of enlightenment of public opinion in this country.

I find the text of his Open Letter excellent and convincing and I am gladly willing to sign it.

I shall be very glad indeed if you give me the opportunity to talk with you about the whole problem (I shall be back in Princeton by the middle of September), and I am gladly willing to help if you see any opportunity for me to do so.

Former Supreme Court justice Owen J. Roberts's open letter appeared in the *New York Times* on 10 October 1945, signed by Einstein, Thomas Mann, and eighteen other eminent figures. By this time, Reves had alerted Einstein to a public policy statement issued by atomic physicists at the facilities in Oak Ridge, Tennessee. This proposal called for an agency under the auspices of the United Nations Security Council that would be solely responsible for regulating all aspects of nuclear energy, including inspections to be carried out in cooperation with those nations with nuclear research and development facilities. Concerned about the inherent weakness of the Security Council's powers, Reves wrote Einstein a letter 27 September outlining arguments against the Oak Ridge proposal.

Drawing on similar reasoning as in his *Anatomy of Peace*, Reves characterized the position of the Oak Ridge scientists as merely another form of "old-fashioned internationalism" based on hopes that a league of sovereign states would find a way to settle all disputes by peaceful diplomacy. "Peace among sovereign powers," he flatly asserted, "is a daydream," citing the authority of Alexander Hamilton, who had warned that "to look for a continuation of harmony between a number of independent, unconnected sovereignties, situated in the same neighborhood, would be to disregard the uniform course of human events, and to set at defiance the accumulated experience of ages" (*The Federalist Papers*, no. 6, cited in N & N 1960, 338). He offered this succinct summary of his position: "There is only one way to prevent an atomic war and that is to prevent war. . . . Analyzing all the wars of history . . . I think it is possible . . . to define the one and only condition in human society that produces war. This is the non-integrated coexistence of sovereign powers. . . . Peace is law. Peace between warring sovereign social units . . . can be achieved only by the integration of these conflicting units into a higher sovereignty . . . by the creation of a world government having direct relations with the individual citizen" (ibid.). These sharply formulated views were thoroughly consistent with those that Einstein had gradually adopted by the early 1930s (see chapters 5 and 6). Similar arguments can also be found in *A Democratic Manifesto* (Reves 1942), a book Einstein later acquired.

Two days later, Einstein forwarded Reves's letter to J. R. Oppenheimer along with a statement strongly supporting its contents.

Letter to J. Robert Oppenheimer, 29 September 1945
Einstein Archives 57-294; N & N 1960, 338–339

Mr. Emery Reves, whom I have known for many years and with whom I have often discussed urgent political problems, has sent me a copy of the statement which you and your colleagues issued for the enlightenment of the public and the government. While I was very much pleased by the candid language and the sincerity of the statement, I was, at the same time, somewhat bewildered by the political recommendations, which I consider inadequate.

The pathetic attempts made by governments to achieve what they consider to be international security have not the slightest effect on the present political structure of the world, nor is it recognized that the real cause of international conflicts is due to the existence of competing sovereign nations. Neither governments nor people seem to have learned anything from the experiences of the past and appear to be unable or unwilling to think that problem through. The conditions existing in the world today force the individual states, out of fear for their own security, to commit acts which inevitably produce war.

At the present high level of industrialization and economic interdependence, it is unthinkable that we can achieve peace without a genuine supranational organization to govern international relations. If war is to be avoided, anything less than such an over-all solution strikes me as illusory.

A few weeks ago, Emery Reves published a short book entitled *The Anatomy of Peace* which, in my opinion, explains the problem as clearly and pertinently as anyone ever has. I have learnt that several men who play an active role in public life are taking steps to make the book known to every American. I urge you and your colleagues to read it and discuss its conclusions. Although it was written before the explosion of the atomic bomb, it contains a solution which is directly applicable to the problem created by this new weapon. I shall be glad to send you a number of copies for distribution or to mail copies directly to you and your colleagues if you send me their addresses. I am convinced that the political part of the statement which you and your colleagues issued would have been formulated differently, had the facts and discussions presented in this book been made known to those who drafted the statement.

I do hope you will forgive my bothering you with this, but the problem is vital and your responsibility great.

Since Oppenheimer had no connection with the statement issued by the Oak Ridge physicists, this particular initiative went aground. By now, however, several eminent figures in American politics and culture, including senators J. William Fulbright and Claude Pepper, as well as Robert Hutchins and Norman Cousins, had come out in public support of a world government as articulated in Reves's *The Anatomy of Peace*.

Soon after conferring with Reves, Einstein helped launch an extremely successful advertising campaign for *Anatomy of Peace* that featured a reproduction of a shorter version of his original letter to Reves of 28 August 1945 (see Plate 23).

LETTER TO EMERY REVES, 29 OCTOBER 1945
ROZELLE, 29

I have read THE ANATOMY OF PEACE with the greatest admiration. Your book is, in my opinion, *the* answer to the present political problem in the world, so drastically precipitated by the release of atomic energy.

It would be most desirable if every political and scientific leader in every country would take a little time to read this book. If this could be brought about, I feel it might avert the disaster of an atomic world war.

Going Public on the Bomb

Somewhat earlier Einstein had already made an important contact with Raymond Gram Swing, a political commentator for ABC radio, who had been one of the first journalists to bring the issue of controlling nuclear armaments before the public. As one of his listeners, Einstein wrote to praise him for his support of world government.

LETTER TO RAYMOND GRAM SWING, 27 AUGUST 1945
EINSTEIN ARCHIVES 57-443; N & N 1960, 346–347

I wish to express my deep gratitude for your systematic endeavor to advise the public about the need for an effective world government.

You were quite correct in pointing out that the occupation of the Pacific islands by American forces alone is a step in the wrong direction, and one that will hinder any future attempts to achieve international security.

But most dangerous of all is the policy of military secrecy and the maintenance of huge organizations which can produce new secret weapons on a national scale. This, too, you have expressed, and very convincingly. You remain one of the few independent persons who have insight into the critical issues of today and exercise great influence on public thought and sentiment. I am astonished that your broadcasts have not had a greater impact on public opinion in the face of the great danger which confronts us and the fateful mistakes of our government.

I hope with all my heart that you may be able to influence the course of events by your courage and sincerity.

Swing's reply led to the first of two interviews which were edited for publication in the *Atlantic Monthly*. The second interview took place two years later at a time when Cold War hostilities had deepened considerably. The two articles that emerged from these interviews were then published together in slightly edited form under the title "Atomic War or Peace" (Einstein 1954, 123–131), thereby obscuring important distinctions between them, in particular due to changes in U.S.-Soviet relations. In the first article, presented below, Einstein was mainly intent on promoting the argument for world government as set forth in Reves's *Anatomy of Peace*, a book he hoped would be touted by other leading intellectuals.

As an experienced journalist, Raymond Swing did his part to dramatize what Einstein had to say. In editing this piece for the *Atlantic Monthly*, he introduced the famous physicist by referring to his exploits in connection with the Manhattan Project. In a headnote Swing wrote: "On August 2, 1939, just a month before the outbreak of World War II, Dr. Einstein wrote a letter which made history" by revealing the possibility of "the construction of bombs . . . extremely powerful bombs." Readers were also informed that the scientific basis for this revelation was also due to the famous physicist: "It was Einstein's daring formula, $E = mc^2$, which led to the concept that atomic energy would someday be unlocked." Having established the physicist's awe-inspiring credentials, Swing braced the reader for the bold thoughts that followed, wherein Einstein "explains how mankind must control atomic power."

On the Atomic Bomb, as Told to Raymond Swing, before 1 October 1945
Atlantic Monthly 176 (November 1945), no. 5, 43–45

The release of atomic energy has not created a new problem. It has merely made more urgent the necessity of solving an existing one. One could say that it has affected us quantitatively, not qualitatively. As long as there are sovereign nations possessing great power, war is inevitable. This does not mean that one can know when war will come but only that one is sure that it will come. This was true even before the atomic bomb was made. What has changed is the destructiveness of war.

I do not believe that the secret of the bomb should be given to the United Nations Organization. I do not believe it should be given to the Soviet Union. Either course would be analogous to a man with capital who, wishing another individual to collaborate with him on an enterprise, starts by giving him half his money. The other man might choose to start a rival enterprise, when what is wanted is his cooperation. The secret of the bomb should be committed to a world government, and the United States should immediately announce its readiness to do so. Such a world government should be established by the United States, the Soviet Union and Great Britain, the only three powers which possess great military strength. The three of them should commit to this world government all of their military resources. The fact that there are only three nations with great military power should make it easier, rather than harder, to establish a world government.

Since the United States and Great Britain have the secret of the atomic bomb and the Soviet Union does not, they should invite the Soviet Union to prepare and present the first draft of a Constitution for the proposed world government. This would help to dispel the distrust of the Russians, which they feel because they know the bomb is being kept a secret chiefly to prevent their having it. Obviously the first draft would not be the final one, but the Russians should be able to feel that the world government will guarantee their security.

It would be wise if this Constitution were to be negotiated by one American, one Briton and one Russian. They would, of course, need advisers, but these advisers should serve only when asked. I believe three men can succeed in preparing a workable Constitution acceptable to all the powers. Were six or seven men, or more, to attempt to do so, they would probably fail. After the three great powers have drafted a Constitution and adopted it, the smaller nations should be invited to join the world government. They should also be free not to join and, though they should feel perfectly secure outside the world government, I am sure they will eventually wish to join. Naturally, they should be entitled to propose changes in the Constitution as drafted by the Big Three. But the Big Three should go ahead and organize the world government, whether or not the smaller nations decide to join.

Such a world government should have jurisdiction over all military matters, and it need have only one other power. That is the power to interfere in countries where a minority is oppressing the majority and, therefore, is creating the kind of instability that leads to war. For example, conditions as they exist today in Argentina and Spain should be dealt with. There must be an end to the concept of non-intervention, for to abandon non-intervention in certain circumstances is part of keeping the peace.

The establishment of a world government should not be delayed until similar conditions of freedom exist in each of the three great powers. While it is true that in the Soviet Union the minority rules, I do not believe that the internal conditions in that country constitute a threat to world peace. One must bear in mind that the people in Russia had not had a long tradition of political education; changes to improve conditions in Russia had to be effected by a minority for the reason that there was no majority capable of doing so. If I had been born a Russian, I believe I could have adjusted myself to the situation.

It should not be necessary, in establishing a world government with a monopoly of authority over military affairs, to change the internal structure of the three great powers. It would be for the three individuals who draft the Constitution to devise ways for collaboration despite the different structures of the countries.

Do I fear the tyranny of a world government? Of course I do. But I fear still more the coming of another war. Any government is certain to be evil to some extent. But a world government is preferable to the far greater evil of wars, particularly when viewed in the context of the intensified destructiveness of war. If such a world government is not established by a process of agreement among nations, I believe it will come anyway, and in a much more dangerous form; for war or wars can only result in one power being supreme and dominating the rest of the world by its overwhelming military supremacy.

Now that we have the atomic secret, we must not lose it, and that is what we would risk doing if we gave it to the United Nations Organization or to the Soviet Union. But, as soon as possible, we must make it clear that we are not keeping the bomb a secret for the sake of maintaining our power but in the hope of establishing peace through world government, and that we will do our utmost to bring this world government into being.

I appreciate that there are persons who approve of world government as the ultimate objective but favor a gradual approach to its establishment. The trouble with taking little steps, one at a time, in the hope of eventually reaching the ultimate goal, is that while such steps are being taken, we continue to keep the bomb without convincing those who do not have the bomb of our ultimate intentions. That of itself creates fear and suspicion, with the consequence that the relations between rival countries deteriorate to a dangerous extent. That is why people who advocate taking a step at a time may think they are approaching world peace, but they actually are contributing by their slow pace to the possibility of war. We have no time to waste in this way. If war is to be averted, it must be done quickly.

Further, we shall not have the secret of the bomb for very long. I know it is being argued that no other country has money enough to spend on the development of the atomic bomb and that, therefore, we are assured of the secret for a long time. But it is a common mistake in this country to measure things by the amount of money they cost. Other countries which have the raw materials and manpower and wish to apply them to the work of developing atomic power can do so; men and materials and the decision to use them, and not money, are all that is needed.

I do not consider myself the father of the release of atomic energy. My part in it was quite indirect. I did not, in fact, foresee that it would be released in my time. I only believed that it was theoretically possible. It became practical through the accidental discovery of chain reaction, and this was not something I could have predicted. It was discovered by Hahn in Berlin, and he himself at first misinterpreted what he discovered. It was Lise Meitner who provided the correct interpretation and escaped from Germany to place the information in the hands of Niels Bohr.

In my opinion, a great era of atomic science cannot be assured by organizing science in the way large corporations are organized. One can organize the application of a discovery already made, but one cannot organize the discovery itself. Only a free individual can make a discovery. However, there can be a kind of organization wherein the scientist is assured freedom and proper conditions of work. Professors of science in American universities, for instance, should be relieved of some of their teaching so as to have more time for research. Can you imagine an organization of scientists making the discoveries of Charles Darwin?

I do not believe that the vast corporations of the United States are suitable to the needs of the times. If a visitor should come to this country from another planet, would he not find it strange that, in this country, private corporations are permitted to wield so much power without having to assume commensurate responsibility? I say this to stress my conviction that the American government must retain control of atomic energy, not because socialism is necessarily desirable but because atomic energy was developed by the government; it would be unthinkable to turn over this property of the people to any individual or group of individuals. As for socialism, unless it is international to the extent of producing a world government which controls all military power, it might lead to wars even more easily than capitalism because it represents an even greater concentration of power.

To give an estimate of when atomic energy might be applied for peaceful, constructive purposes is impossible. All that we know now is how to use a fairly large quantity of uranium. The use of small quantities sufficient, say to operate a car or an airplane, is thus far impossible, and one cannot predict when it will be accomplished. No doubt, it will be achieved, but no one can say when. Nor can one predict when materials more common than uranium can be used to supply atomic energy. Presumably, such materials would be among the heavier elements of high atomic weight and would be relatively scarce due to their lesser stability. Most of these materials may already have disappeared through radioactive disintegration. So, though the release of atomic energy can be, and no doubt will be, a great boon to mankind, this may not come about for some time.

I myself do not have the gift of explanation which would be needed to persuade large numbers of people of the urgency of the problems that now face the human race. Hence, I should like to commend someone who has this gift of explanation: Emery Reves, whose book *Anatomy of Peace* is intelligent, clear, brief, and, if I must use the absurd term, dynamic on the topic of war and need for world government.

Since I do not foresee that atomic energy will prove to be a boon within the near future, I have to say that, for the present, it is a menace. Perhaps it is well that it should be. It may intimidate the human race into bringing order to its international affairs, which, without the pressure of fear, undoubtedly would not happen.

While the political import of Einstein's text was clear, it was far from accurate when it came to the historical facts, indicating that Einstein knew very little about the scientific events that preceded work on the Manhattan Project. Lise Meitner was already in Sweden when she learned about Otto Hahn's experimental results and interpreted these as due to nuclear fission. She did not flee to Bohr with this information, but rather conveyed it to him just before she and Otto Frisch published their work in *Nature*. Moreover, the possibility of setting off a chain reaction was first contemplated only somewhat later by Enrico Fermi.

Needless to say, these slips passed without comment, as Einstein's name had by now become indissolubly linked with nuclear research. On the other hand, his actual political agenda became even more transparent when on 27 October 1945 the *New York Times* published an abridged version of the preceding text that omitted the paragraph referring to Reves's book. A few days later, Einstein wrote to protest this oversight. The version presented here follows the text in the *New York Times*, which differs somewhat from that in N & N 1960.

RECOMMENDED READING, 30 OCTOBER 1945
LETTER TO THE *NEW YORK TIMES*, 1 NOVEMBER 1945; N & N 1960, 352

... I regret that in your story you have failed to quote the paragraph in which I commend to the people the reading and study of a short book by Emery Reves entitled "The Anatomy of Peace."

To draw the attention of the American public to this book was one of the main reasons I wrote the article. The vast and complex problem of the urgent need for the establishment of a world-wide governmental organization to prevent an atomic war cannot be sufficiently explained in a short article and I am afraid that some of my statements, without further explanation, may surprise people and be misunderstood by them.

Therefore, I think it in [the] public interest that the attention of the people be drawn to this book, which I believe to be the clearest and most complete analysis of the problem we are facing today, and in which can be found all the reasons for the arguments in my article.

Seasoned political analysts did not think much of Einstein's argument that the Big Three powers should enter into a world government. In his invited response in the *Atlantic Monthly* of January 1946, former undersecretary of state Sumner Welles characterized Einstein as a prominent voice within a band of "idealists" who sought to abandon the United Nations at the very moment when the new organization promised a fresh start for the future. Regarding Einstein's claim that it would *not* be necessary for the great powers to alter their respective forms of government in order to draw up a constitution for a world government, Welles called this a "wholly impossible" proposal.

In a letter to Einstein, Reves called Welles's article "silly" and thought that Thomas Finletter, a future secretary of the Air Force, whom he characterized as a "staunch believer in world government," would respond to Welles in a forthcoming issue (letter from Reves, 4 January 1946). Einstein drafted the following reply, which was, however, not published until after Einstein's death.

REPLY TO SUMNER WELLES'S ARTICLE ON THE ATOMIC
BOMB AND WORLD GOVERNMENT, JANUARY 1946
EINSTEIN ARCHIVES 28-720; N & N 1960, 352–353

The importance of the issue obliges me to reply, even though briefly, to the recent remarks of Sumner Welles. Eventualities which depend solely upon the decisions of men should never, in advance, be labeled "impossible." In the sphere of human activity, everything depends on the strength of men's convictions. These convictions must be based upon clear understanding of the prevailing objective conditions which have been affected by the unexpectedly swift and radical technological development of weapons.

Surely, no one can doubt that a war among the great powers would lead to the destruction of a large part of the world's population, cities, and industrial resources. No thoughtful person can fail to be convinced that there is no conceivable cause which could justify so great a sacrifice.

Hence, I should like to raise the following question: Is it really a sign of unpardonable naïveté to suggest that those in power decide among themselves that future conflicts must be settled by constitutional means rather than by the senseless sacrifice of great numbers of human lives? Once such a firm decision has been reached, nothing will be "impossible." And a second decision must necessarily follow, which is, to make certain that no individual nation is able to use its own independent military resources for the purpose of forcing its will upon other nations.

A "sophisticated" person might well comment: We have been working toward the small goal by means of small, patient steps, which, in view of human psychology, is the only possible method. But I, the so-called "idealist," regard this attitude as a fatal illusion. There is no *gradual* way to secure peace. As long as nations have no real security against aggression, they will, inevitably, continue to prepare for war. And, as history has proven conclusively, preparation for war always leads to actual war. When the North American Colonies united and created a central government in

Washington, it came about not through a slow process but through a resolute and creative act.

Only such a resolute and creative act can provide a possible solution to the present perilous situation in which the nations of the world find themselves. If we fail to take such action, murderous conflicts are bound to develop, which will bring about unimaginable destruction and, eventually, result in the oppression of all by a single power.

Just a month earlier, Einstein presented a similar critique of conventional political wisdom in a stirring speech at the fifth Nobel anniversary dinner held at the Hotel Astor in New York.

THE WAR IS WON, BUT THE PEACE IS NOT, 10 DECEMBER 1945
EINSTEIN 1954, 115–117

Physicists find themselves in a position not unlike that of Alfred Nobel. Alfred Nobel invented the most powerful explosive ever known up to his time, a means of destruction par excellence. In order to atone for this, in order to relieve his human conscience he instituted his awards for the promotion of peace and for achievements of peace. Today, the physicists who participated in forging the most formidable and dangerous weapon of all times are harassed by an equal feeling of responsibility, not to say guilt. And we cannot desist from warning, and warning again, we cannot and should not slacken in our efforts to make the nations of the world, and especially their governments, aware of the unspeakable disaster they are certain to provoke unless they change their attitude toward each other and toward the task of shaping the future. We helped in creating this new weapon in order to prevent the enemies of mankind from achieving it ahead of us, which, given the mentality of the Nazis, would have meant inconceivable destruction and the enslavement of the rest of the world. We delivered this weapon into the hands of the American and the British people as trustees of the whole of mankind, as fighters for peace and liberty. But so far we fail to see any guarantee of peace, we do not see any guarantee of the freedoms that were promised to the nations in the Atlantic Charter. The war is won, but the peace is not. The great powers, united in fighting, are now divided over the peace settlements. The world was promised freedom from fear, but in fact fear has increased tremendously since the termination of the war. The world was promised freedom from want, but large parts of the world are faced with starvation while others are living in abundance. The nations were promised liberation and justice. But we have witnessed, and are witnessing even now, the sad spectacle of "liberating" armies firing into populations who want their independence and social equality, and supporting in those countries, by force of arms, such parties and personalities as appear to be most suited to serve vested interests. Territorial questions and arguments of power, obsolete though they are, still prevail over the essential demands of common welfare and justice. . . .

The picture of our postwar world is not bright. So far as we, the physicists, are concerned, we are no politicians and it has never been our wish to meddle in politics. But we know a few things that the politicians do not know. And we feel the duty to speak up and to remind those responsible that there is no escape into easy comforts, there is no distance ahead for proceeding little by little and delaying the necessary changes into an indefinite future, there is no time left for petty bargaining. The situation calls for a courageous effort, for a radical change in our whole attitude, in the entire political concept. May the spirit that prompted Alfred Nobel to create his great institution, the spirit of trust and confidence, of generosity and brotherhood among men, prevail in the minds of those upon whose decisions our destiny rests. Otherwise, human civilization will be doomed.

Roadblock at the United Nations

With Einstein's appointment as chair of the Emergency Committee of Atomic Scientists (ECAS), he began actively campaigning for an international framework for the control of nuclear energy. The following interview with Michael Amrine, originally published in the Sunday magazine section of the *New York Times*, was also reprinted by the ECAS for distribution in its fund-raising activities.

The version presented here follows the text in the *New York Times*, which differs slightly from that reprinted in N & N 1960.

The Real Problem Is in the Hearts of Men
New York Times Magazine, 23 June 1946; N & N 1960, 383–388

Many persons have inquired concerning a recent message of mine that "a new type of thinking is essential if mankind is to survive and move to higher levels."

Often in evolutionary processes a species must adapt to new conditions in order to survive. Today the atomic bomb has altered profoundly the nature of the world as we knew it, and the human race consequently finds itself in a new habitat to which it must adapt its thinking.

In the light of new knowledge, a world authority and an eventual world state are not just *desirable* in the name of brotherhood, they are *necessary* for survival. In previous ages a nation's life and culture could be protected to some extent by the growth of armies in national competition. Today we must abandon competition and secure cooperation. This must be the central fact in all our considerations of international affairs; otherwise we face certain disaster. Past thinking and methods did not prevent world wars. Future thinking *must* prevent wars.

Modern war, the bomb, and other discoveries or inventions, present us with revolutionary circumstances. Never before was it possible for one nation to make war on another without sending armies across borders. Now, with rockets and atomic bombs no center of population on the earth's surface is secure from surprise destruction in a single attack.

America has a temporary superiority in armament, but it is certain that we have no lasting secret. What nature tells one group of men, she will tell in time to any other group interested and patient enough in asking the questions. But our temporary superiority gives this nation the tremendous responsibility of leading mankind's effort to surmount the crisis.

Being an ingenious people, Americans find it hard to believe there is no foreseeable defense against atomic bombs. But this is a basic fact. Scientists do not even know of any field which promises us any hope of adequate defense. The military-minded cling to old methods of thinking and one Army department has been surveying possibilities of going underground and in wartime placing factories in places like Mammoth Cave. Others speak of dispersing our population centers into "linear" or "ribbon" cities.

Reasonable men with these new facts to consider refuse to contemplate a future in which our culture would attempt to survive in ribbons or in underground tombs. Neither is there reassurance in proposals to keep a hundred thousand men alert along the coasts scanning the sky with radar. There is no radar defense against the V-2, and should a "defense" be developed after years of research, it is not humanly possible for any defense to be perfect. Should one rocket with atomic warhead strike Minneapolis, that city would look almost exactly like Nagasaki. Rifle bullets kill men, but atomic bombs kill cities. A tank is a defense against a bullet but there is no defense in science against the weapon which can destroy civilization.

Our defense is not in armaments, nor in science, nor in going underground. Our defense is in law and order.

Henceforth, every nation's foreign policy must be judged at every point by one consideration: Does it lead us to a world of law and order or does it lead us back toward anarchy and death? I do not believe that we can prepare for war and at the same time prepare for a world community. When humanity holds in its hand the weapon with which it can commit suicide, I believe that to put more power into the gun is to increase the probability of disaster. . . .

We are still making bombs and the bombs are making hate and suspicion. We are keeping secrets and secrets breed distrust. I do not say we should now turn the secret of the bomb loose in the world, but are we ardently seeking a world in which there will be no need for bombs or secrets, a world in which science and men will be free?

While we distrust Russia's secrecy and she distrusts ours we walk together to certain doom.

The basic principles of the Acheson-Lilienthal Report are scientifically sound and technically ingenious, but as Mr. Baruch wisely said, it is a problem not of physics but of ethics. There has been too much emphasis on legalisms and procedure; it is easier to denature plutonium than it is to denature the evil spirit of man.

The United Nations is the only instrument we have to work with in our struggle to achieve something better. But we have used U.N. and U.N. form and procedure to outvote the Russians on some occasions when the Russians were right. Yes, I do not think it is possible for any nation to be right all the time or wrong all the time. In all negotiations, whether over Spain, Argentina, Palestine, food, or atomic energy, so long as we rely on procedure and keep the threat of military power, we are attempting to use old methods in a world which is changed forever.

No one gainsays that the United Nations Organization at times gives great evidence of eventually justifying the desperate hope that millions have in it. But time is not given to us in solving the problems science and war have brought. Powerful forces in the political world are moving swiftly toward crisis. When we look back to the end of the war it does not seem ten months—it seems ten years ago! Many leaders express well the need for world authority and an eventual world government, but actual planning and action to this end have been appallingly slow. . . . Meanwhile, men high in government propose defense or war measures which would not only compel us to live in a universal atmosphere of fear but would cost untold billions of dollars and ultimately destroy our American free way of life—even before a war. . . .

Before the raid on Hiroshima, leading physicists urged the War Department not to use the bomb against defenseless women and children. The war could have been won without it. The decision was made in consideration of possible future loss of American lives—and now we have to consider possible loss in future atomic bombings of *millions of lives*. The American decision may have been a fatal error, for men accustom themselves to thinking a weapon which was used once can be used again.

Had we shown other nations the test explosion at Alamogordo, New Mexico, we could have used it as an education for new ideas. It would have been an impressive and favorable moment to make considered proposals for world order to end war. Our renunciation of this weapon as too terrible to use would have carried great weight in negotiations and made convincing our sincerity in asking other nations for a binding partnership to develop these newly un-leashed powers for good.

The old type of thinking can raise a thousand objections of "re-alism" against this simplicity. But such thought ignores the *psy-chological realities*. All men fear atomic war. All men hope for benefits from these new powers. Between the realities of man's true desires and the realities of man's danger, what are the obso-lete "realities" of protocol and military protection?

During the war many persons fell out of the habit of doing their own thinking, for many had to do simply what they were told to do. Today, lack of interest would be a great error, for there is much the average man can do about this danger.

This nation held a great debate concerning the menace of the Axis, and again today we need a great chain reaction of awareness and communication. Current proposals should be discussed in the light of the basic facts, in every newspaper, in schools, churches, in town meetings, in private conversations, and neighbor to neighbor. Merely reading about the bomb promotes knowledge in the mind, but only talk between men promotes feeling in the heart.

Not even scientists completely understand atomic energy, for each man's knowledge is incomplete. Few men have ever seen the bomb. But all men if told a few facts can understand that this bomb and the danger of war is a very real thing, and not something far away. It directly concerns every person in the civilized world. We cannot leave it to generals, Senators, and diplomats to work out a solution over a period of generations. Perhaps five years from now several nations will have made bombs and it will be too late to avoid disaster.

Ignoring the realities of faith, good-will, and honesty in seeking a solution, we place too much faith in legalisms, treaties, and mechanisms. We must begin through the U.N. Atomic Energy Commission to work for binding agreement, but America's decision will not be made over a table in the United Nations. Our representatives in New York, in Paris, or in Moscow depend ultimately on decisions made in the village square.

To the village square we must carry the facts of atomic energy. From there must come America's voice.

This belief of physicists promoted our formation of the Emergency Committee of Atomic Physicists, with headquarters at Princeton, N.J., to make possible a great national campaign for education on these issues, through the National Committee on Atomic Information. Detailed planning for world security will be easier when negotiators are assured of public understanding of our dilemmas.

Then our American proposals will be not merely documents about machinery, the dull, dry statements of a government to other governments, but the embodiment of a message to humanity from a nation of human beings.

Science has brought forth this danger, but the real problem is in the minds and hearts of men. We will not change the hearts of other men by mechanisms, but by changing *our* hearts and speaking bravely.

We must be generous in giving to the world the knowledge we have of the forces of nature, after establishing safeguards against abuse.

We must be not merely willing but actively eager to submit ourselves to binding authority necessary for world security.

We must realize we cannot simultaneously plan for war and peace.

When we are clear in heart and mind—only then shall we find courage to surmount the fear which haunts the world.

The Acheson-Lilienthal Report on the international control of atomic energy, published three months before the Baruch Plan, called on all countries to renounce any intention of developing more atomic weapons.

One year later, the situation looked far bleaker for Einstein and the ECAS. By this time, Einstein was urging the United States to initiate a series of reforms of the United Nations that would eventually enable it to function as the framework for an international government. In his "Open Letter to the General Assembly," he forcefully argued for the viability of such reforms, with or without the cooperation of the Soviet Union. Einstein prefaced his remarks with the following note: "As I see it, this is the way for the nations of the world to break the vicious circle which threatens the continued existence of mankind, as no other situation in human history has ever done."

Open Letter to the General Assembly of the United Nations
United Nations World 1 (October 1947), no. 8, 13–14;
Einstein 1956, 156–160

We are caught in a situation in which every citizen of every country, his children, and his life's work, are threatened by the terrible insecurity which reigns in our world today. The progress of technological development has not increased the stability and the welfare of humanity. Because of our inability to solve the problem of international organization, it has actually contributed to the dangers which threaten peace and the very existence of mankind.

The delegates of fifty-five governments, meeting in the second General Assembly of the United Nations, undoubtedly will be aware of the fact that during the last two years—since the victory over the Axis powers—no appreciable progress has been made either toward the prevention of war or toward agreement in specific fields such as control of atomic energy and economic cooperation in the reconstruction of war-devastated areas.

The UN cannot be blamed for these failures. No international organization can be stronger than the constitutional powers given it, or than its component parts want it to be. As a matter of fact, the United Nations is an extremely important and useful institution *provided* the peoples and governments of the world realize that it is merely a transitional system toward the final goal, which is the establishment of a supra-national authority vested with sufficient legislative and executive powers to keep the peace. The present impasse lies in the fact that there is no sufficient, reliable supra-national authority. Thus the responsible leaders of all governments are obliged to act on the assumption of eventual war. Every step motivated by that assumption contributes to the general fear and distrust and hastens the final catastrophe. However strong national armaments may be, they do not create military security for any nation nor do they guarantee the maintenance of peace.

There can never be complete agreement on international control and the administration of atomic energy or on general disarmament until there is a modification of the traditional concept of national sovereignty. For as long as atomic energy and armaments are considered a vital part of national security no nation will give more than lip service to international treaties. Security is indivisible. It can be reached only when necessary guarantees of law and enforcement obtain everywhere, so that military security is no longer the problem of any single state. There is no compromise possible between preparation for war, on the one hand, and preparation of a world society based on law and order on the other.

Every citizen must make up his mind. If he accepts the premise of war, he must reconcile himself to the maintenance of troops in strategic areas like Austria and Korea; to the sending of troops to Greece and Bulgaria; to the accumulation of stockpiles of uranium by whatever means; to universal military training, to the progressive limitation of civil liberties. Above all, he must endure the consequences of military secrecy which is one of the worst scourges of our time and one of the greatest obstacles to cultural betterment.

If on the other hand every citizen realizes that the only guarantee for security and peace in this atomic age is the constant development of a supra-national government, then he will do everything in his power to strengthen the United Nations. It seems to me that every reasonable and responsible citizen in the world must know where his choice lies.

Yet the world at large finds itself in a vicious circle since the UN powers seem to be incapable of making up their minds on this score. The Eastern and Western blocs each attempt frantically to strengthen their respective power positions. Universal military training, Russian troops in Eastern Europe, United States control over the Pacific Islands, even the stiffening colonial policies of the Netherlands, Great Britain and France, atomic and military secrecy—are all part of the old familiar jockeying for position.

The time has come for the UN to strengthen its moral authority by bold decisions. First, the authority of the General Assembly must be increased so that the Security Council as well as all other bodies of the UN will be subordinated to it. As long as there is a conflict of authority between the Assembly and the Security Council, the effectiveness of the whole institution will remain necessarily impaired.

Second, the method of representation at the UN should be considerably modified. The present method of selection by government appointment does not leave any real freedom to the appointee. Furthermore, selection by governments cannot give the peoples of the world the feeling of being fairly and proportionately represented. The moral authority of the UN would be considerably enhanced if the delegates were elected directly by the people. Were they responsible to an electorate, they would have much more freedom to follow their consciences. Thus we could hope for more statesmen and fewer diplomats.

Third, the General Assembly should remain in session throughout the critical period of transition. By staying constantly on the job, the Assembly could fulfill two major tasks: first, it could take the initiative toward the establishment of a supra-national order; second, it could take quick and effective steps in all those danger areas (such as currently exist on the Greek border) where peace is threatened.

The Assembly, in view of these high tasks, should not delegate its powers to the Security Council, especially while that body is paralyzed by the shortcomings of the veto provisions. As the only body competent to take the initiative boldly and resolutely, the UN must act with utmost speed to create the necessary conditions for international security by laying the foundations for a real world government.

Of course there will be opposition. It is by no means certain that the U.S.S.R.—which is often represented as the main antagonist to the idea of world government—would maintain its opposition if an equitable offer providing for real security were made. Even assuming that Russia is now opposed to the idea of world government, once she becomes convinced that world government is nonetheless in the making her whole attitude may change. She may then insist on only the necessary guarantees of equality before the law so as to avoid finding herself in perennial minority as in the present Security Council.

Nevertheless, we must assume that despite all efforts Russia and her allies may still find it advisable to stay out of such a world government. In that case—and only after all efforts have been made in utmost sincerity to obtain the cooperation of Russia and her allies—the other countries would have to proceed alone. It is of the utmost importance that this partial world government be very strong, comprising at least two-thirds of the major industrial and economic areas of the world. Such strength in itself would make it possible for the partial world government to abandon military secrecy and all the other practices born of insecurity.

Such a partial world government should make it clear from the beginning that its doors remain wide open to any non-member—particularly Russia—for participation on the basis of complete equality. In my opinion, the partial world government should accept the presence of observers from non-member governments at all its meetings and constitutional conventions.

In order to achieve the final aim—which is one world, and not two hostile worlds—such a partial world government must never act as an alliance against the rest of the world. The only real step toward world government is world government itself.

In a world government the ideological differences between the various component parts are of no grave consequence. I am convinced that the present difficulties between the U.S.A. and the U.S.S.R. are not due primarily to ideological differences. Of course, these ideological differences are a contributing element in an already serious tension. But I am convinced that even if the U.S.A. and Russia were both capitalist countries—or communist, or monarchist, for that matter—their rivalries, conflicting interests, and jealousies would result in strains similar to those existing between the two countries today.

The UN now and world government eventually must serve one single goal—the guarantee of the security, tranquillity, and the welfare of all mankind.

A year earlier, Sumner Welles had sharply criticized Einstein's contention that ideological and systemic differences posed no grave obstacle to world government. This time a rebuttal came from the Soviet camp in the form of an open letter entitled "About Certain Fallacies of Professor Albert Einstein." It was signed by four leading members of the Soviet Academy of Sciences—Sergei Vavilov, A. N. Frumkin, A. F. Joffe, and N. N. Semyonov—and appeared in the English-language *New Times* (Moscow) on 26 November 1947. Although respectful in tone, the Soviets argued that Einstein's appeal for a world government merely echoed the interests of the capitalist monopolies that could only function in the framework of "world markets, world sources of raw materials and regions for investment capital" (Open Letter to Dr. Einstein, *Bulletin of the Atomic Scientists*, February 1948, 34).

In his sober reply Einstein pleaded for mutual understanding while dismissing the substance of the Soviet critique as merely a reflection of the deep estrangement that separated intellectuals in the world's two great superpowers. In contending that his critics' position amounted to anarchy in the sphere of international politics, Einstein echoed the language of Reves's *Anatomy of Peace*.

After its initial publication, this text received widespread circulation through the ECAS.

A Reply to the Soviet Scientists, December 1947
Bulletin of the Atomic Scientists 4 (February 1948), no. 2, 35–37; Einstein 1956, 169–175

Four of my Russian colleagues have published a benevolent attack upon me in an open letter carried by the *New Times*. I appreciate the effort they have made and I appreciate even more the fact that they have expressed their point of view so candidly and straightforwardly. To act intelligently in human affairs is only possible if an attempt is made to understand the thoughts, motives, and apprehensions of one's opponent so fully that one can see the world through his eyes. All well-meaning people should try to contribute as much as possible to improving such mutual understanding. It is in this spirit that I should like to ask my Russian colleagues and any other reader to accept the following answer to their letter. It is the reply of a man who anxiously tries to find a feasible solution without having the illusion that he himself knows "the truth" or "the right path" to follow. If in the following I shall express my views somewhat dogmatically, I do it only for the sake of clarity and simplicity.

Although your letter, in the main, is clothed in an attack upon the non-socialistic foreign countries, particularly the United States, I believe that behind the aggressive front there lies a defensive mental attitude which is nothing else but the trend towards an almost unlimited isolationism. The escape into isolationism is not difficult to understand if one realizes what Russia has suffered at the hands of foreign countries during the last three decades—the German invasions with planned mass murder of the civilian population, foreign interventions during the civil war, the systematic campaign of calumnies in the western press, the support of Hitler as an alleged tool to fight Russia. However understandable this desire for isolation may be, it remains no less disastrous to Russia and to all other nations; I shall say more about it later on.

The chief object of your attack against me concerns my support of "world government." I should like to discuss this important problem only after having said a few words about the antagonism between socialism and capitalism; for your attitude on the significance of this antagonism seems to dominate completely your views on international problems. If the socio-economic problem is considered objectively, it appears as follows: technological development has led to increasing centralization of the economic mechanism. It is this development which is also responsible for the fact that economic power in all widely industrialized countries has become concentrated in the hands of relatively few. These people, in capitalist countries, do not need to account for their actions to the public as a whole; they must do so in socialist countries in which they are civil servants similar to those who exercise political power.

I share your view that a socialist economy possesses advantages which definitely counterbalance its disadvantages whenever the management lives up, at least to some extent, to adequate standards. No doubt, the day will come when all nations (as far as such nations still exist) will be grateful to Russia for having demonstrated, for the first time, by vigorous action the practical possibility of planned economy in spite of exceedingly great difficulties. I also believe that capitalism, or, we should say, the system of free enterprise will prove unable to check unemployment, which will become increasingly chronic because of technological progress, and unable to maintain a healthy balance between production and the purchasing power of the people.

On the other hand we should not make the mistake of blaming capitalism for all existing social and political evils, and of assuming that the very establishment of socialism would be able to cure all the social and political ills of humanity. The danger of such a belief lies, first, in the fact that it encourages fanatical intolerance on the part of all the "faithfuls" by making a possible social method into a type of church which brands all those who do not belong to it as traitors or as nasty evildoers. Once this stage has been reached, the ability to understand the convictions and actions of the "unfaithfuls" vanishes completely. You know, I am sure, from history how much unnecessary suffering such rigid beliefs have inflicted upon mankind.

Any government is in itself an evil insofar as it carries within it the tendency to deteriorate into tyranny. However, except for a very small number of anarchists, everyone of us is convinced that civilized society cannot exist without a government. In a healthy nation there is a kind of dynamic balance between the will of the people and the government which prevents its degeneration into tyranny. It is obvious that the danger of such deterioration is more acute in a country in which the government has authority not only over the armed forces but also over all the channels of education and information as well as over the economic existence of every single citizen. I say this merely to indicate that socialism as such cannot be considered the solution to all social problems but merely as a framework within which such a solution is possible.

What has surprised me most in your general attitude, expressed in your letter, is the following aspect: You are such passionate opponents of anarchy in the economic sphere, and yet equally passionate advocates of anarchy, e.g., unlimited sovereignty, in the sphere of international politics. The proposition to curtail the sovereignty of individual states appears to you in itself reprehensible, as a kind of violation of a natural right. In addition, you try to prove that behind the idea of curtailing sovereignty the United States is hiding her intention of economic domination and exploitation of the rest of the world without going to war. You attempt to justify this indictment by analyzing in your fashion the individual actions of this government since the end of the last war. You attempt to show that the Assembly of the United Nations is a mere puppet show controlled by the United States and hence the American capitalists.

Such arguments impress me as a kind of mythology; they are not convincing. They make obvious, however, the deep estrangement among the intellectuals of our two countries which is the result of a regrettable and artificial mutual isolation. If a free personal exchange of views should be made possible and should be encouraged, the intellectuals, possibly more than anyone else, could help to create an atmosphere of mutual understanding between the two nations and their problems. Such an atmosphere is a necessary prerequisite for the fruitful development of political cooperation. . . .

Is it not true, however, that we have stumbled into a state of international affairs which tends to make every invention of our minds and every material good into a weapon and, consequently, into a danger for mankind?

This question brings us to the most important matter, in comparison to which everything else appears insignificant indeed. We all know that power politics, sooner or later, necessarily leads to war, and that war, under present circumstances, would mean a mass destruction of human beings and material goods, the dimensions of which are much, much greater than anything that has ever before happened in history.

Is it really unavoidable that, because of our passions and our in-herited customs, we should be condemned to annihilate each other so thoroughly that nothing would be left over which would deserve to be conserved? Is it not true that all the controversies and differences of opinion which we have touched upon in our strange exchange of letters are insignificant pettiness compared to the danger in which we all find ourselves? Should we not do everything in our power to eliminate the danger which threatens all nations alike?

If we hold fast to the concept and practice of unlimited sover-eignty of nations it only means that each country reserves the right for itself of pursuing its objectives through warlike means. Under the circumstances, every nation must be prepared for that possibil-ity; this means it must try with all its might to be superior to any-one else. This objective will dominate more and more our entire public life and will poison our youth long before the catastrophe is itself actually upon us. We must not tolerate this, however, as long as we still retain a tiny bit of calm reasoning and human feelings.

This alone is on my mind in supporting the idea of "World Government," without any regard to what other people may have in mind when working for the same objective. I advocate world government because I am convinced that there is no other possible way of eliminating the most terrible danger in which man has ever found himself. The objective of avoiding total destruction must have priority over any other objective.

I am sure you are convinced that this letter is written with all the seriousness and honesty at my command; I trust you will ac-cept it in the same spirit.

In the face of this Russian intransigence, some felt that the time had come to issue an ultimatum. A British document, signed by Bertrand

Russell, T. S. Eliot, Baron Robert Vansittart, and Gordon Lang, head of the organization Crusade for World Government, brought this issue to a head by calling on the British and American governments to begin rearming if the Soviets continued to refuse to cooperate. Einstein took a far more sanguine view of the situation and strongly opposed such alarmism.

LETTER TO M.P. HENRY C. USBORNE, 9 JANUARY 1948
EINSTEIN ARCHIVES 58-922; N & N 1960, 463–464

1. I believe that the statement issued by Gordon Lang, Lord Vansittart, etc., is harmful indeed and that a statement opposing it is necessary. (It seems that Bertrand Russell's signature was not authorized by him.)

2. I agree with what Cord Meyer says about the Baruch proposals. It is not feasible to abolish one single weapon as long as war itself is not abolished. This can be done only by establishing effective world government.

3. The Russians would have been justified in their rejection of the Baruch proposals had they explained their rejection by genuine considerations and had it been accompanied by constructive and reasonable proposals of their own. Their actual counterproposal of general disarmament without supranational control cannot be taken seriously; and they must have known this.

4. It has become clear to me that the Soviets will categorically reject any supranational institution which would involve inspection of their country by foreigners or which would impose any limitations on their sovereignty. They will adhere to this position as long as they have any hope that this policy may successfully prevent the establishment of an effective supranational organization. This is manifest, for example, in the Open Letter addressed to me by four Russian scientists and recently published in the magazine *New Times*. This Open Letter must be regarded as a semiofficial statement.

5. The following difficult question remains: How can one work vigorously for world government without creating the risk that the result will be a coalition directed against Soviet Russia? It is precisely this question that makes it so difficult for the atomic scientists in this country to reach agreement on a specific and immediate policy.

6. Nevertheless, I myself believe that we should take a vigorous position in favor of world government, even at the risk of temporarily driving Soviet Russia into self-imposed isolation; and, naturally, everything should be done in such a way as to make it easy and attractive for the Russians to change their isolationist attitude. I believe that if this were done intelligently (rather than in clumsy Truman style!), Russia would co-operate once she realized that she was no longer able to prevent world government anyhow.

7. Matters would be considerably simplified if the United States were to adopt an unequivocal and unwavering position in favor of a supranational solution of the security problem. I believe that, in this connection, some criticism from the other side of the Atlantic, especially from England, would prove useful.

You may quote everything I have said in this letter, providing it is made clear that I am speaking only as an individual.

Cord Meyer Jr. was a staff member of the U.S. delegation to the United Nations organizational conference and an advocate of world government. His article "Peace Is Still Possible" appeared in the *Atlantic Monthly* of October 1947.

In this early stage of the Cold War, Einstein's approach to the critical impasse discussed above bears a striking resemblance to the strategy he advocated after the United States entered the First World War (letter to

Heinrich Zangger, 21 August 1917, chapter 1). Then he proposed to meet German military aggression with a strong international body comprised of nations dedicated to peace; here he indicated that Russian unilateralism could be overcome by means of a similarly constituted world government.

When a twenty-minute color animation film ("Where Will You Hide?") was produced in 1948, illustrating the idea that no one would be safe after another war with weapons of total destruction, Einstein wrote the following comment, uncharacteristically in English: "Somebody, after having seen this film, may say to you: This representation of our situation may be right, but the idea of world government is not realistic. You may answer him: If the idea of world government is not realistic, then there is only *one* realistic view of our future: holesale [*sic*] destruction of man by man" (Einstein Archives 28-817). A month later, in response to a question about whether world government was practicable immediately, he wrote: "It is downright foolish to pose the question.... It is like with an unavoidable operation. The probability that the patient will survive decreases every day that it is postponed" (Einstein Archives 28-825.1).

Controlling Weapons of Mass Destruction

On 23 September 1949 the Truman administration announced that the Soviet Union had detonated an atomic bomb, thereby ending the United States' monopoly on nuclear weapons. In the wake of this announcement, Einstein was approached by the Reverend Abraham J. Muste, who hoped to gain his support for an effort to stop the mounting arms race. As the leading figure in the American Workers Party, Muste had once been closely allied with American Trotskyites, but after meeting Trotsky personally in 1936 he left the movement to become an outspoken pacifist. He was appointed executive secretary of the Fellowship of Reconciliation in 1940. Einstein dismissed Muste's proposal as unrealistic and ineffectual.

LETTER TO A. J. MUSTE, 31 OCTOBER 1949
EINSTEIN ARCHIVES 58-574; N & N 1960, 517–518

There is no *purely mechanical* solution for the security problem. Also, security cannot be attained through armament but only on the basis of a give-and-take relationship which would make the creation and maintenance of a policy of peaceful co-operation desirable to both parties. Once this course is taken, it would mean the beginning of the development of mutual confidence which alone can make disarmament possible.

We should do our utmost to convince our fellow citizens that it is practically impossible for the Soviets to accept the Baruch Plan. . . .

Temporary suspension of the production of atom bombs is, in my opinion, ineffective. On the one hand, nobody abroad would really believe it; on the other, such a single measure would be of little significance as long as the armaments race continues. An especially bad policy, in my opinion, is our economic boycott of Eastern Europe. To give you an example: The United States Government has prevented the shipment of goods (machinery) to Czechoslovakia, which that country had bought in the United States and paid for. One can easily imagine what bitterness such a policy must create in a small, impoverished and war-ravaged country. The American public hardly ever hears of such happenings.

The concatenation is this: No peace without disarmament; no disarmament without confidence; no confidence without mutual and effective economic relations. I cannot help feeling that since the death of President Roosevelt our foreign policy has proceeded in the wrong direction, and there seems to be little prospect at the moment for a shift toward a more reasonable policy.

Freedom of research and publication, and preservation of civil liberties, can be secured only in an atmosphere of genuine peace. Otherwise these freedoms will slowly, but irretrievably, be lost. I do not expect very much even from a concerted effort on the part of the scientists since we are faced with a political and economic, rather than a technical, problem. The American physicists, individually and collectively, have again and again expressed their conviction that it is impossible to attain security through armaments; they have, unfortunately, been unsuccessful in convincing others. . . .

Plans to develop the "super," Edward Teller's name for the hydrogen bomb, now went ahead rapidly (Rhodes 1985, 382–408). On 30 January 1950, only a day before Truman announced that the United States would undertake an all-out effort to produce the H-bomb, Muste again appealed to Einstein to help put the brakes on this decision, citing numerous other American religious leaders who joined him in doing so. He pleaded to Einstein in a telegram that "people must have opportunities to ponder and discuss this life-and-death issue, if [the] United States is to remain [a] democratic nation" (N & N 1960, 520). Einstein responded immediately, characterizing Muste's proposal as "quite impracticable," given the arms race already underway. "The only possible solution," he added, "would be an honest attempt to work out a reasonable agreement with Soviet Russia and, beyond this, for security on a supranational basis" (ibid.).

Two weeks later, Einstein's views became widely known when he appeared on the NBC television program *Today with Mrs. Roosevelt* to discuss the government's decision to pursue a crash program to produce hydrogen bombs. The former first lady also invited a number of other guests to speak on her program, including J. R. Oppenheimer, who had recently resigned as director of the Manhattan Project, and David Lilienthal, head of the Atomic Energy Commission. Einstein's remarks were prerecorded in his Princeton home two days before the broadcast (see Plate 21).

Statement for Appearance on *Today with Mrs. Roosevelt*,
10 February 1950
Einstein Archives 28-870 (fragment); N & N 1960, 520–522

I am grateful to you, Mrs. Roosevelt, for the opportunity to express my convictions on this most important political question.

The belief that it is possible to achieve security through armaments on a national scale is, in the present state of military technology, a disastrous illusion. In the United States, this illusion has been strengthened by the fact that this country was the first to succeed in producing an atomic bomb. This is why people tended to believe that this country would be able to achieve permanent and decisive military superiority which, it was hoped, would deter any potential enemy and thus bring about the security, so intensely sought by us as well as by the rest of the world. The maxim we have followed these last five years has been, in short, security through superior force, whatever the cost.

This technology as well as psychological orientation in military policy has had its inevitable consequences. Every action related to foreign policy is governed by one single consideration: How should we act in order to achieve the utmost superiority over the enemy in the event of war? The answer has been: Outside the United States, we must establish military bases at every possible, strategically important point of the globe as well as arm and strengthen economically our potential allies. And inside the United States, tremendous financial power is being concentrated in the hands of the military; youth is being militarized; and the loyalty of citizens, particularly civil servants, is carefully supervised by a police force growing more powerful every day. People of independent political thought are harassed. The public is subtly indoctrinated by the radio, the press, the schools. Under the pressure of military secrecy, the range of public information is increasingly restricted.

The arms race between the United States and the Soviet Union, initiated originally as a preventive measure, assumes hysterical proportions. On both sides, means of mass destruction are being perfected with feverish haste and behind walls of secrecy. And now the public has been advised that the production of the hydrogen bomb is the new goal which will probably be accomplished. An accelerated development toward this end has been solemnly proclaimed by the President. If these efforts should prove successful, radioactive poisoning of the atmosphere and, hence, annihilation of all life on earth will have been brought within the range of what is technically possible. The weird aspect of this development lies in its apparently inexorable character. Each step appears as the inevitable consequence of the one that went before. And at the end, looming ever clearer, lies general annihilation.

Is there any way out of this impasse created by man himself? All of us, and particularly those who are responsible for the policies of the United States and the Soviet Union, must realize that, although we have vanquished an external enemy, we have proved unable to free ourselves from the war mentality. We shall never achieve real peace as long as every step is taken with a possible future conflict in view, especially since it becomes ever clearer that such a war would spell universal annihilation. The guiding thought in all political action should therefore be: What can we do in the prevailing situation to bring about peaceful coexistence among all nations? The first goal must be to do away with mutual fear and distrust. Solemn renunciation of the policy of violence, not only with respect to weapons of mass destruction, is without doubt necessary. Such renunciation, however, will be effective only if a supranational judicial and executive agency is established at the same time, with power to settle questions of immediate concern to the security of nations. Even a declaration by a number of nations that they would collaborate loyally in the realization of such a "restricted world government" would considerably reduce the imminent danger of war.

In the last analysis the peaceful coexistence of peoples is primarily dependent upon mutual trust and, only secondarily, upon institutions such as courts of justice and the police. This holds true for nations as well as for individuals. And the basis of trust is a loyal relationship of give-and-take.

And what about international control? Well, it may be useful as a police measure but cannot be considered a prime factor. In any event, it may be wise not to overestimate its importance. The example of Prohibition comes to mind and gives one pause.

A year earlier, during the celebration of his seventieth birthday, Einstein had been asked what weapons he thought might be used in the event of a third world war. His alleged reply: "I do not know how the Third World War will be fought, but I can tell you what they will use in the Fourth—rocks!" (Alfred Werner interview, "Einstein at Seventy," *Liberal Judaism* 16 (April–May 1949), 4–12). Precisely to forestall this possibility, Einstein later joined forces with Bertrand Russell to mount a campaign repudiating nuclear weapons. This led to the Russell-Einstein Manifesto, the last major political document to carry the physicist's famous name (see chapter 10).

CHAPTER 9

Soviet Russia, Political Economy, and Socialism, 1918–1952

Having now covered Einstein's views on several major political issues from the Great War to the Cold War era, we will in the present chapter both complement and amplify the previous eight by presenting a set of his writings that deal with socialism, Soviet communism, and political economy. These texts show not only how Einstein's ideas evolved over three decades but also how he refused to reduce political matters to black and white terms. His political philosophy remained throughout this time one in which individual rights stood in a delicate balance with the larger interests of society as a whole.

Before the onset of the Great Depression, Einstein appears to have given relatively scant thought to matters of political economy; moreover, his rejection of the proposal for a "nourishment army" in 1918 indicates that he had little sympathy for a welfare state. During the 1920s his knowledge of developments in the Soviet Union came largely from books written by intellectuals like Bertrand Russell and Emil Gumbel, both of whom had briefly visited the country. His ambivalent attitude toward Bolshevism at this time is apparent from his remarks in the last two documents of the opening section, though his unequivocal concern for the plague-ridden Soviet population was already evident in his support of attempts in October 1919 to lift the Allied blockade of foodstuffs and medicine on Russia (see CPAE 9, Doc. 141).

Around 1930 Einstein began to outline his tentative thoughts on political economy as set forth in the three texts presented in the section "Reflections on Economics at the Onset of the Great Depression." These constitute his first serious attempt to diagnose the crisis plaguing modern capitalist economies, though they reveal no particularly striking insights into the problem. Nor do they provide any evidence that Einstein was tempted to give up free enterprise in favor of a rigidly planned economy, least of all at the price of basic freedoms. While he may have felt a certain attraction to Bolshevism and class warfare in 1920—as reflected in a letter he wrote to Max and Hedwig Born—by 1930 he had seen and heard about enough violence. Pacifism had become his watchword, and his approach to economics and social justice desired "nothing but the good of humanity and the most harmonious possible scheme of human existence."

Einstein's experiences with the Geneva Disarmament Conference showed him that the inertia of governments and their agencies is far too great to accomplish anything radically new that would help promote world peace. His brief flirtation with an international council of distinguished pacifists (the "wise men") was partly a reaction to the impotence of government representatives to accomplish anything noteworthy through conventional diplomacy and mediation. Einstein's encounters with Communist tactics during the early 1930s only reinforced his erstwhile skepticism and fervent elitism. As a staunch antifascist, he came under strong pressure to join hands with leading leftists like Henri Barbusse and Willi Münzenberg. Both felt free to use Einstein's name to advertise causes like the 1932 International Congress against Imperialist Wars, which Münzenberg staged in Amsterdam at the behest of Moscow authorities. A glimpse into Einstein's private correspondence at this time reveals not only the depth of his misgivings regarding the activities of the Comintern but also his willingness to dissociate himself from causes he believed were tarnished by Communist influence. Indeed, his disillusionment with Barbusse and associates provided the impetus for Einstein's continued efforts to form a small coalition of independent intellectuals who could stand above party politics.

This elitist endeavor came to naught after 1932, when Einstein emerged as one of the early voices warning of the imminent collapse of European civilization. It was during the midst of the fight against fascism in the early 1940s that he once again began to weigh the pros and cons of a planned economy. The first two shorter texts in the section "Freedom and Socialism Reconsidered" can be regarded almost as preliminary sketches for the views he set out in "Why Socialism?" Whereas many of the writings in this chapter are barely known, "Why Socialism?" constitutes one of Einstein's best known political essays, having been reprinted countless times since it first appeared in the *Monthly Review* in 1949. Properly contextualized, it offers a remarkably clear synthesis of Einstein's mature political thought, one that transcends familiar discourse in the ongoing debate about capitalism versus socialism.

Einstein was never much of a debater, nor did he have any real desire to philosophize about politics, a point apparently lost on some of his leading contemporaries, like Sidney Hook. In the final section of this chapter, spanning the period 1934 to 1952, we see Einstein offering a wide variety of opinions about the quality of life in the Soviet Union. Hook, who drew a sharp line between democratic socialism and Soviet Marxism, grew increasingly dismayed by Einstein's unwillingness to join his anti-Communist crusade. Their exchanges suggest two very different mindsets, and in Einstein's case the need to distinguish between political principles and political praxis.

Early Thoughts on Socialism and the Soviet Experiment

During the First World War, the Viennese engineer and philosopher Josef Popper-Lynkeus called for the creation of a compulsory national labor service that would produce and distribute the minimum of goods and services needed for each individual's material and cultural existence.

Fearing the cost in economic compulsion and bureaucracy, Einstein opposed the idea.

To the Society "A Guaranteed Subsistence for All," 12 December 1918
Einstein Archives 32-754; CPAE 7, Doc. 16

I too believe Popper's suggestion that a secured minimum level of subsistence for every citizen is the most important goal toward which we can strive. But I can by no means agree with the idea of a "nourishment army" because I am convinced such a thing would produce highly uneconomical results. Everybody should, in case of disability or unemployment, have the right to get the absolutely essential things one needs to live from public soup kitchens. I am convinced that the effort of striving for an improved existence as a wage earner combined with the disdain people have toward those who are not gainfully employed are sufficiently strong psychological forces to ensure the healthy development of economic life. Experience seems to show that dire need and bitter worries act to paralyze rather than promote the desire to work.

Faith in the "automatic regulation of the problems of distribution through supply and demand" was a position that Einstein later came to regret, writing in 1944 that he had earlier believed "that the great goal [of improving the lot of humankind] might be achieved at a lesser cost in compulsion, organization, and bureaucracy" (draft preface for a booklet by Yisrael Doryon, Einstein Archives 28-602).

Only a short time before disgruntled nationalists briefly terrorized Berlin and threatened to topple the Weimar Republic in the Kapp Putsch, Einstein felt sure that the radical left was gaining momentum.

LETTER TO HEDWIG AND MAX BORN, 27 JANUARY 1920
EINSTEIN ARCHIVES 8-144; CPAE 9, DOC. 284

... The political situation is developing consistently in favor of the Bolsheviks. It seems that the Russians' considerable external achievements are gathering an irresistible momentum in relation to the increasingly untenable position of the West; particularly our position. But before this can happen, streams of blood will have to flow; the forces of reaction are also growing more violent all the time. Nicolai is being attacked and insulted so much that he is no longer able to lecture, not even in the Charité [clinic at the University of Berlin]. Once again I have had to intercede for him in public ... By the way, I must confess to you that the Bolsheviks do not seem so bad to me, however ridiculous their theories. It would be really interesting just to have a look at the thing at close quarters. At any rate, their message seems to be very effective, for the weapons the Allies used to destroy the German army melt away in Russia like snow in the spring sun. Those fellows have gifted politicians at the top. I recently read a brochure by Radek—one has to hand it to him, the man knows his business.

Einstein's most recent intercession on Nicolai's behalf had occurred the previous day ("In Support of Georg Nicolai," chapter 1).

In all likelihood it was Radek 1918 ("The Development of Socialism from Utopia to a Science") that Einstein read. Karl Radek was associated with both Lenin and Trotsky in Zurich during the war and accompanied the Bolshevik leaders on the sealed train that took them out of Switzerland to lay the groundwork for the October Revolution. After the Kaiser's abdication in November 1918, Radek collaborated with the far left Spartacists in their abortive attempt to seize power in January 1919. He was arrested soon after the assassination of the Spartacist leaders, Rosa Luxemburg and Karl Liebknecht, and deported to Moscow. In 1920 he was appointed to the executive committee of the Communist International.

In 1922 Emil J. Gumbel prepared a German translation of a number of Bertrand Russell's political writings from the war years under the title *Politische Ideale*, for which Einstein wrote the foreword.

FOREWORD TO GERMAN VERSION OF BERTRAND RUSSELL'S
POLITICAL IDEALS
RUSSELL 1922, 5

It is welcome that the lucid comments of the great English mathematician are being made available to the German public. It is not a shaky professor who speaks to us in the equivocal tones of "on the one hand" and "on the other," but one of the decisive, straightforward individuals, who are independent of the period into which they are by coincidence born. Implacable rigor and a warm human sensibility guide him on his path. He is unconcerned by the consequences which follow from his remarks. He allows himself to be stripped of his academic chair without posing as a martyr and goes to prison as a result of his antimilitaristic propaganda.

He wishes to dismantle military force completely and recommends consistent education in organized passive popular resistance against oppressive foreign military coercion. This solution will no longer appear utopian to those who lived through the Kapp putsch.

Russell also deals with the socio-political problem. Driven by a burning interest in the progress of human society, he travels through Bolshevik Russia to learn. His ideal is the development of the freely creative powers of individuals within a social order, in which fear that basic needs cannot be met is reduced without succumbing to the worst enemy of socialist efforts, a hypertrophic bureaucracy.

One may agree or disagree with some of Russell's opinions. It is in any case a pleasure to acquaint oneself with the thoughts of a penetrating and truly noble contemporary on such matters as stir all serious individuals today.

Let everyone come to terms for himself with the great Englishman.

In 1925 Einstein was contacted by the journalist Isaac Don Levine, who requested an introductory note for a document collection entitled *Letters from Russian Prisons*. A Russian-born Jew, Levine immigrated to the United States in 1911 and later worked for various American newspapers. During the civil war that followed the Russian Revolution he returned to his native country to cover events. The document collection, published on behalf of the International Committee of Political Prisoners and gathered with the assistance of the anarchist Alexander Berkman, contained correspondence of inmates of the early Soviet gulag, as well as reprints of affidavits on political persecution in the Soviet Union. The volume, which appeared in New York and London in 1925, harshly condemned the methods used by the Bolsheviks to suppress political opposition. The following note by Einstein was one of twenty-two introductory comments secured by Levine. Its title in Einstein 1929 is "On a Document Collection on the Russian Revolution."

ON A DOCUMENT COLLECTION FROM RUSSIAN PRISONS, 1925
EINSTEIN ARCHIVES 28-029; EINSTEIN 1929, 20

If you study these accounts as a reader in a peaceful, well-regulated system of government, don't imagine that those around you are different and better than those who conduct a regime of terror in Russia. Shudder to view this tragedy of human history where one murders out of fear that one will be murdered. It is the best, the most altruistic who are tortured and killed because their political influence is feared—but not just in Russia.

All serious men owe a debt of gratitude to the editor of these documents. He will help to reverse this dreadful fate. After the publication of these documents the rulers of Russia will have to change their methods if they wish to continue their effort to gain moral credibility with civilized nations. They will lose all sympathy if they cannot show through a great and courageous act of liberation that they do not need to rely on bloody terror to lend support to their political ideals.

A more favorable account of conditions in Russia was provided by Emil Gumbel, who published his views of the country after travels there in winter 1925–26 (Gumbel 1927). Einstein wrote a short prefatory note for the book, in which he expressed his admiration: "As few could have done, you have made remarkably objective observations from many perspectives" (letter to Gumbel, 23 October 1926, quoted in Gumbel 1991, 84). His own mixed feelings about events in Russia are revealed in the following statement solicited by the German League of Human Rights. In the cover letter, Einstein requests that "either everything or nothing [of the statement] be published" (Einstein Archives 47-470).

ON THE FIFTH ANNIVERSARY OF LENIN'S DEATH, 6 JANUARY 1929
EINSTEIN ARCHIVES 47-471; EINSTEIN 1929, 20–21

In Lenin I honor a man, who in total sacrifice of his own person has committed his entire energy to realizing social justice. I do not find his methods advisable. One thing is certain, however: men like him are the guardians and renewers [*Erneuerer*] of mankind's conscience.

Reflections on Economics at the Onset of the Great Depression

In the decades following, Einstein continued to take a keen interest in economic and political developments in the Soviet Union (see "Principles versus Praxis," the closing section of this chapter). His later reflections on these matters were deeply affected by the Great Depression and the Second World War. Regarding the former, Einstein put various economic proposals to paper that came to light in 1934 when Rudolf Kayser included these texts in *Mein Weltbild*.

Thoughts on the World Economic Crisis, ca. 1930
Einstein Archives 28-120; Einstein 1954, 87–91

If there is anything that can give a layman in the sphere of economics the courage to express an opinion on the nature of the alarming economic difficulties of the present day, it is the hopeless confusion of opinions among the experts. What I have to say is nothing new and does not pretend to be anything more than the expression of the opinion of an independent and honest man who, unburdened by class or national prejudices, desires nothing but the good of humanity and the most harmonious possible scheme of human existence. If in what follows I write as if I were sure of the truth of what I am saying, this is merely done for the sake of an easier mode of expression; it does not proceed from unwarranted self-confidence or a belief in the infallibility of my somewhat simple intellectual conception of problems which are in reality uncommonly complex.

As I see it, this crisis differs in character from past crises in that it is based on an entirely new set of conditions, arising out of the rapid progress in methods of production. Only a fraction of the available human labor in the world is now needed for the production of the total amount of consumption goods necessary to life. Under a completely laissez-faire economic system, this fact is bound to lead to unemployment.

For reasons which I do not propose to analyze here, the majority of people are compelled to work for the minimum wage on which life can be supported. If two factories produce the same sort of goods, other things being equal, that factory will be able to produce them more cheaply which employs fewer workmen—i.e., makes the individual worker work as long and as hard as human nature permits. From this it follows inevitably that, with methods of production as they are today, only a portion of the available labor can be used. While unreasonable demands are made on this portion, the remainder is automatically excluded from the process of production. This leads to a fall in sales and profits. Businesses go smash, which further increases unemployment and diminishes confidence in industrial concerns and therewith public participation in the mediating banks; finally the banks become insolvent through the sudden withdrawal of accounts and the wheels of industry therewith come to a complete standstill.

The crisis has also been attributed to other causes which we will now consider.

Over-production. We have to distinguish between two things here—real over-production and apparent over-production. By real over-production I mean a production so great that it exceeds the demand. This may perhaps apply to motor cars and wheat in the United States at the present moment, although even that is doubtful. By "over-production" people usually mean a condition in which more of one particular article is produced than can, in existing circumstances, be sold, in spite of a shortage of consumption goods among consumers. This I call apparent over-production. In this case it is not the demand that is lacking but the consumers' purchasing power. Such apparent over-production is only another word for a crisis and therefore cannot serve as an explanation of the latter; hence people who try to make over-production responsible for the present crisis are merely juggling with words.

Reparations. The obligation to pay reparations lies heavy on the debtor nations and their economies. It compels them to go in for dumping and so harms the creditor-nations too. This is beyond dispute. But the appearance of the crisis in the United States, in spite of the high tariff-wall, proves that this cannot be the principal cause of the world crisis. The shortage of gold in the debtor countries due to reparations can at most serve as an argument for putting an end to these payments; it cannot provide an explanation of the world crisis.

Erection of new tariff-walls. Increase in the unproductive burden of armaments. Political insecurity owing to latent danger of war. All these things make the situation in Europe considerably worse without really affecting America. The appearance of the crisis in America shows that they cannot be its principal causes.

The dropping-out of the two powers, China and Russia. Also this blow to world trade cannot make itself very deeply felt in America and therefore cannot be the principal cause of the crisis.

The economic rise of the lower classes since the War. This, supposing it to be a reality, could only produce a scarcity of goods, not an excessive supply.

I will not weary the reader by enumerating further contentions which do not seem to me to get to the heart of the matter. Of one thing I feel certain: this same technical progress which, in itself, might relieve mankind of a great part of the labor necessary to its subsistence, is the main cause of our present misery. Hence there are those who would in all seriousness forbid the introduction of technical improvements. This is obviously absurd. But how can we find a more rational way out of our dilemma?

If we could somehow manage to prevent the purchasing power of the masses, measured in terms of goods, from sinking below a certain minimum, stoppages in the industrial cycle such as we are experiencing today would be rendered impossible.

The logically simplest but also most daring method of achieving this is a completely planned economy, in which consumption goods are produced and distributed by the community. That is essentially what is being attempted in Russia today. Much will depend on what results this forced experiment produces. To hazard a prophecy here would be presumption. Can goods be produced as economically under such a system as under one which leaves more freedom to individual enterprise? Can this system maintain itself at all without the terror that has so far accompanied it, to which none of us westerners would care to expose himself? Does not such a rigid, centralized economic system tend toward protectionism and toward resistance to advantageous innovations? We must take care, however, not to allow these misgivings to become prejudices which prevent us from forming an objective judgment.

My personal opinion is that those methods are in general preferable which respect existing traditions and habits so far as that is in any way compatible with the end in view. Nor do I believe that a sudden transference of economy into governmental management would be beneficial from the point of view of production; private enterprise should be left its sphere of activity, in so far as it has not already been eliminated by industry itself by the device of cartelization.

There are, however, two respects in which this economic freedom ought to be limited. In each branch of industry the number of working hours per week ought so to be reduced by law that unemployment is systematically abolished. At the same time minimum wages must be fixed in such a way that the purchasing power of the workers keeps pace with production.

Further, in those industries which have become monopolistic in character through organization on the part of the producers, prices must be controlled by the state in order to keep the issue of capital within reasonable bounds and prevent the artificial strangling of production and consumption.

In this way it might perhaps be possible to establish a proper balance between production and consumption without too great a limitation of free enterprise and at the same time to stop the intolerable tyranny of the owners of the means of production (land and machinery) over the wage-earners, in the widest sense of the term.

Einstein's comments on production and purchasing power of 7 November 1931 were written in the form of a letter to the editor of the *Berliner Tageblatt*, but were not published there.

Production and Purchasing Power, 7 November 1931
Einstein Archives 28-161; Einstein 1954, 91–92

... I do not believe that the remedy for our present difficulties lies in a knowledge of productive capacity and consumption, because this knowledge is likely, in the main, to come too late. Moreover, the trouble in Germany seems to me to be not hypertrophy of the machinery of production but deficient purchasing power in a large section of the population, which has been cast out of the productive process through the rationalization of industry.

The gold standard has, in my opinion, the serious disadvantage that a shortage in the supply of gold automatically leads to a contraction of credit and also of the amount of currency in circulation, to which contraction prices and wages cannot adjust themselves sufficiently quickly.

The natural remedies of our troubles are, in my opinion, as follows:

(1) A statutory reduction of working hours, graduated for each department of industry, in order to get rid of unemployment, combined with the fixing of minimum wages for the purpose of adjusting the purchasing power of the masses to the amount of goods available.

(2) Control of the amount of money in circulation and of the volume of credit in such a way as to keep the price level steady, abolishing any monetary standard.

(3) Statutory limitation of prices for such articles as have been practically withdrawn from free competition by monopolies or the formation of cartels.

Einstein's proposals were largely restricted to the sphere of monetary policy in contrast to John Maynard Keynes's call for government intervention in the labor market, a strategy which would emerge as a driving force behind the New Deal. In the following article, written in response to a query from a Swedish architect, H. Cederström, Einstein again took the more conservative monetarist tack.

PRODUCTION AND WORK, 22 SEPTEMBER 1932
EINSTEIN ARCHIVES 28-192; EINSTEIN 1954, 92–93

The fundamental trouble seems to me to be the almost unlimited freedom of the labor market combined with extraordinary progress in the methods of production. To satisfy the needs of the world today nothing like all the available labor is wanted. The result is unemployment and unhealthy competition among the workers, both of which reduce purchasing power and thereby put the whole economic system intolerably out of gear.

I know Liberal economists maintain that every economy in labor is counterbalanced by an increase in demand. But, to begin with, I don't believe that; and even if it were true, the above-mentioned factors would always operate to force the standard of living of a large portion of the human race down to an unnaturally low level.

I also share your conviction that steps absolutely must be taken to make it possible and necessary for the younger people to take part in the productive process. Further, that the older people ought to be excluded from certain sorts of work (which I call "unqualified" work), receiving instead a certain income, as having by that time done enough work of a kind accepted by society as productive.

I, too, am in favor of abolishing large cities, but not of settling people of a particular type, e.g., old people, in particular towns. Frankly, the idea strikes me as horrible.

I am also of the opinion that fluctuations in the value of money must be avoided, by substituting for the gold standard a standard based on certain classes of goods selected according to the conditions of consumption—as Keynes, if I am not mistaken, long ago proposed. With the introduction of this system one might consent to a certain amount of "inflating," as compared with the present monetary situation, if one could believe that the state would really make a rational use of the windfall thus accruing to it.

The weaknesses of your plan lie, so it seems to me, in the sphere of psychology, or rather, in your neglect of it. It is no accident that capitalism has brought with it progress not merely in production but also in knowledge. Egoism and competition are, alas, stronger forces than public spirit and sense of duty. In Russia, they say, it is impossible to get a decent piece of bread . . . Perhaps I am over-pessimistic concerning state and other forms of communal enter-prise, but I expect little good from them. Bureaucracy is the death of any achievement. I have seen and experienced too many dread-ful warnings, even in comparatively model Switzerland.

I am inclined to the view that the state can only be of real use to industry as a limiting and regulative force. It must see to it that competition among the workers is kept within healthy limits, that all children are given a chance to develop soundly, and that wages are high enough for the goods produced to be consumed. But it can exert a decisive influence through its regulative function if its measures are framed in an objective spirit by independent experts.

Intellectual Elitism and Political Idealism

As noted in chapter 4, the famous correspondence between Einstein and Freud, published under the title *Why War?* in 1933, revealed a number of common bonds, not least their mutual commitment to paci-fism. Freud even suggested that this was not a matter of conventional politics but rather that pacifism was for both him and Einstein akin to a congenital condition that stemmed from their special place within the course of human evolution.

Einstein may have resisted that particular conclusion; nevertheless, he clearly accepted the elitism behind it (Scheideler 2002). Freud em-phasized that "men are divided into the leaders and the led." Since "the second class constitutes the vast majority, they need a high command to make decisions for them, to which decisions they usually bow without demur . . . men should be at greater pains than heretofore to form a superior class of independent thinkers, unamenable to intimidation and fervent in the quest of truth, whose function it would be to guide the masses dependent on their lead" (N & N 1960, 199–200).

When he wrote these words, Freud could be sure they would meet with Einstein's approval as he had earlier received the following private letter from him.

Letter to Sigmund Freud, 1931–1932
Einstein Archives 32-558; N & N 1960, 186–187

I greatly admire your passion to ascertain the truth—passion that has come to dominate all else in your thinking. You have shown with irresistible lucidity how inseparably the aggressive and destructive instincts are bound up in the human psyche with those of love and the lust for life. At the same time, your convincing arguments make manifest your deep devotion to the great goal of the internal and external liberation of man from the evils of war. This was the profound hope of all those who have been revered as moral and spiritual leaders beyond the limits of their own time and country, from Jesus to Goethe and Kant. Is it not significant that such men have been universally recognized as leaders, even though their desire to affect the course of human affairs was quite ineffective?

I am convinced that almost all great men who, because of their accomplishments, are recognized as leaders even of small groups share the same ideals. But they have little influence on the course of political events. It would almost appear that the very domain of human activity most crucial to the fate of nations is inescapably in the hands of wholly irresponsible political rulers.

Political leaders or governments owe their power either to the use of force or to their election by the masses. They cannot be regarded as representative of the superior moral or intellectual elements in a nation. In our time, the intellectual elite does not exercise any direct influence on the history of the world; the very fact of its division into many factions makes it impossible for its members to co-operate in the solution of today's problems. Do you not share the feeling that a change could be brought about by a free association of men whose previous work and achievements offer a guarantee of their ability and integrity? Such a group of international scope, whose members would have to keep contact with each other through constant interchange of opinions, might gain a significant and wholesome moral influence on the solution of political problems if its own attitudes, backed by the signatures of its concurring members, were made public through the press. Such an association would, of course, suffer from all the defects that have so often led to degeneration in learned societies; the danger that such a degeneration may develop is, unfortunately, ever present in view of the imperfections of human nature. However, and despite those dangers, should we not make at least an attempt to form such an association in spite of all dangers? It seems to me nothing less than an imperative duty!

Once such an association of intellectuals—men of real stature—has come into being, it might then make an energetic effort to enlist religious groups in the fight against war. The association would give moral power for action to many personalities whose good intentions are today paralyzed by an attitude of painful resignation. I also believe that such an association of men, who are highly respected for their personal accomplishments, would provide important moral support to those elements in the League of Nations who actively support the great objective for which that institution was created.

I offer these suggestions to you, rather than to anyone else in the world, because your sense of reality is less clouded by wishful thinking than is the case with other people and since you combine the qualities of critical judgment, earnestness and responsibility.

Einstein was also on friendly terms with a leading French leftist intellectual, Henri Barbusse, whose war novel *Le Feu* created a minor literary sensation. Barbusse served as one of the principal organizers of a peace conference scheduled to open in Geneva in June 1932 while the General Disarmament Conference was in session (see chapter 4). Among its many sponsors were to be such luminaries as Mme. Sun Yat-sen, Theodore Dreiser, Upton Sinclair, John Dos Passos, Heinrich Mann, Maxim Gorki, George Bernard Shaw, H. G. Wells, Romain Rolland, and Paul Langevin. When Barbusse first wired Einstein asking for his support, he got a stiff rejection: "I would never take part in such a congress, which would be pathetic in its impotence . . . like holding a congress to keep volcanoes from erupting or to increase the rainfall in the Sahara" (20 April 1932, N & N 1960, 176).

Soon afterward, however, Einstein warmed to the idea, hoping that the congress would serve to unify and mobilize the antiwar movement which was then focused on Japanese aggression in Asia. He shared the widespread view on the left that Japan's attack on Manchuria was encouraged by those who sought to undermine the Soviet Union. To Barbusse he wrote: "Ever since Japan embarked on its Manchurian adventure, it has been clear to me that it was supported by powerful, invisible allies," and he further presumed that "they are the same forces which are sabotaging the disarmament effort." As plans for the congress unfolded, Barbusse continually bombarded Einstein with messages. On 18 May he sent him a letter along with the text of an appeal that sharply criticized Japan while heaping lavish praise on the Soviet Union (N & N 1960, 178). Einstein refused to sign it for the following reasons.

Letter to Henri Barbusse, 6 June 1932
Einstein Archives 34-542; N & N 1960, 178–179

On my return from England I received your letter enclosing the draft of an appeal. Because of the glorification of Soviet Russia which it includes, I cannot bring myself to sign it. I have of late tried very hard to form a judgment of what is happening there, and I have reached some rather somber conclusions.

At the top there appears to be a personal struggle in which the foulest means are used by power-hungry individuals acting from purely selfish motives. At the bottom there seems to be complete suppression of the individual and of freedom of speech. One wonders what life is worth under such conditions.

This, however, does not cause me to consider the intrigues of Japan and the powers behind her any less damnable than you do. I have on various occasions hinted at the possibility of an international economic boycott against Japan, only to find that nothing could be achieved, obviously because of the powerful private economic interests that are involved! If you could bring yourself to rephrase the first paragraph of your appeal along more objective lines, omitting any glorification of conditions in the Soviet Union, I would certainly sign it.

Barbusse soon realized that he had no time to wait for the various sponsors' signatures. He thus informed Einstein that the appeal had already been published under just two names: those of Einstein and Rolland. As for negative reports on conditions in Russia, Barbusse rejected these as capitalist propaganda, urging Einstein to visit the Soviet Union in order to form an independent judgment. He received this amicable rejoinder.

LETTER TO HENRI BARBUSSE, 17 JUNE 1932
EINSTEIN ARCHIVES 34-546; N & N 1960, 179–180

You are undoubtedly aware that I agree with you on all essential points. Everything must be done to prevent any external threat to the development of Russia. In my criticism of the appeal I merely sought to avoid glorification of the internal situation in Russia. I do not wish to discuss the matter at length at this time, but I am rather convinced that if Barbusse happened to be in Russia he would be somewhere in prison or exile, if indeed his life had been spared.

In any event, in the interest of the extraordinarily important cause of peace I should like to urge that, at the Geneva Congress, all peace efforts be supported with great energy, but without any expression in favor of Bolshevism. Otherwise ugly controversies will be inevitable among the most sincere and most devoted people, and your project will probably have even less practical success than it, unfortunately, anyhow has.

It is my conviction that the cause of peace can be effectively served only if the principle of the inviolability of life and the individual is not made the subject of political controversies. We must never forget that the most courageous fighters against militarism come from a religious group, from among the Quakers.

Only a short time before the congress was scheduled to take place, the authorities in Geneva decided to withdraw permission, forcing the organizers to scramble for another venue. They eventually settled on Amsterdam, where the International Congress against Imperialist Wars opened on 27 August. Einstein came under considerable pressure to attend, receiving not only a barrage of mail from Barbusse but even a message from Willi Münzenberg, Berlin's "red press baron," who was one of the major orchestrators behind the scenes. Five days before it opened, Barbusse wired Einstein an urgent message imploring him to attend as "absence would be interpreted [as a] vote of nonconfidence" (N & N 1960, 180). Einstein scribbled a message on the back of the telegram which was read to the participants of the congress.

STATEMENT FOR THE AMSTERDAM PEACE CONGRESS,
26 AUGUST 1932
N & N 1960, 180–181

When Japan invaded Manchuria, the conscience of the civilized world was not strong enough to prevent this injustice. The economic interests of the war industries proved more powerful than the peoples' desire for justice. Now it has become clear to everyone that one of the purposes of the Japanese adventure was to weaken Russia by military attack and to obstruct its economic development.

All supporters of a healthy development of international legal institutions, regardless of their political or economic ideology, must do all in their power to insure that brute force and the unrestrained lust for profits be replaced by institutions which will make just and considerate decisions. Everyone who stands by idly while our culture is so gravely threatened must share in the guilt.

When, in the leading countries of the world, the desire for justice is strong enough, justice will be achieved. I hope that this congress will help to mobilize public opinion in such a way as to compel governments of the great powers to take the measures necessary to avert the impending disaster.

Einstein had sensed all along that the long arm of Moscow, operating through the Comintern, was trying to exploit this International Congress against Imperialist Wars as an instrument of propaganda. Now that Einstein's name appeared on the congress letterhead as one of its 26 sponsors, anti-Communists henceforth had an easy target. They were, however, *not* aware of the stance he took when afterward invited to join the German Committee against Imperialist Wars, an offshoot of the congress. The following letter was addressed to its founder and chairman, Felix Boenheim.

LETTER TO THE GERMAN COMMITTEE AGAINST IMPERIALIST WARS, 29 SEPTEMBER 1932
EINSTEIN ARCHIVES 50-439; N & N 1960, 181–182

I have carefully considered your invitation to become a member of your committee, but I feel unable to accept this invitation. I have long since come to regret my membership in the International

Committee which prepared the Amsterdam Congress, although I was in complete accord with the congress about the aims it pursued. I objected, however, to the fact that the congress was entirely under Russian-Communist domination. Also, its resolution is styled in the phraseology customary to the Communist movement. Prominent Social Democrats were excluded from the congress committee. Militant pacifists pleaded with me not to participate in the congress in view of its intolerant political attitude. Had I known in advance about the political position of the congress leadership, I should never have allowed myself to be elected to the Amsterdam Committee; it is important for me to retain a position of political neutrality. Failure to do so might jeopardize my chances of serving the cause of militant pacifism.

Another committee member, the French writer Victor Margueritte, sent Einstein a copy of his sharply worded letter of resignation. Five months earlier, Margueritte had supported the stance taken by Lord Ponsonby and Einstein in protesting the timid negotiations at the Geneva Disarmament Conference (see chapter 4). In the present situation Einstein sent him an understanding reply.

LETTER TO VICTOR MARGUERITTE, 19 OCTOBER 1932
EINSTEIN ARCHIVES 34-556; N & N 1960, 183

You are quite right . . . Barbusse is a fine man, but unfortunately a poor performer. He allowed himself to be so completely taken in by the Bolshevists that the congress lost its suprapartisan character. I am sure that Romain Rolland must also have been greatly dissatisfied with the leadership of the congress but finds it perhaps difficult to withdraw gracefully. I believe that the cause of peace can best be served by creating a militant pacifist organization composed of eminent artists and scholars. Such a group, which would undoubtedly exert great influence, could be assembled if the project were handled with sufficient skill.

Margueritte responded enthusiastically to this suggestion, leading Einstein to propose that he contact Paul Langevin about the prospects for creating such an organization under the latter's leadership. Apparently Langevin took a serious interest in the matter, for one month later Einstein wrote his friend Maurice Solovine in Paris that he had been in touch with Langevin "about the formation of an international association of leading intellectuals, of dependable pacifist orientation, that should seek, through the press, to exert political influence in matters of disarmament, security, etc. Langevin should be the central figure of such a group, because he is a man not only of good will but also of good political sense" (20 November 1932, N & N 1960, 183).

On the same day he wrote to Chaim Weizmann in London in a similar vein.

LETTER TO CHAIM WEIZMANN, 20 NOVEMBER 1932
EINSTEIN ARCHIVES 33-423; N & N 1960, 183–184

. . . I believe you are entirely right in your remarks about the danger of war and the futility of efforts made by political leaders. For some time I have attempted to help assemble an effective group of independent individuals whose prestige derives from their intellectual achievements and whose objectivity and honesty are unquestioned. The purpose of the group would be to support or influence all efforts directed toward international co-operation, disarmament and security by issuing public announcements on the important questions of the day. Coudenhove would not qualify for this group because his reputation is not based on achievements in a nonpolitical field; furthermore, his political orientation—for example, his sharp attack on Russia—might jeopardize the effectiveness of such a group. It is indeed doubtful whether any professional politician should be an official member of the organization . . .

Count R. N. Coudenhove-Kalergi was head of the Pan-European Union. In July, Einstein had declined his invitation to act as a sponsor for this organization because of its position on Russia.

Langevin's initiative apparently withered after Hitler came to power in January 1933. Writing to a Frenchwoman the following July, Einstein reported that he had received no news from him for some time. "It appears to be exceedingly difficult," he noted, "to find people who have achieved eminence solely by virtue of their moral qualities. That is why we have tried to form a group of some thirty persons who, in addition to their moral qualities, have gained [a] world-wide reputation through their intellectual achievements" (1 July 1933, N & N 1960, 184).

By this time, Einstein was embroiled in the public controversy occasioned by his resignation from the Prussian Academy (see "Response to the Declaration . . . ," chapter 6). While the right-wing press labeled him an anti-German agitator, those on the left were heartened by his harsh words for the Nazi regime. One of these was the Communist wheeler-dealer Willi Münzenberg, a member of the Reichstag who barely escaped arrest after it was set afire in late February. Münzenberg fled to Paris, where he set up new headquarters in order to put his propaganda machine back in operation. From there he launched the World Committee for the Victims of German Fascism, naming Einstein as the organization's honorary president. The physicist asked that his name be withdrawn, but Münzenberg's people apparently convinced him to stay on until they had completed the task at hand.

The plan was to compile massive documentation on events in Germany since Hitler's assumption of power in *The Brown Book of the Hitler Terror and the Burning of the Reichstag*, which appeared in numerous languages on 31 August 1933. The timing for its release was crucial, as the Nazis just then opened the trial of a group of Communists, led by the Bulgarian Georgi Dimitrov, who stood accused of conspiring to burn down the Reichstag. Münzenberg, following plans laid in Moscow, used the Brown Book to help launch a countertrial in London aimed at proving to the world that the Nazis themselves were the arsonists. The Brown Book also contained a chapter on the persecution of Jews in Hitler's Germany, ending with a verbatim citation of Einstein's statement to the International League for Combating Anti-Semitism (see "Statement on Conditions in Germany," chapter 6; Braunbuch, 241–242).

Asked to what degree he had been involved in the publication of the Brown Book, Einstein declared the following.

Statement on Willi Münzenberg's Brown Book, early September 1933
Einstein Archives 28-250

About half a year ago a committee was formed in Paris to support German political refugees. At the written urging of the committee I joined the honorary steering committee. Accordingly I had no real influence on the direction of this committee and was not informed that it contemplated political activity in addition to its charitable function. I also knew nothing of the publication of the Brown Book, which occurred without any participation on my part.

This only as a statement of the truth. Credit is due to those who really did the work.

Some six years later, on the occasion of Einstein's sixtieth birthday, the American public was given a chance to ask him questions about a variety of topics, including political matters. Although Einstein's days as a spokesman for militant pacifism were long behind him, the old ideal of an elite group of morally incorruptible leaders still burned brightly within. Asked to describe such an organization, he proposed some basic principles for how such an advisory body might function.

Thoughts on Forming a Council of the Wise, 14 March 1939
Einstein Archives 28-473

If such a corporation could be formed, which by virtue of its intellectual and moral qualities represented a kind of conscience for humanity, such a body could exert through its resolutions a beneficial and, in the course of time, even decisive influence on the shape of social and economic relationships in the world.

This idea is, of course, not new, and it has occupied me a great deal. I believe that one should make all effort to support its realization should a path open to do so. The difficulties that stand in the way are very significant, however, and are likely to prevent the realization of such a plan.

Assuming, however, that such a body were successfully formed, consisting of perhaps twenty permanent members, the question arises as to how vacancies are to be filled in such a way that the degeneration of quality is prevented to the degree that this is possible. I believe that the only course of action that could come into consideration would be an election by the members of the corporation itself, as otherwise external influences would very quickly lead to a leveling effect. This is the old principle of the Academy. Experience shows, nevertheless, that this principle in no way excludes the possibility of degeneration; I just do not know a better one.

Still more difficult appears to me the initial formation of the corporation. I believe this could only take place spontaneously through the union of the right people who undertake this on their own initiative. I do not know how one could consciously bring this about or induce it to happen. Of course one can try to engage serious people to keep this idea in mind and reflect upon it continuously.

The function of the corporation would consist in taking public positions as a collective on problems of vital interest to the human community.

Freedom and Socialism Reconsidered

Einstein was long aware of the essential tension between freedom and socialism, an issue he addressed on a number of occasions over the years. As can be seen from the three texts below, his views with regard to this issue remained fundamentally unchanged throughout the 1940s, culminating with the argument set forth in "Why Socialism?" There he begins by suggesting that the pros and cons of socialism cannot and should not be seen in purely economic terms; indeed, he stresses that "economic science . . . can throw little light on the socialist society of the future." Moreover, the larger problem of framing an ideal political system falls outside the range of science itself, being a matter of human values. "Scientific socialism" is thus a misnomer, since "socialism is directed towards a social-ethical end."

Already in the following essay, which appeared in 1940, Einstein confronted the issue of social and individual values head on in the form of an adversary who questions the value of human life. Science and logic alone, he emphasized, cannot overcome the deep sense of cultural crisis and even nihilism that afflicted the age (echoing the message in "Ten Fateful Years," chapter 6). In analyzing the social preconditions for human freedom, he here anticipates many of the key issues he would later develop in "Why Socialism?" Particularly noteworthy is the value Einstein places on a proper balance between "outward" and "inward" freedom, as presented in the final paragraph. His inclination toward cultural elitism here bears comparison with the preceding text, "Thoughts on Forming a Council of the Wise."

The article was presumably solicited by Ruth Nanda Anshen, a well-known figure in New York literary circles who edited numerous works in fields ranging from physics and biology to philosophy, education, psychology, and aesthetics. In the original version in Anshen's edition, there are two parts to the article, which is entitled "Freedom and Science." The following is the second part that was reprinted verbatim in Einstein 1954 under the title "On Freedom."

On Freedom, ca. 1940
Anshen 1940, 92–93; Einstein 1954, 31–32

I know that it is a hopeless undertaking to debate about fundamental value judgments. For instance, if someone approves, as a goal, the extirpation of the human race from the earth, one cannot refute such a viewpoint on rational grounds. But if there is agreement on certain goals and values, one can argue rationally about the means by which these objectives may be attained. Let us, then, indicate two goals which may well be agreed upon by nearly all who read these lines.

1. Those instrumental goods which should serve to maintain the life and health of all human beings should be produced by the least possible labour of all.

2. The satisfaction of physical needs is indeed the indispensable precondition of a satisfactory existence, but in itself it is not enough. In order to be content, men must also have the possibility of developing their intellectual and artistic powers to whatever extent accords with their personal characteristics and abilities.

The first of these two goals requires the promotion of all knowledge relating to the laws of nature and the laws of social processes, that is, the promotion of all scientific endeavour. For scientific endeavour is a natural whole, the parts of which mutually support one another in a way which, to be sure, no one can anticipate. However, the progress of science presupposes the possibility of unrestricted communication of all results and judgments—freedom of expression and instruction in all realms of intellectual endeavour. By freedom I understand social conditions of such a kind that the expression of opinions and assertions about general and particular matters of knowledge will not involve dangers or serious disadvantages for him who expresses them. This freedom of communication is indispensable for the development and extension of scientific knowledge, a consideration of much practical import. In the first instance it must be guaranteed by law. But laws alone cannot secure freedom of expression; in order that every man may present his views without penalty there must be a spirit of tolerance in the entire population. Such an ideal of external liberty can never be fully attained but must be sought unremittingly if scientific thought, and philosophical and creative thinking in general, are to be advanced as far as possible.

If the second goal, that is, the possibility of the spiritual development of all individuals, is to be secured, a second kind of outward freedom is necessary. Man should not have to work for the achievement of the necessities of life to such an extent that he has neither time nor strength for personal activities. Without this second kind of outward liberty, freedom of expression is useless for him. Advances in technology would provide the possibility of this kind of freedom if the problem of a reasonable division of labor were solved.

The development of science and of the creative activities of the spirit in general requires still another kind of freedom, which may be characterized as inward freedom. It is this freedom of the spirit which consists in the independence of thought from the restrictions of authoritarian and social prejudices as well as from unphilosophical routinizing and habit in general. This inward freedom is an infrequent gift of nature and a worthy objective for the individual. Yet the community can do much to further this achievement, too, at least by not interfering with its development.

Thus schools may interfere with the development of inward freedom through authoritarian influences and through imposing on young people excessive spiritual burdens; on the other hand, schools may favour such freedom by encouraging independent thought. Only if outward and inner freedom are constantly and consciously pursued is there a possibility of spiritual development and perfection and thus of improving man's outward and inner life.

Two years later Einstein took up the issue of freedom and socialism directly in an unpublished text that was solicited by the Jewish Telegraphic Agency of New York. In the words of the agency's managing director, the question posed "has become acute in view of the emphasis Churchill is giving during the election campaign to the thesis that a victory of the Labor Party would mean the establishment of a totalitarian system" (letter from Jacob Landau, 27 June 1945). In spite of the dire prophecy, Labor won in a landslide on 5 July.

Einstein's argumentation closely parallels the one later presented in "Why Socialism?" Just as in that published piece, his critique of capitalism ends by sharply distinguishing between a social system based on a planned economy and true socialism, which Einstein identified with social justice and solidarity. Thus he regarded socialism as an ideal that human beings should strive to attain, an ideal that "only constant political struggle and vigilance can create and maintain."

Is There Room for Individual Freedom in a Socialist State?, ca. July 1945
Einstein Archives 28-661

I will gladly answer your question: is there room for individual freedom in a socialist state?

This question preoccupies people at the present time, and for good reason because to a considerable extent the answer shapes one's political position. The question sounds simple, but it contains imprecise expressions that make it misleading.

First we should ask: what is individual freedom? It is a condition in which every individual acts in accordance with his personal wishes and decisions without restriction by others. Obviously such a condition cannot exist in any society, not even one in which total anarchy reigns; for in such circumstances all are in constant danger of being robbed of nourishment and their very existence by their fellow human beings. Everyone is then slave to the general insecurity.

Individual freedom in a community is thus only possible in a restricted sense, but it is also absolutely necessary for a worthy human existence. One should strive to attain it to the degree that this is compatible with the need to protect the security of the individual and to provide essential economic needs. In other words: the first priority is a secure life, and then comes fulfillment of the need for freedom. To recognize this is not to underestimate the high value of individual freedom. One sees, however, that this freedom can only be achieved by forgoing unrestricted freedom. In fact, the question of freedom cannot even be sensibly posed independent of the question of securing the conditions for physical existence. It was through neglect of this commonplace that liberalism forfeited a great deal of its popular appeal.

What then is a socialist state? The idea of a socialist state emerged as the counterpart to the capitalist states that exist almost everywhere. What is a capitalist state? It is a state in which the principal means of production, such as farmland, real estate in the cities, the supply of water, gas, and electricity, public transportation, as well as the larger industrial plants are owned by a minority of the citizenry. Productivity is geared toward making a profit for the owners rather than providing the population with a uniform distribution of essential goods. This propertied minority dominates the rest of the population by dictating working conditions and controlling jobs in accordance with its interests. It dominates public opinion through its influence on schools, the press, government, and legislation. The harshness of this situation is sometimes tempered or at least disguised by a democratic form of government in which all citizens are guaranteed formal equality.

A state can be characterized as "socialistic" when the principal means of production are owned collectively and are administered by individuals responsible to them and who are paid by the state.

What role does individual freedom play in such a state? Is it not inevitable that in such a state political and economic power is concentrated in the hands of a relatively small number of people? Won't this lead to an even harsher tyranny than is the case when the means of production are privately held by a minority?

My answer to this is that freedom, in any case, is only possible by constantly struggling for it. A citizenry that is politically indifferent will always end up enslaved no matter what form its constitution and legal institutions take. I am convinced, however, that in a state with a socialist economy the prospects are better for the average individual to achieve the maximum degree of freedom that is compatible with the well-being of the community.

The reasoning behind this is as follows: in a soundly administered socialist society work is undertaken not for the profit of a propertied minority but for the satisfaction of the needs of all. The problem of finding a reasonably equitable distribution of the workload can, in my view, only be solved in a planned economy and not in a free-market system in which entrepreneurs are compelled to reduce the number of workers while maximizing their productivity as much as possible. Chronic unemployment grows as new labor-saving machinery is developed. This leads to rising unemployment and economic insecurity which means a loss of freedom, insofar as freedom is dependent on economic factors.

Socializing the means of production, to be sure, does not yet constitute socialism, though it serves as a precondition. Socialism also requires that concentrated power be under the effective control of the citizenry, so that the planned economy benefits the entire population, and that everyone is free to rise to the more important positions in accordance with his or her natural qualifications. Only constant political struggle and vigilance can create and maintain such a condition.

Thus, the conditions for the struggle to attain individual freedom for the majority are more favorable in a socialist state than in an economic system based on private property.

The preceding texts provide key elements of the argument in "Why Socialism?" They nevertheless skirt the real heart of his position, which he lays out below beginning with the passage: "Man is, at one and the same time, a solitary being and a social being." Themes already raised in his correspondence with Freud (chapter 4) here again come to the fore—the obstinacy of human nature and what Thorstein Veblen had called the "predatory phase" of human development.

Einstein's article helped launch the journal *Monthly Review*, which commemorates its founding every year by republishing "Why Socialism?" in its May issue. Its editors took special delight in doing so in May 2000 when they could call attention to how *Time* magazine had overlooked this aspect of his political biography: "Neither Einstein's advocacy of socialism nor the FBI's increased surveillance of him during the McCarthy era (much less the role of *MR*)—is discussed in *Time's* 'Person of the Century' issue. Einstein, we are certain, would not have been in the least surprised by such silences and distortions in the establishment press."

The last paragraph of the essay was excluded in Einstein 1954.

Why Socialism?

Monthly Review, an Independent Socialist Magazine 1 (May 1949),
no. 1, 9–15; Einstein 1954, 151–158

Is it advisable for one who is not an expert on economic and social issues to express views on the subject of socialism? I believe for a number of reasons that it is.

Let us first consider the question from the point of view of scientific knowledge. It might appear that there are no essential methodological differences between astronomy and economics: scientists in both fields attempt to discover laws of general acceptability for a circumscribed group of phenomena in order to make the interconnection of these phenomena as clearly understandable as possible. But in reality such methodological differences do exist. The discovery of general laws in the field of economics is made difficult by the circumstance that observed economic phenomena are often affected by many factors which are very hard to evaluate separately. In addition, the experience which has accumulated since the beginning of the so-called civilized period of human history has—as is well known—been largely influenced and limited by causes which are by no means exclusively economic in nature. For example, most of the major states of history owed their existence to conquest. The conquering peoples established themselves, legally and economically, as the privileged class of the conquered country. They seized for themselves a monopoly of the land ownership and appointed a priesthood from among their own ranks. The priests, in control of education, made the class division of society into a permanent institution and created a system of values by which the people were thenceforth, to a large extent unconsciously, guided in their social behavior.

But historic tradition is, so to speak, of yesterday; nowhere have we really overcome what Thorstein Veblen called "the predatory phase" of human development. The observable economic facts belong to that phase, and even such laws as we can derive from them are not applicable to other phases. Since the real purpose of

socialism is precisely to overcome and advance beyond the predatory phase of human development, economic science in its present
state can throw little light on the socialist society of the future.

Second, socialism is directed toward a social-ethical end. Science, however, cannot create ends and, even less, instill them in
human beings; science, at most, can supply the means by which to
attain certain ends. But the ends themselves are conceived by personalities with lofty ethical ideals and—if these ends are not still-
born, but vital and vigorous—are adopted and carried forward by
those many human beings who, half unconsciously, determine the
slow evolution of society.

For these reasons, we should be on our guard not to overestimate science and scientific methods when it is a question of
human problems; and we should not assume that experts are the
only ones who have a right to express themselves on questions affecting the organization of society.

Innumerable voices have been asserting for some time now that
human society is passing through a crisis, that its stability has
been gravely shattered. It is characteristic of such a situation that
individuals feel indifferent or even hostile toward the group, small
or large, to which they belong. In order to illustrate my meaning,
let me record here a personal experience. I recently discussed with
an intelligent and well-disposed man the threat of another war,
which in my opinion would seriously endanger the existence of
mankind, and I remarked that only a supra-national organization
would offer protection from that danger. Thereupon my visitor,
very calmly and coolly, said to me: "Why are you so deeply opposed to the disappearance of the human race?"

I am sure that as little as a century ago no one would have
so lightly made a statement of this kind. It is the statement of a
man who has striven in vain to attain an equilibrium within himself and has more or less lost hope of succeeding. It is the expression of a painful solitude and isolation from which so many
people are suffering in these days. What is the cause? Is there a
way out?

It is easy to raise such questions, but difficult to answer them with any degree of assurance. I must try, however, as best I can, although I am very conscious of the fact that our feelings and strivings are often contradictory and obscure and that they cannot be expressed in easy and simple formulas.

Man is, at one and the same time, a solitary being and a social being. As a solitary being, he attempts to protect his own existence and that of those who are closest to him, to satisfy his personal desires, and to develop his innate abilities. As a social being, he seeks to gain the recognition and affection of his fellow human beings, to share in their pleasures, to comfort them in their sorrows, and to improve their conditions of life. Only the existence of these varied, frequently conflicting strivings accounts for the special character of a man, and their specific combination determines the extent to which an individual can achieve an inner equilibrium and can contribute to the well-being of society. It is quite possible that the relative strength of these two drives is, in the main, fixed by inheritance. But the personality that finally emerges is largely formed by the environment in which a man happens to find himself during his development, by the structure of the society in which he grows up, by the tradition of that society, and by its appraisal of particular types of behavior. The abstract concept "society" means to the individual human being the sum total of his direct and indirect relations to his contemporaries and to all the people of earlier generations. The individual is able to think, feel, strive, and work by himself; but he depends so much upon society—in his physical, intellectual, and emotional existence—that it is impossible to think of him, or to understand him, outside the framework of society. It is "society" which provides man with food, clothing, a home, the tools of work, language, the forms of thought, and most of the content of thought; his life is made possible through the labor and the accomplishments of the many millions past and present who are all hidden behind the small word "society."

It is evident, therefore, that the dependence of the individual upon society is a fact of nature which cannot be abolished—just as in the case of ants and bees. However, while the whole life process of ants and bees is fixed down to the smallest detail by rigid, hereditary instincts, the social pattern and interrelationships of human beings are very variable and susceptible to change. Memory, the capacity to make new combinations, the gift of oral communication have made possible developments among human beings which are not dictated by biological necessities. Such developments manifest themselves in traditions, institutions, and organizations; in literature; in scientific and engineering accomplishments; in works of art. This explains how it happens that, in a certain sense, man can influence his life through his own conduct, and that in this process conscious thinking and wanting can play a part.

Man acquires at birth, through heredity, a biological constitution which we must consider fixed and unalterable, including the natural urges which are characteristic of the human species. In addition, during his lifetime, he acquires a cultural constitution which he adopts from society through communication and through many other types of influences. It is this cultural constitution which, with the passage of time, is subject to change and which determines to a very large extent the relationship between the individual and society. Modern anthropology has taught us, through comparative investigation of so-called primitive cultures, that the social behavior of human beings may differ greatly, depending upon prevailing cultural patterns and the types of organization which predominate in society. It is on this that those who are striving to improve the lot of man may ground their hopes: human beings are *not* condemned, because of their biological constitution, to annihilate each other or to be at the mercy of a cruel, self-inflicted fate.

If we ask ourselves how the structure of society and the cultural attitude of man should be changed in order to make human life as satisfying as possible, we should constantly be conscious of the fact that there are certain conditions which we are unable to modify. As mentioned before, the biological nature of man is, for all practical purposes, not subject to change. Furthermore, technological and demographic developments of the last few centuries have created conditions which are here to stay. In relatively densely settled populations with the goods which are indispensable to their continued existence, an extreme division of labor and a highly-centralized productive apparatus are absolutely necessary. The time—which, looking back, seems so idyllic—is gone forever when individuals or relatively small groups could be completely self-sufficient. It is only a slight exaggeration to say that mankind constitutes even now a planetary community of production and consumption.

I have now reached the point where I may indicate briefly what to me constitutes the essence of the crisis of our time. It concerns the relationship of the individual to society. The individual has become more conscious than ever of his dependence upon society. But he does not experience this dependence as a positive asset, as an organic tie, as a protective force, but rather as a threat to his natural rights, or even to his economic existence. Moreover, his position in society is such that the egotistical drives of his make-up are constantly being accentuated, while his social drives, which are by nature weaker, progressively deteriorate. All human beings, whatever their position in society, are suffering from this process of deterioration. Unknowingly prisoners of their own egotism, they feel insecure, lonely, and deprived of the naïve, simple, and unsophisticated enjoyment of life. Man can find meaning in life, short and perilous as it is, only through devoting himself to society.

The economic anarchy of capitalist society as it exists today is, in my opinion, the real source of the evil. We see before us a huge community of producers the members of which are unceasingly striving to deprive each other of the fruits of their collective labor—not by force, but on the whole in faithful compliance with legally established rules. In this respect, it is important to realize that the means of production—that is to say, the entire productive capacity that is needed for producing consumer goods as well as additional capital goods—may legally be, and for the most part are, the private property of individuals.

For the sake of simplicity, in the discussion that follows I shall call "workers" all those who do not share in the ownership of the means of production—although this does not quite correspond to the customary use of the term. The owner of the means of production is in a position to purchase the labor power of the worker. By using the means of production, the worker produces new goods which become the property of the capitalist. The essential point about this process is the relation between what the worker produces and what he is paid, both measured in terms of real value. Insofar as the labor contract is "free," what the worker receives is determined not by the real value of the goods he produces, but by his minimum needs and by the capitalists' requirements for labor power in relation to the number of workers competing for jobs. It is important to understand that even in theory the payment of the worker is not determined by the value of his product.

Private capital tends to become concentrated in few hands, partly because of competition among the capitalists, and partly because technological development and the increasing division of labor encourage the formation of larger units of production at the expense of the smaller ones. The result of these developments is an oligarchy of private capital the enormous power of which cannot be effectively checked even by a democratically organized political society. This is true since the members of legislative bodies are selected by political parties, largely financed or otherwise influenced by private capitalists who, for all practical purposes, separate the electorate from the legislature. The consequence is that the representatives of the people do not in fact sufficiently protect the interests of the underprivileged sections of the population. Moreover, under existing conditions, private capitalists inevitably control, directly or indirectly, the main sources of information (press, radio, education). It is thus extremely difficult, and indeed in most cases quite impossible, for the individual citizen to come to objective conclusions and to make intelligent use of his political rights.

The situation prevailing in an economy based on the private ownership of capital is thus characterized by two main principles: first, means of production (capital) are privately owned and the owners dispose of them as they see fit; second, the labor contract is free. Of course, there is no such thing as a *pure* capitalist society in this sense. In particular, it should be noted that the workers, through long and bitter political struggles, have succeeded in securing a somewhat improved form of the "free labor contract" for certain categories of workers. But taken as a whole, the present-day economy does not differ much from "pure" capitalism.

Production is carried on for profit, not for use. There is no provision that all those able and willing to work will always be in a position to find employment; an "army of unemployed" almost always exists. The worker is constantly in fear of losing his job. Since unemployed and poorly paid workers do not provide a profitable market, the production of consumers' goods is restricted, and great hardship is the consequence. Technological progress frequently results in more unemployment rather than in an easing of the burden of work for all. The profit motive, in conjunction with competition among capitalists, is responsible for an instability in the accumulation and utilization of capital which leads to increasingly severe depressions. Unlimited competition leads to a huge waste of labor, and to that crippling of the social consciousness of individuals which I mentioned before.

This crippling of individuals I consider the worst evil of capitalism. Our whole educational system suffers from this evil. An exaggerated competitive attitude is inculcated into the student, who is trained to worship acquisitive success as a preparation for his future career.

I am convinced there is only *one* way to eliminate these grave evils, namely through the establishment of a socialist economy, accompanied by an educational system which would be oriented toward social goals. In such an economy, the means of production are owned by society itself and are utilized in a planned fashion. A planned economy, which adjusts production to the needs of the community, would distribute the work to be done among all those able to work and would guarantee a livelihood to every man, woman, and child. The education of the individual, in addition to promoting his own innate abilities, would attempt to develop in him a sense of responsibility for his fellow men in place of the glorification of power and success in our present society.

Nevertheless, it is necessary to remember that a planned economy is not yet socialism. A planned economy as such may be accompanied by the complete enslavement of the individual. The achievement of socialism requires the solution of some extremely difficult socio-political problems: how is it possible, in view of the far-reaching centralization of political and economic power, to prevent bureaucracy from becoming all-powerful and overweening? How can the rights of the individual be protected and therewith a democratic counterweight to the power of bureaucracy be assured?

Clarity about the aims and problems of socialism is of greatest significance in our age of transition. Since, under present circumstances, free and unhindered discussion of these problems has come under a powerful taboo, I consider the foundation of this magazine to be an important public service.

——————————————— ✿ ———————————————

Einstein's closing comments clearly allude to the postwar political atmosphere in the United States, a major topic in chapter 10. To gain a better appreciation of how he saw the ideological rift that led up to the Cold War, however, we return full circle to the theme addressed in this chapter's opening section by documenting Einstein's subsequent views on the Russian brand of socialism.

Principles versus Praxis

Joseph Wood Krutch, a leading social and literary critic, was on the staff of *The Nation* from 1924 to 1952. In August and September 1934, *The Nation* published a series of four articles by him addressing the question "Was Europe a Success?" Afterward, the magazine's editors invited "a number of persons whose opinions it was thought would be pungent and interesting ... to comment on them." Einstein was joined by Bertrand Russell, Aldous Huxley, H. L. Mencken, and James Burnham. The version presented here follows the text in *The Nation*, which differs slightly from that in Einstein 1956.

Was Europe a Success?
The Nation 139 (3 October 1934), no. 3613, 373;
Einstein 1956, 181–182

Your articles "Was Europe a Success?" must make a great impression on thoughtful people. The humanitarian ideal of Europe appears indeed to be unalterably bound up with the free expression of opinion, to some extent with the free-will of the individual, with the effort toward objectivity in thought without consideration of mere utility, and with the encouragement of differences in the realm of mind and taste. I agree with you that these requirements and ideals comprise the nature of the European spirit. One cannot establish with reason the worth of these values and maxims, for they are matters of fundamental principle in the approach to life and are points of departure which can only be affirmed or denied by emotion. I only know that I affirm them with my whole soul, and would find it intolerable to belong to a society which consistently denied them.

I do not share the pessimism of those who believe that full intellectual growth is dependent on the foundation of open or concealed slavery. That may be true for eras of primitive technical development, where the production of the necessaries of life requires physical work by a majority of the people to the point of total exhaustion. In our time of high technical development, with a reasonably equitable division of labor and adequate provisions for all, the individual would have both time and strength to participate receptively and productively in the finest intellectual and artistic efforts his abilities and inclinations allowed. Unfortunately nothing approaching such conditions exist in our society. But everyone devoted to the specific European ideals will do his utmost to achieve aims of whose desirability and practicability an increasing number of right-minded persons are convinced.

You ask if it is justifiable to set aside for a time the principles of individual freedom in deference to the high endeavor to improve economic organization. A fine and shrewd Russian scholar very skillfully defended this point of view to me in comparing the success of compulsion and terror—at least at the outset—in a functioning Russian Communism with the failure of German Social Democracy after the war. He did not convince me. No purpose is so high that unworthy methods in achieving it can be justified in my eyes. Violence sometimes may have cleared away obstructions quickly, but it never has proved itself creative.

No longer an outspoken pacifist, Einstein nevertheless distanced himself from the use of violence, except in cases of self-defense. This was his well-known position in early 1937 when he was approached by the political philosopher Sidney Hook, who sought his support for an international commission that would investigate the charges brought against Leon Trotsky by authorities in Moscow. Hook's former mentor, John Dewey, had agreed to serve on this commission, and he appealed to Einstein to support this cause. "There is no one in the world," Hook contended, "whose moral authority is greater than yours on the issues involved" (letter from Hook, 22 February 1937). Einstein answered this request the next day.

LETTER TO SIDNEY HOOK, 23 FEBRUARY 1937
EINSTEIN ARCHIVES 34-735; HOOK, 463

According to my view there is no doubt that every accused person should be given an opportunity to establish his innocence. This certainly holds true for Trotsky.

The only question that can be raised is *how* this should be done. That question is pertinent because Trotsky is an active and skilled political personality who is sure to seek an opportunity for the effective propagation of his political goals in public. I believe that a *public* procedure [trial] would serve Trotsky's purposes to the highest degree. There is on the other hand the question whether such a public procedure, conducted in this country, would really further the ends of justice. For it is questionable whether sufficiently competent judges can be found in view of the great difficulty of assembling authentic material evidence. I fear that the only consequence will be the achievement of an effective propaganda for Trotsky's cause, without the possibility of reaching a well-founded verdict.

That is why the public character of the proposed undertaking seems to me to be mistaken. If a number of intelligent jurists were to investigate the case privately, and if, after they were successful in reaching a truly convincing conclusion, went public with it, I would welcome such an approach with much enthusiasm. For it would serve the ends of justice and at the same time avoid the side effects whose harm under the circumstances could far outweigh the value of any positive results.

Hook afterward visited Einstein in Princeton, hoping to persuade him to change his mind, but to no avail. He was later dismayed to learn that Einstein had privately entertained the possibility that the Moscow trials of the 1930s might be a legitimate undertaking, an opinion he expressed in the following undated letter to Max Born. Einstein apparently took a keen interest in the 1937 trial of Karl Radek and sixteen others accused of betraying the party; his library contained a copy of the German edition of the official Soviet report (*Pressebericht 1937*).

Letter to Max Born, 1937
Einstein Archives 8-199; Born/Einstein 2005, Doc. 73

... There are increasing signs that the Russian trials are not faked, but that there is a plot by those who look upon Stalin as a stupid reactionary who has betrayed the ideas of the revolution. Though we find it difficult to imagine this kind of internal thing those who know Russia best are all more or less of the same opinion. I was firmly convinced to begin with that it was a case of a dictator's despotic acts, based on lies and deception, but this was a delusion ...

Hook contacted Einstein a few years later, in June 1940, to remind him of a conversation that Einstein once had with the revisionist Marxist philosopher Eduard Bernstein, a fellow member of the short-lived New Fatherland Association during the First World War. In 1928 Bernstein told Hook that he had shown Einstein the manuscript of Friedrich Engels's *Dialectics and Nature* in order to gain his opinion of its scientific value. Einstein allegedly informed Bernstein that the book was of no importance either for contemporary physics or for the history of the subject.

A movement was then underway to rehabilitate Engels's text, and Hook expressed concern that "Stalinists are claiming both here and abroad, that the only part of the manuscript which Bernstein showed you was the section on electricity and not the sections on Dialectic, Forms of Motion, Heat, etc. which they claim are very profound" (letter from Hook, 15 June 1940). Einstein replied immediately: "Eduard Bernstein made the entire manuscript available to me and my stated opinion referred to the whole manuscript. I am firmly convinced that if Engels himself could see the great importance being attached to his modest effort after so many years, he would find this ridiculous" (letter to Hook, 17 June 1940).

During the war years, Einstein voiced strong sympathy not only for the Russian people but also for the Soviet government and its leaders. "Why," he asked in September 1942 "is there no really serious effort to assist Russia in her dire need?" The answer, he said, lay in class hostility to the Soviet system: "[The U.S. has] a government controlled to a large degree by financiers the mentality of whom is near to the fascist frame of mind. If Hitler were not a lunatic he could easily have avoided the hostility of the Western powers" (letter to Frank Kingdon, 3 September 1942; see Einstein's letter to Michele Besso in chapter 6 for a similar sentiment).

The tide of war was beginning to turn, however. By the time that Einstein delivered the following address over the telephone, the Red Army had proven its indispensability to Allied efforts by halting the Nazi juggernaut before Stalingrad.

Address to Jewish Council on Russian War Relief, 25 October 1942
Einstein Archives 28-571; N & N 1960, 322–324

I consider this an occasion of great importance. As friends of human progress, as Americans, and not least as Jews, we are most keenly interested in giving our utmost to the struggle of the Russian people for freedom.

Let us be clear at the outset. For many years our press has misled us about the efforts and achievements of the Russian people and their government. But today, everybody knows that Russia has promoted and continues to promote the advancement of science with the same zeal as our own country. Moreover, her conduct of the war has made obvious her great achievements in all industrial and technical fields. From rudimentary beginnings, the rate of her development in the last twenty-five years has been so extraordinary that it is virtually without parallel in history. But it would be quite unfair to mention only those accomplishments which are chiefly due to improvements in organization. We must particularly emphasize the fact that the Russian Government has labored more honestly and unequivocally to promote international security than any of the other great powers. Her foreign policy was consistently directed toward this goal until shortly before the outbreak of war, actually up to the time the other powers brusquely excluded her from the European concert, in the days of Czechoslovakia's betrayal. Thus was she driven into the unfortunate pact with Germany, for, by then, it had become obvious that attempts were being made to turn the force of the German attack eastward. Russia, unlike the Western Powers, supported the legitimate government of Spain, offered assistance to Czechoslovakia and was never guilty of increasing the power of the German and Japanese adventurers. Russia, in short, cannot be accused of disloyalty in the field of foreign policy. It seems reasonable, then, to look forward to her powerful and loyal cooperation in devising an effective scheme of supra-national security, provided, of course, that she encounters the same degree of seriousness and good will in the other great powers.

Now, a comment on the domestic affairs of Russia. It is undeniable that a policy of severe coercion exists in the political sphere. This may, in part, be due to the necessity of breaking the power of the former ruling class, of protecting the country against foreign aggression and of converting a politically inexperienced, culturally backward people, deeply rooted in the traditions of their past, into a nation well organized for productive work. I do not presume to pass judgement in these difficult matters; but, in the unity of the Russian people against a powerful enemy from without and in the limitless sacrifice and exemplary self-denial of every single individual, I see proof of a strong and universal will to defend what they have won. We should also remember that the achieve-

ment of economic security for the individual and the utilization of
the country's productive powers for the common good must necessarily have entailed certain sacrifices of personal freedom, a
freedom which may have relatively little meaning unless accompanied by a measure of economic security.

Then, let us consider Russia's extraordinary success in fostering
the intellectual life of her people. Enormous quantities of the best
books are distributed and are eagerly read and studied; and this in
a country where until twenty-five years ago cultural education
was restricted to the very privileged few. It is difficult even to conceive of such revolutionary changes.

Finally, let me mention a fact of particular importance to us
Jews. In Russia the equality of all national and cultural groups is
not merely nominal but is actually practiced. "Equal objectives
and equal rights together with equal obligations to society" is not
merely an empty slogan but a practice which is realized in everyday life.

So much for Russia as she appears today. Now let us briefly
consider her present significance to the United States and the
Western Powers. Suppose she too had succumbed to the German
hordes, as has almost the whole Continent of Europe. How would
this have affected the situation of England and the United States? I
am sure it requires very little imagination to realize that we should
be in a very bad way. In fact, I believe that, without Russia, the
German bloodhounds would have already achieved their goal, or
would achieve it very soon.

Thus, it is no more than a dictate of self-preservation that we
help Russia in whatever way we can, to the utmost limit of our resources. And, quite apart from this selfish interest, we and our
children owe a great debt of gratitude to the Russian people for
having experienced such immense losses and suffering. If we wish
to retain our self-respect as human beings, we must be conscious
of their great sacrifice every hour of our lives.

Let us act accordingly.

After the war, Einstein felt that the Soviet Union played a regressive role in foreign affairs. In a radio broadcast from July 1947 he said that "Russia has become xenophobic; she is obsessed by the utopia of isolationism" (N & N 1960, 416). Although critical of the policies of the Truman administration, he disagreed with those who thought that American capitalism was primarily to blame for the tensions of the Cold War. After the publication of his "Reply to the Soviet Scientists" (see chapter 8), he faced sharp criticism from a reader who thought his liberal views ignored the conflict between capitalism and socialism. To this Einstein replied as follows.

LETTER TO JOHN DUDZIC, 8 MARCH 1948
EINSTEIN ARCHIVES 58-108; N & N 1960, 468–469

I indeed fully agree with your critical remarks with respect to America's foreign policy since the death of Roosevelt. Your criticisms of the liberal viewpoint, however, are less convincing. The meaning of the term *liberal* has become so watered down as to cover the most diverse views and attitudes. Your criticism should not be directed, for example, against Henry Wallace, who is, without doubt, a liberal. On the other hand, when you speak of socialism, it really seems to me that socialism, as I understand it, does not exist anywhere today.

You say that socialism by its very nature rejects the remedy of war. I do not believe that. I can easily imagine that two socialist states might fight a war against each other.

I know very well that a world government may have both good and bad qualities. Nonetheless, it is the only conceivable machinery which can prevent war. I do not believe that a world government would be just in all its decisions; but with technology at its present level, even a poor world government is preferable to none, since our first goal must be to avoid total destruction through war.

I am far from blaming the Soviet Union for the injustices and barbarism of our age. But I am convinced of one thing: If you knew the Soviet Union as well as you know the United States, you would be no less bitter in your judgment of conditions there than you now are about conditions in this country. What you do not eat does not taste bitter . . .

On 15 May 1950 Einstein received another letter from Sidney Hook, who informed him that he had been asked to review the recently published collection of Einstein's writings, *Out of My Later Years*. Hook went on to query Einstein about two passages that he found difficult to reconcile. The first of these came from "Was Europe a Success?" (see above), where Einstein wrote: "The humanitarian ideal of Europe . . . I affirm . . . with my whole soul, and would find it intolerable to belong to a society which consistently denied them." Hook then cited the following passage from "Atomic War or Peace" (the first part of which is presented in original form in chapter 8 as "On the Atomic Bomb, as Told to Raymond Swing"): "One must bear in mind that the people in Russia had not had a long tradition of political education; changes to improve conditions in Russia had to be effected by a minority for the reason that there was no majority capable of doing so. If I had been born a Russian, I believe I could have adjusted myself to the situation." After pointing out other apparent ambiguities in Einstein's political writings, he asked him to clarify these inconsistencies; Einstein responded the following day.

LETTER TO SIDNEY HOOK, 16 MAY 1950
EINSTEIN ARCHIVES 59-1018; HOOK, 476–477

... The two statements you mentioned do not contradict each other. In the first . . . I profess to intellectual and moral individualism. In the second statement . . . I try to seek understanding for the necessity of the Russian revolution and recognize . . . that under the circumstances prevailing in Russia at that time this revolution could only have been undertaken successfully by a resolute minority. It was natural, under the conditions, for a Russian who had the welfare of the people at heart to cooperate with and submit to this minority because the immediate goals could not have been achieved otherwise. It cannot be doubted that for an independent individual this meant a painful *temporary* renunciation of his personal independence. But I believe that I myself would have deemed it my duty to make this temporary sacrifice (as the lesser evil).

However, with this I do not mean to say that I do approve of the direct and indirect interference by the Soviet Government in intellectual and artistic matters. Such interference seems to me objectionable, harmful, and even ridiculous. As far as the centralization of political power and the limitations of the freedom of action for the individual are concerned, I am of the opinion that these restrictions should not exceed the limit demanded by exterior security, inner stability, and the necessities resulting from a planned economy. An outsider is hardly able to judge the facts and possibilities. In any case it cannot be doubted that the achievements of the Soviet Regime are considerable in the fields of education, public health, social welfare, and economics, and that the people as a whole have greatly gained by these achievements.

Needless to say, Hook was both unimpressed and unpersuaded by this response. Yet if Einstein counseled open-mindedness with regard to the Soviet Union, he had no sympathy at all for the dogmatic brand of Marxism it promulgated as official state doctrine. As he was well aware, his theory of relativity was harshly criticized by Soviet Marxists during the 1930s, an attack that reached its high point in 1952 when I. V. Kuznetsov denounced Einstein's theory as absurd and beyond repair (Vucinich, 222–224). Interestingly enough, soon after the cult of Stalin had passed, younger Russians began to read Einstein's humanistic writings, drawn in part by the "ethical satisfaction" he found in Dostoevsky (ibid., 181). Einstein's private opinion of doctrinaire Marxists can be gleaned from the following two stanzas.

The Wisdom of Dialectical Materialism, 1952
Einstein Archives 28-948

Through sweat and effort beyond compare
To arrive at a small grain of truth?
A fool is he who toils to find
What we simply ordain as the Party line.

And those who dare to express doubt
Will quickly find their skulls bashed in
And thus we educate as never before
Bold spirits to embrace harmony.

On the same page he scrawled a sarcastic "Inscription for the Marx-Engels Institute: In the realm of truth-seekers there is no human authority. He who attempts to play the ruler there will run afoul of the laughter of the gods."

The message is unambiguous: Einstein shunned ideologues, whether on the left or the right. By the same token, he had the deepest sympathy for all those who spoke out against tyranny and in favor of human freedom. Indeed, for the last twenty-five years of his life he was an indefatigable advocate of civil liberties and a staunch defender of those who put their lives in jeopardy to advance human rights.

CHAPTER 10

Political Freedom and the Threat of Nuclear War, 1931–1955

Throughout his life Einstein recognized the inherent conflict between the freedom of the individual and the growing power of modern states and social organizations. Thus it is no accident that those whom he admired most in political life were men like Gandhi who risked everything to challenge the authority of the repressive political system under which they lived. During the waning days of the Weimar Republic, two such courageous figures were the Heidelberg mathematician, Emil Julius Gumbel, and the editor of the *Weltbühne*, Carl von Ossietzky. Both did much to publicize the attitudes and behavior of those who were intent on toppling the "system," a standard right-wing euphemism for Germany's fledgling democracy, and both paid a heavy price for this even before the Nazis came to power. For Einstein, the attacks against them in 1931–1932 were among the several factors that signaled the beginning of the end of his relationship with Germany, in sharp contrast to the attitude he expressed earlier, in particular with regard to academic youth (see "Interview in *Neue Zürcher Zeitung*," chapter 4).

Einstein's transition from bird of passage to stalwart partisan of the "American way of life" came quite naturally given the sense of urgency he felt not only about the fate of his Jewish brethren but of European civilization in general. Only after the war did he begin actively to voice his views, always in a diplomatic tone, regarding the evils of racism in the United States. His principal reservations, however, focused on American militarism during the postwar era.

As the Cold War conflict deepened, Einstein's confidence in American foreign policy quickly came to an end. On numerous occasions he likened the political atmosphere in the United States to the one he had witnessed in Germany during the First World War, as well as the waning years of the Weimar Republic. Although he remained largely aloof from the public arena, his fame inevitably brought adoration as well as notoriety. His support for Henry Wallace, the estranged New Dealer who ran against Truman in 1948, led to accusations that he was a fellow traveler, one of many "Moscow dupes." A heated exchange with the political philosopher Sidney Hook, a staunch anti-Communist, reveals how Einstein chose to distance himself from Bolshevist-style socialism while blaming the policies of the Truman administration for exacerbating the conflict between the world's two superpowers. In discerning symptoms of cultural decay within American society, he bemoaned the steady erosion of individual freedoms in an atmosphere dominated by propaganda and fear.

Heading the list of such anxieties was the threat of nuclear holocaust, particularly after the Russians detonated their first atomic bomb in 1949. As documented in chapter 8, Einstein's reputation as the "grandfather of the bomb" was firmly implanted in the public's imagination immediately after the war. In Japan, the issue resurfaced in 1952 when the government finally released photographs of the devastation wrought by the atomic bombs that destroyed Hiroshima and Nagasaki. Thirty years after his triumphant visit to Japan, Einstein was contacted once again by an editor from *Kaizo* requesting his reaction to the photographs. His reply prompted new correspondence with Hook and the pacifist Seiei Shinohara, leading to a clarification of his views on pacifism as well as the efficacy of Gandhi's nonviolent resistance in the future.

Fear of an international Communist conspiracy intensified in the early 1950s, leading to a sharp crackdown on those whose activities were viewed as betraying a lack of firm allegiance to the American political system. Although Einstein and those close to him were spared the ignominy of having to testify before HUAC or Senator McCarthy's Permanent Investigations Subcommittee, in 1953 he lashed out publicly against such inquisitorial methods, urging those called to Washington to refuse to testify on the grounds that the hearings constituted an infringement of their First Amendment rights. The fallout after an open letter to William Frauenglass appeared in the *New York Times* brought an amusing response from Bertrand Russell and did much to enhance Einstein's reputation as an advocate of human rights. His public speech on this topic one year later covered relatively familiar ground, whereas his private letter to Norman Thomas provides an illuminating picture of how Einstein gauged the threat to American civil liberties posed by domestic Communists.

During the last months of his life Einstein was directly involved in a new political effort aimed at overcoming the impasse that separated East and West. With the advent of the hydrogen bomb in 1952, the threat of nuclear disaster loomed larger than ever. This circumstance prompted Bertrand Russell to make a worldwide effort to avert catastrophe by means of a widely publicized statement issued in his and Einstein's names. The Russell-Einstein Manifesto vividly describes the unimaginable potential destruction unleashed by bombs 2,500 times more powerful than the one that fell over Hiroshima. Stressing the need to bridge the ideological gap separating Communists and anti-Communists, the document called for the governments of the world "to acknowledge publicly, that their purposes cannot be furthered by a world war" and "to find peaceful means for the settlement of all matters of dispute between them" (N & N 1960, 635). Although he did not live to see it publicized, just by lending his signature to this document Einstein helped bring public attention to these pressing political problems.

Alongside this initiative, he also hoped to promote another cause long dear to his heart: the reconciliation of Jews and Arabs in the newly founded state of Israel. While the message he hoped to convey to the Israeli people was, unfortunately, never delivered, some sense of what he wanted to say can be gleaned from that portion of the text he was able to complete before his death.

Voices of Freedom and Dissent

E. J. Gumbel had long been a thorn in the side of the Heidelberg faculty and student body, both of which tended toward the right end of the political spectrum (see A. Brenner, 90–114). In the 1930 Reichstag elections, over 25 percent of Heidelberg voters cast their ballots for the National Socialists. Soon afterward the NS-League of Students began disrupting Gumbel's lectures, forcing the university to undertake protective measures. The case drew national attention in 1931 when radical elements in Heidelberg tried to have him ousted from the university, thereby posing a test of academic freedom in the Weimar Republic. On 27 January 1931 the NS-League organized a public protest against Gumbel that ended in a brawl. This served as a prelude to the university's decision in the summer of 1932 to dismiss Gumbel in the face of renewed pressure from right-wing demagogues.

The statistician Gumbel occasionally wrote for *Die Weltbühne*, whose readership learned from its editor Carl von Ossietzky that Gumbel was "a socialist, republican, and pacifist who never mixed Marx with algebraic formulas." Ossietzky well understood the motives of the brown shirts who opposed the professor: they demanded Gumbel's removal from the university not because his political views constituted a threat to the established political order but rather because "they will neither forgive nor forget [his] chronicle of political murders in the first German Republic" (Ossietzky, 150–151).

Einstein gave his assessment of the situation in a letter to Gustav Radbruch, Social Democratic politician and professor of law at Heidelberg, that was excerpted in the leading Berlin daily. The excerpt was imbedded in an article entitled "Kowtow to the Mob," written by

Rudolf Olden, its political editor, who in the same year served as a defense attorney in Ossietzky's trial for treason.

Red Terror by Way of Fascism: The Gumbel Case
Berliner Tageblatt, 9 April 1931; Einstein 1933, 29

The conduct of the academic youth against Gumbel, by failing to live up to the ideals of justice, tolerance, and truth, offers one of the saddest aspects of our time. Professor Gumbel's only offense has been to fight against political murder and in so doing he has maintained high ethical standards. What is to become of a people who brutally harass such a contemporary and whose leaders offer no resistance to the base mob? . . .

What is so terrible is the way inexperienced youth is being misled for self-seeking reasons. If it goes further in this direction, we shall arrive at a reign of red terror by way of a fascist regime of tyranny.

Einstein and Gumbel had known each other since the war years when they belonged to the New Fatherland Association. During the postwar era they joined its successor organization, the German League of Human Rights, which held a meeting in Berlin on 27 April 1931 to support Gumbel's cause. The theme of the rally was "Reactionary Politics in the University." Shortly before, the League had published Gumbel's latest book (Gumbel 1931), in which he demonstrated the correlation between increased political violence and the resurgence of the Nazis. As one of the invited speakers, Einstein addressed a capacity crowd of more than a thousand extemporaneously and succinctly. (Some of the speakers are depicted in Plate 11.) After urging the audience not only to applaud Gumbel, but also to read him more, Einstein pointed out that "I myself have learned much from this book, and I daresay, you too could learn something from it" (Hobohm 1931, 122).

The following day he put his thoughts to paper.

ON ACADEMIC FREEDOM: THE GUMBEL CASE, 28 APRIL 1931
EINSTEIN ARCHIVES 28-151; EINSTEIN 1954, 28–30

Numerous are the academic chairs, but rare are wise and noble teachers. Numerous and large are the lecture halls, but far from numerous the young people who genuinely thirst for truth and justice. Numerous are the wares that nature produces by the dozen, but her choice products are few.

We all know that, so why complain? Was it not always thus and will it not always thus remain? Certainly, and one must take what nature gives as one finds it. But there is also such a thing as a spirit of the times, an attitude of mind characteristic of a particular generation, which is passed on from individual to individual and gives its distinctive mark to a society. Each of us has to do his little bit toward transforming this spirit of the times.

Compare the spirit which animated the youth in our universities a hundred years ago with that prevailing today. They had faith in the amelioration of human society, respect for every honest opinion, the tolerance for which our great minds had lived and fought. In those days men strove for a larger political unity, which at the time was called Germany. It was the students and the teachers at the universities in whom these ideals were alive.

Today also there is an urge toward social progress, toward tolerance and freedom of thought, toward a larger political unity, which we today call Europe. But the students at our universities have ceased as completely as their teachers to embody the hopes and ideals of the people. Anyone who looks at our times soberly and dispassionately must admit this.

We are assembled today to take stock of ourselves. The external reason for this meeting is the Gumbel case. This apostle of justice has written about unexpiated political crimes with devoted industry, high courage, and exemplary fairness, and has done the community a signal service by his books. And this is the man whom the students and a good many of the faculty of his university are today doing their best to expel.

Political passion cannot be allowed to go to such lengths. I am convinced that every man who reads Mr. Gumbel's books with an open mind will get the same impression from them as I have. Men like him are needed if we are ever to build up a healthy political society.

Let every man judge by himself, by what he has himself read, not by what others tell him.

If that happens, this Gumbel case, after an unedifying beginning, may still do good.

In October 1934 Einstein wrote to Jane Addams, the founder of Hull House in Chicago and a recipient of the Nobel Peace Prize, suggesting that she recommend Ossietzky for the prize the following year. She did so, joining a galaxy of European émigrés, including Thomas and Heinrich Mann, Romain Rolland, Lion Feuchtwanger, and Arnold Zweig. At first reluctant to join this campaign, Einstein feared that the Nazi regime might make Ossietzky's life even more miserable should they learn that he was among those who nominated him. He eventually changed his mind, however, and addressed the following letter to the Nobel Committee.

Nominating Carl von Ossietzky for Nobel Peace Prize, 27 October 1935
Einstein Archives 34-706; N & N 1960, 266

Formally speaking, I have no right to propose a candidate for the Nobel Prize. But under the conditions now prevailing, my conscience dictates that I address to you this letter.

In awarding this prize, the Nobel Committee has a unique opportunity to accomplish an act of great historical significance, an act whose repercussions would most likely contribute to a solution of the peace problem. This could be accomplished only by awarding the prize to a man who, by his actions and his agony, is more deserving of it than any other living person: Carl von Ossietzky. To award the Peace Prize to him would instill new life in the cause of pacifism in the very country which, because of the circumstances now prevailing there, constitutes the gravest threat to world peace. Moreover, such a gesture would arouse the conscience of all well-meaning people the world over and inspire them to work for the establishment of a secure international order.

In November, the Nobel Committee finally decided not to award the Peace Prize to anyone in 1935, but the following year Ossietzky received this coveted honor, much to the embarrassment of the Nazi regime. Hitler responded by refusing to allow any German citizen to accept a Nobel Prize in the future. By this time, the pacifist's health had deteriorated to the point that he was removed from Esterwegen concentration camp and placed under guard in a Berlin hospital. Carl von Ossietzky spent the remainder of his days under arrest there. He died of tuberculosis on 4 May 1938. Einstein spoke in his honor eight years later at a Nobel Foundation dinner.

CARL VON OSSIETZKY, 10 DECEMBER 1946
EINSTEIN ARCHIVES 28-722; EINSTEIN 1956, 241

Only one who has spent the years following the First World War in Germany can fully understand how hard a battle it was that a man like Ossietzky had to fight. He knew that the tradition of his countrymen, bent on violence and war, had not lost its power. He knew how difficult, thankless, and dangerous a task it was, to preach sanity and justice to his countrymen who had been hardened by a rough fate and the demoralizing influence of a long war. In their blindness they repaid him in hatred, persecution, and slow destruction; to heed him and to act accordingly would have meant their salvation and would have been a true relief for the whole world.

It will be to the eternal fame of the Nobel Foundation that it bestowed its high honor on this humble martyr, and that it is resolved to keep alive his memory and the memory of his work. It is also wholesome for mankind today, since the fatal illusion against which he fought has not been removed by the outcome of the last war. The abstention from the solution of human problems by brute force—is the task today as it was then.

Among all his colleagues in France, Einstein felt closest to Paul Langevin, a friendship deepened by their mutual efforts to strengthen Franco-German relations during the early 1920s (chapter 2). Soon after the Nazis seized power, Einstein wrote Langevin, describing the danger he perceived for democracy in Europe: "A group of armed bandits in Germany has successfully silenced the responsible segments of the population and imposed a kind of revolution from below which will soon succeed in destroying or paralyzing everything that is civilized in society. That which, today, threatens our cultural values will, in a few years, become a grave military menace unless the countries still living under a parliamentary system eventually decide to take vigorous action" (5 May 1933, N & N 1960, 220–221).

Like Einstein, Langevin instinctively understood that the key to European peace lay in harmonious relations between France and Germany. His death so soon after peace was finally restored brought back sorrowful memories that Einstein openly shared with the French public in the journal that Langevin had coedited since its founding in 1939.

In Memoriam Paul Langevin

La Pensée (May–June 1947), no. 12, 13–14; Einstein 1956, 231–232

The news of Paul Langevin's death dealt me a greater blow than most of the events of these fateful years, so fraught with disappointment. Why should this have been the case? Was his not a long life, crowded with fruitful creative work—the life of a man in harmony with himself? Was he not widely revered for his keen insight into intellectual problems, universally beloved for his devotion to every good cause, for his understanding kindness toward all creatures? Is there not a certain satisfaction in the fact that natural limits are set to the life of the individual, so that at its conclusion it may appear as a work of art?

The sorrow brought on by Paul Langevin's passing has been so particularly poignant because it has given me a feeling of being left utterly alone and desolate. There are so very few in any one generation, in whom clear insight into the nature of things is joined with an intense feeling for the challenge of true humanity and the capacity for militant action. When such a man departs, he leaves a gap that seems unbearable to his survivors.

Langevin was endowed with unusual clarity and agility in scientific thought, together with a sure intuitive vision for the essential points. It was a result of these qualities that his lectures exerted a crucial influence on more than one generation of French theoretical physicists ... Yet the burden of responsibility which he was always ready to assume circumscribed his own research work, so that the fruits of his labors emerge in the publications of other scientists to a greater extent than in his own.

It appears to me as a foregone conclusion that he would have developed the Special Theory of Relativity, had that not been done elsewhere; for he had clearly perceived its essential aspects. Another admirable thing is that he fully appreciated the significance of De Broglie's ideas—from which Schrödinger subsequently developed the methods of wave mechanics—even before these ideas had become consolidated into a consistent theory. I vividly recall the pleasure and warmth with which he told me about it— and I also remember that I followed his remarks but hesitantly and doubtfully.

All his life Langevin suffered from an awareness of the deficiencies and inequities of our social and economic institutions. Yet he believed firmly in the power of reason and knowledge. So pure in heart was he that he was convinced all men should be ready for complete personal renunciation, once they had seen the light of reason and justice. Reason was his creed—a creed that was to bring not only light but also salvation. His desire to promote the happier life for all men was perhaps even stronger than his craving for pure intellectual enlightenment. Thus it was that he devoted much of his time and vital energy to political enlightenment. No one who appealed to his social conscience ever went away from him empty-handed. Thus it was too that the very moral grandeur of his personality earned him the bitter enmity of many of the more humdrum intellectuals. He in turn understood them all and in his kindness never harbored resentment against anyone.

I can only give expression to my gratitude for having personally known this man of purity and illumination.

Outspoken American Citizen

Having long pleaded for America to take a more active part in world affairs, Einstein showed that he was prepared to play his part. In June 1940, following an examination on his application for American citizenship in Trenton, New Jersey, he took part in a radio broadcast for the Immigration and Naturalization Service (INS). The interview was conducted by an official from the State Department, to whom he repeated his "Political Manifesto" of March 1933 (see chapter 6) and strongly affirmed his faith in American democracy.

I Am an American, 22 June 1940
Einstein Archives 29-092; N & N 1960, 312–314

... I must tell you that I do not think words alone will solve humanity's present problems. The sound of bombs drowns out men's voices. In times of peace I have great faith in the communication of ideas among thinking men, but today, with brute force dominating so many millions of lives, I fear that the appeal to man's intellect is fast becoming virtually meaningless ...

Making allowances for human imperfections, I do feel that, in America, the development of the individual and his creative powers is possible, and that, to me, is the most valuable asset in life. In some countries men have neither political rights nor the opportunity for free intellectual development. But for most Americans such a situation would be intolerable. In this country, it has been generations since men were subject to the humiliating necessity of unquestioning obedience. Here, human dignity has been developed to a point where people would find it almost impossible to endure life under a system in which the individual is only a slave of the state and has neither a voice in his government nor any control over his own way of life ...

I gather from what I have seen of Americans since I came here that they are not suited, either by temperament or tradition, to live under a totalitarian system. I believe that many of them would find life not worth living under such circumstances. Hence, it is all the more important for them to see to it that these liberties be preserved and protected ...

Science has gone a long way toward helping man to free himself from the burden of hard labor; yet, science itself is not a liberator. It creates means, not goals. It is up to men to utilize those means to achieve reasonable goals. When men are engaged in war and conquest, the tools of science become as dangerous as a razor in the hands of a child. We must not condemn man because his inventiveness and patient conquest of the forces of nature are being exploited for false and destructive purposes. Rather, we should remember that the fate of mankind hinges entirely upon man's moral development . . .

Anyone who seeks to affect the course of events must have the gift of being able to exert direct influence on men and their activities . . . Intellectuals often lack the gift of impressing their audiences. Among the outstanding American statesmen, Woodrow Wilson probably provides the clearest example of an intellectual. Yet not even Wilson seems to have mastered the art of dealing with men. At first glance, his greatest contribution, the League of Nations, appears to have failed. Still, despite the fact that the League was crippled by his contemporaries and rejected by his own country, I have no doubt that Wilson's work will one day emerge in more effective form. Only then will the stature of that great innovator be fully recognized . . .

I am convinced that an international political organization is not only a possibility but an absolute necessity; life on our planet will otherwise become intolerable. The League of Nations failed because its members were unwilling to surrender any part of their sovereignty and because the League itself did not have any executive power. Similarly, a world-state which does not control all the resources of its member states will not be able to ensure peace.

Excessive nationalism is a state of mind which is artificially induced by the prevalent obsession of nations that they must, at all times, be prepared for war. If the danger of war were eliminated, nationalism would soon disappear. Further, I do not accept the thesis that the unequal geographical distribution of raw materials must necessarily lead to war. So long as a nation has access to the resources of other countries which it needs for its industrial development, it will be able to develop its economy adequately. This is clearly demonstrated by the economies of nations such as Switzerland, Finland, Denmark, and Norway, which before the

war were among the most prosperous countries in Europe ...
One of the most important functions of an international organization, thus, would be to guarantee the unhampered marketing of
raw materials.

Concerning the possibility of creating an international organization, I am far from being optimistic. I merely intended to suggest certain possibilities which might prevent human existence
from becoming totally intolerable. Concerning the formation of
an international organization, there is probably general agreement that we now seem even further from this goal than we were
ten years ago. We would hardly have suffered such a setback
if, ten years ago, the democracies had evidenced the same solidarity and readiness for sacrifice they exhibit now in the face of
a grave emergency. Solidarity, foresight and the will to sacrifice
are, however, most effective *before* an emergency has actually
arisen ...

I believe that America will prove that democracy is not merely a
form of government based on a sound Constitution but is, in fact,
a way of life tied to a great tradition, the tradition of moral
strength. Today more than ever, the fate of the human race depends upon the moral strength of human beings.

Three months later he took the oath of U.S. citizenship (see Plate 17).

After the war, the INS no longer found Einstein an appropriate
spokesman for American values. He had figured as one of the famous
names quoted in the agency's booklet *Gateway to Citizenship*, first
published in 1943. In the section on "Freedom of Expression" one
finds the familiar Einstein aphorism from his "Political Manifesto" of
ten years earlier in slightly variant form: "Political liberty implies liberty to express one's political opinion orally and in writing, and a toler-
ant respect for any and every individual opinion." When the INS

published the 1953 edition it contained exactly the same collection of quotations, except for this one, which was deleted.

Even during those years when he identified closely with the political culture of the United States, Einstein voiced outspoken criticism of the many manifestations of racism that permeated American life.

ON POLITICAL FREEDOM IN THE U.S.A., 1945
EINSTEIN ARCHIVES 28-627

When I decided twelve years ago to take up permanent residence in America, I was no longer a stranger to this country. After all, I had, for a number of years following an invitation from the California Institute of Technology, spent several months here. Today I am happy to be a citizen of this country and am pleased to explain why . . .

America is a democratic country not only by virtue of its laws and institutions but also because of the mentality of its inhabitants. No one abases himself before another; everyone is conscious of the fact that he should feel and show respect for his fellow creatures. No one takes himself too seriously, and a person would look ridiculous were he to advance himself in pompous fashion. Apart from this democratic mentality, there is a sense of humor that is already instilled in children in a delightful way. As regards conduct toward others, people would be truly democratic were it not for the still present dark shadow of racial prejudices, particularly toward Negroes. I believe that each individual must work within his or her circle to eradicate this shameful evil . . .

Although fully aware of anti-Semitic currents in American society, Einstein felt these latent prejudices paled in significance compared to the endemic racism faced by blacks in all walks of life. In Einstein 1956 the following article is entitled "The Negro Question."

A Message to My Adopted Country
Pageant 1 (January 1946), no. 12, 36–37; Einstein 1956, 132–134

I am writing as one who has lived among you in America only a little more than ten years. And I am writing seriously and warningly. Many readers may ask: "What right has he to speak out about things which concern us alone, and which no newcomer should touch?"

I do not think such a standpoint is justified. One who has grown up in an environment takes much for granted. On the other hand, one who has come to this country as a mature person may have a keen eye for everything peculiar and characteristic. I believe he should speak out freely on what he sees and feels, for by so doing he may perhaps prove himself useful.

What soon makes the new arrival devoted to this country is the democratic trait among the people. I am not thinking here so much of the democratic political constitution of this country, however highly it must be praised. I am thinking of the relationship between individual people and of the attitude they maintain toward each other.

In the United States everyone feels assured of his worth as an individual. No one humbles himself before another person or class. Even the great difference in wealth, the superior power of a few, cannot undermine this healthy self-confidence and natural respect for the dignity of one's fellow-man.

There is, however, a somber point in the social outlook of Americans. Their sense of equality and human dignity is mainly limited to men of white skins. Even among these there are prejudices of which I as a Jew am clearly conscious; but they are unimportant in comparison with the attitude of the "Whites" toward their fellow-citizens of darker complexion, particularly toward Negroes. The more I feel an American, the more this situation pains me. I can escape the feeling of complicity in it only by speaking out.

Many a sincere person will answer me: "Our attitude towards Negroes is the result of unfavorable experiences which we have had by living side by side with Negroes in this country. They are not our equals in intelligence, sense of responsibility, reliability."

I am firmly convinced that whoever believes this suffers from a fatal misconception. Your ancestors dragged these black people from their homes by force; and in the white man's quest for wealth and an easy life they have been ruthlessly suppressed and exploited, degraded into slavery. The modern prejudice against Negroes is the result of the desire to maintain this unworthy condition.

The ancient Greeks also had slaves. They were not Negroes but white men who had been taken captive in war. There could be no talk of racial differences. And yet Aristotle, one of the great Greek philosophers, declared slaves inferior beings who were justly subdued and deprived of their liberty. It is clear that he was enmeshed in a traditional prejudice from which, despite his extraordinary intellect, he could not free himself.

A large part of our attitude toward things is conditioned by opinions and emotions which we unconsciously absorb as children from our environment. In other words, it is tradition—besides inherited aptitudes and qualities—which makes us what we are. We but rarely reflect how relatively small as compared with the powerful influence of tradition is the influence of our conscious thought upon our conduct and convictions.

It would be foolish to despise tradition. But with our growing self-consciousness and increasing intelligence we must begin to control tradition and assume a critical attitude toward it, if human relations are ever to change for the better. We must try to recognize what in our accepted tradition is damaging to our fate and dignity—and shape our lives accordingly.

I believe that whoever tries to think things through honestly will soon recognize how unworthy and even fatal is the traditional bias against Negroes.

What, however, can the man of good will do to combat this deeply rooted prejudice? He must have the courage to set an example by word and deed, and must watch lest his children become influenced by this racial bias.

I do not believe there is a way in which this deeply entrenched evil can be quickly healed. But until this goal is reached there is no greater satisfaction for a just and well-meaning person than the knowledge that he has devoted his best energies to the service of the good cause.

That is precisely what I have tried to do in writing this.

Cold War Reverberations

During the Second World War, the American government funded scientific research as never before, most famously in the U.S. Army's Manhattan Project. After the war, the army tried to maintain control of nuclear research through the May-Johnson Bill only to encounter steadfast resistance from the scientific community (on the efforts of Szilard and Einstein, see chapter 8). The army's plans went up in smoke in the spring of 1946, when President Truman announced his support for the McMahon Bill, which authorized the creation of a civilian-controlled Atomic Energy Commission. Still, by means of an amendment proposed by Republican senator Arthur H. Vandenberg, the army maintained its influence through a permanent Military Liaison Committee to the AEC, a feature that led to a split among leading atomic scientists (Lanouette, 357).

This struggle soon spawned a general debate over military versus civilian control of large-scale scientific research projects. The following essay was prompted by an earlier article by Louis N. Ridenour, "Military Support of American Science, a Danger?" and appeared under the title "Should the Scientists Resist Military Intrusion?" Besides Einstein, a number of prominent individuals addressed the issue, including Aldous Huxley, Vannevar Bush, and Norbert Wiener.

The Military Mentality
The American Scholar 16 (Summer 1947), no. 3, 353–354;
Einstein 1956, 212–214

It seems to me that the decisive point in the situation under discussion lies in the fact that the problem before us cannot be viewed as an isolated one. First of all, one may pose the following question: From now on institutions for learning and research will more and more have to be supported by grants from the state, since, for various reasons, private sources will not suffice. Is it at all reasonable that the distribution of the funds raised for these purposes from the taxpayer should be entrusted to the military? To this question every prudent person will certainly answer: "No!" For it is evident that the difficult task of the most beneficent distribution should be placed in the hands of people whose training and life's work give proof that they know something about science and scholarship.

If reasonable people nevertheless favor military agencies for the distribution of a major part of the available funds, the reason for this lies in the fact that they subordinate cultural concerns to their general political outlook. We must then focus our attention on these practical political viewpoints, their origins and their implications. In doing so we shall soon recognize that the problem here under discussion is but one of many, and can only be fully estimated and properly adjudged when placed in a broader framework.

The tendencies we have mentioned are something new for America. They arose when, under the influence of the two World Wars and the consequent concentration of all forces on a military goal, a predominantly military mentality developed, which with the almost sudden victory became even more accentuated. The characteristic feature of this mentality is that people place the importance of what Bertrand Russell so tellingly terms "naked power" far above all other factors which affect the relations between peoples. The Germans, misled by Bismarck's successes in particular, underwent just such a transformation of their mentality—in consequence of which they were entirely ruined in less than a hundred years.

I must frankly confess that the foreign policy of the United States since the termination of hostilities has reminded me, sometimes irresistibly, of the attitude of Germany under Kaiser Wilhelm II, and I know that, independent of me, this analogy has most painfully occurred to others as well. It is characteristic of the military mentality that non-human factors (atom bombs, strategic bases, weapons of all sorts, the possession of raw materials, etc.) are held essential, while the human being, his desires and thoughts—in short, the psychological factors—are considered as unimportant and secondary. Herein lies a certain resemblance to Marxism, at least insofar as its theoretical side alone is kept in view. The individual is degraded to a mere instrument; he becomes "human materiel." The normal ends of human aspiration vanish with such a viewpoint. Instead, the military mentality raises "naked power" as a goal in itself—one of the strangest illusions to which men can succumb.

In our time the military mentality is still more dangerous than formerly because the offensive weapons have become much more powerful than the defensive ones. Therefore it leads, by necessity, to preventive war. The general insecurity that goes hand in hand with this results in the sacrifice of the citizen's civil rights to the supposed welfare of the state. Political witch-hunting, controls of all sorts (e.g., control of teaching and research, of the press, and so forth) appear inevitable, and for this reason do not encounter that popular resistance, which, were it not for the military mentality, would provide a protection. A reappraisal of all values gradually takes place insofar as everything that does not clearly serve the utopian ends is regarded and treated as inferior.

I see no other way out of prevailing conditions than a far-seeing, honest and courageous policy with the aim of establishing security on supranational foundations. Let us hope that men will be found, sufficient in number and moral force, to guide the nation on this path so long as a leading role is imposed on her by external circumstances. Then problems such as have been discussed here will cease to exist.

But the specter of a "military mentality" did not so readily recede. Some four years later, Einstein wrote a friend that "people here . . . are well on the way to surpassing the Germans in military sentiments" (letter to Michele Besso, 12 December 1951).

On 30 March 1948 Einstein was quoted in the *New York Times* as endorsing the presidential campaign of Henry Wallace, a man he called "clear, honest, and unassuming." This endorsement, more specifically intended for the soon-to-be-published book, *Toward World Peace*, caught the eye of the ever-vigilant Sidney Hook, who fired off a letter to Einstein in order to inquire about the accuracy of this report. If so, Hook wrote, "I think it is nothing short of disastrous for the cause of genuine peace and cultural freedom with which your name has been until now indissolubly associated" (Hook 1987, 469). After presenting a litany of complaints about Wallace's recent public pronouncements, the former Trotskyite came to the heart of the matter: "Politically, Wallace today is a captive of the Communist party whose devious work in other countries you are familiar with much better than most scientists. His speeches are written by fellow-travelers, his line is indistinguishable from that of *Pravda* and the *Daily Worker*. It expresses from the first to last the illogic of appeasement" (ibid.). A photograph taken in autumn 1947 showing Einstein with Wallace and in the company of Frank Kingdon of the Progressive Citizens of America and Paul Robeson would only have further reinforced Hook's suspicions (see Plate 25).

Recalling a conversation they had in Princeton a year earlier, Hook reminded Einstein that he had dismissed the Russians as "a half savage horde" who posed no real threat to the United States. That view, he countered, overlooked the aggressive nature of the Soviet regime: "With the help of its Fifth Column [the Russians] can take all of Europe and Asia which they will undoubtedly do if Wallace's policy prevails" (Hook 1987, 470).

Einstein found none of this at all persuasive and said so.

LETTER TO SIDNEY HOOK, 3 APRIL 1948
EINSTEIN ARCHIVES 58-300; HOOK, 471

I must openly confess that I was very astonished by your letter. What I have really done was to recommend warmly Wallace's book and—in one sentence—paid tribute to Wallace as a man who is above all the petty interests. I have furthermore never spoken of the Russians as a "half-savage horde," for this is not my way of expression. I believe one could say this with some justification of every nation on earth; but I don't like the expression.

In my opinion your views are far from objective. If you ask yourself who, since the termination of the war, has threatened his opponent to a higher degree by direct action—the Russians, the Americans or the Americans, the Russians? The answer is, in my opinion, not doubtful and is accurately given in Wallace's book. It is, furthermore, not doubtful that the military strength of the U.S.A. is at present much greater than that of Soviet Russia. It would therefore be sheer madness if the Russians would seek war. On the other side I have heard influential people in this country pleading for "preventive war" even before the last war was finished.

I am not blind to the serious weaknesses of the Russian system of government and I would not like to live under such government. But it has, on the other hand, great merits and it is difficult to decide whether it would have been possible for the Russians to survive by following softer methods. If you should be interested in my opinion you may read my answer to a few Russian scientists which I am enclosing.

An oral discussion seems to me not promising because I see from your letter the rigidity and one-sidedness of your judgment.

─────────── ❧ ───────────

Einstein's closing rebuff was a response to Hook's hope that he might be able to talk with the famous physicist again. After reading the "Open Letter from Four Soviet Physicists" and Einstein's reply, Hook was disappointed to see that this exchange was irrelevant to the issues he had tried to raise. He nevertheless wrote a lengthy letter to Einstein on 3 April disputing virtually every point the latter had made in his "Reply to Soviet Scientists" (chapter 8).

Hook wrote Einstein again a year later on learning that the latter was a sponsor of the Cultural and Scientific Congress for World Peace held in late March 1949 at the Waldorf-Astoria Hotel in New York. Hook complained that the program committee for the Congress had refused to let him present a paper that argued against the pernicious influence of "national" or "class" or "party" truths in science. This event, in many ways reminiscent of the 1932 Amsterdam Peace Congress, made Einstein once again the target of anti-Communist watchdogs. He related his own misgivings about the Congress and an earlier one in Wroclaw, Poland in a letter to his friend, the mathematician Jacques Hadamard.

Letter to Jacques Hadamard, 7 April 1949
Einstein Archives 12-061; N & N 1960, 512

I was deeply moved by the address you delivered over the radio on the occasion of my seventieth birthday.

In answer to your cable I must frankly confess that, in view of my experience with the first congress of this kind in Wroclaw last August, and from what I have observed concerning the recent congress in New York, I have the strong impression that this kind of procedure does not really serve the cause of international understanding. The reason is simply that it is more or less a Soviet enterprise and everything is managed accordingly. This in itself would not be so bad if the Russians and the men coming from the countries affiliated with Russia were really free to express their personal opinions rather than having to express what is currently the official Russian point of view. Therefore, most people have the impression that these gatherings constitute "Soviet propaganda," particularly since the speakers from the Western countries are so selected that they do not disturb the over-all pattern of the meetings. So the result of these meetings is that they tend to sharpen the silly controversies and polemics which characterize the international situation of today.

You may be certain that I would gladly lend my name to any endeavor to bring about an honest discussion of the possibilities for reaching understanding and international security.

If Einstein's patience with Soviet propaganda had worn thin, he was also deeply disappointed by the repressive measures adopted by the American government against those who refused to fall in line with its Cold War mentality. A young German refugee, Gerhard Nellhaus, informed him about the case of Robert Michener, a Quaker who was sentenced to ten years in prison for refusing to be inducted into the military service. Einstein offered this assessment of the issues at stake in this case.

Letter to Gerhard Nellhaus, 20 March 1951
Einstein Archives 60-683; N & N 1960, 542–543

What we have here is the old conflict between conscience and law. During the Nuremberg trials, the various governments adopted the position that immoral acts cannot be excused on the plea that they were committed on government orders. What constitutes an immoral act can be determined only by one's own individual judgment and conscience. The attitude that moral law takes precedence over secular law is very much in line with people's general sense of right and wrong.

Whenever a person disobeys the law because of his moral convictions, the government considers him a rebel who has breached the law and must be punished. Hence it makes little sense in such cases for the individual in question to appeal to the very officials who are duty-bound to enforce the existing laws.

The conscientious objector is a revolutionary. In deciding to disobey the law he sacrifices his personal interests to the most important cause of working for the betterment of society. In matters of crucial significance this is often the only way to bring about social progress; this is particularly true when the prevailing balance of power precludes the successful utilization of normal legal and political institutions. It was in this sense that the Fathers of the American Constitution specifically acknowledged the people's right to revolution.

Revolution without the use of violence was the method by which Gandhi brought about the liberation of India. It is my belief that the problem of bringing peace to the world on a supranational basis will be solved only by employing Gandhi's method on a large scale.

Einstein strongly supported the stance of the Allied Powers at Nuremberg regarding war crimes, which made him all the more critical of the harsh treatment accorded to conscientious objectors by authorities in the United States. In putting one of his private thoughts regarding this matter to paper, he nicely captured the double standard underlying the government's policy in an epigram from 1951: "There is a curious inconsistency in a government which punishes aliens for *not* following their conscience in a given conflict, while penalizing its own citizens for *following* their conscience in the same kind of conflict. Apparently such a government holds the conscience of its own citizens in lower esteem than that of aliens" (N & N 1960, 542).

In the public arena, Einstein spoke out forcefully against the trend toward government intervention in the affairs of scientists. One example that directly affected the quality and integrity of scientific work in the United States were new restrictions on the admission of foreigners on ideological grounds as introduced in the McCarran-Walter Immigration Act of June 1952. Requested by the *Bulletin of the Atomic Scientists* to present his view on the new American visa policy, Einstein chose to generalize his response.

Symptoms of Cultural Decay
Bulletin of the Atomic Scientists 8 (October 1952), no. 7, 217–218; Einstein 1954, 166–167

The free, unhampered exchange of ideas and scientific conclusions is necessary for the sound development of science as it is in all spheres of cultural life. In my opinion, there can be no doubt that the intervention of political authorities of this country in the free exchange of knowledge between individuals has already had significantly damaging effects. First of all, the damage is to be seen in the field of scientific work proper, and, after a while, it will become evident in technology and industrial production.

The intrusion of the political authorities into the scientific life of our country is especially evident in the obstruction of the travels of American scientists and scholars abroad and of foreign scientists seeking to come to this country. Such petty behavior on the part of a powerful country is only a peripheral symptom of an ailment which has deeper roots.

Interference with the freedom of the oral and written communication of scientific results, the widespread attitude of political distrust which is supported by an immense police organization, the timidity and anxiety of individuals to avoid everything which might cause suspicion and which could threaten their economic position—all these are only symptoms, even though they reveal more clearly the threatening character of the illness.

The real ailment, however, seems to me to lie in the attitude which was created by the World War and which dominates all our actions; namely, the belief that we must in peacetime so organize our whole life and work that in the event of war we would be sure of victory. This attitude gives rise to the belief that one's freedom and indeed one's existence are threatened by powerful enemies.

This attitude explains all of the unpleasant facts which we have designated above as symptoms. It must, if it does not rectify itself, lead to war and to very far-reaching destruction. It finds its expression in the budget of the United States.

Only if we overcome this obsession can we really turn our attention in a reasonable way to the real political problem which is, "How can we contribute to make the life of man on this diminishing earth more secure and more tolerable?"

It will be impossible to cure ourselves of the symptoms we have mentioned and many others if we do not overcome the deeper ailment which is affecting us.

Two years later, Einstein was gloomier than ever about the prospects for America's scientific community. It was within the context of Cold War restraints on academic freedom that he issued a famous letter to the editor that caused considerable uproar in the press.

ON INTELLECTUAL FREEDOM
THE REPORTER 11 (18 NOVEMBER 1954), NO. 9, 8; N & N 1960, 613

You have asked me what I thought about your articles concerning the situation of scientists in America. Instead of trying to analyze the problem I should like to express my feeling in a short remark: If

I would be a young man again and had to decide how to make my living, I would not try to become a scientist or scholar or teacher. I would rather choose to be a plumber or a peddler in the hope to find that modest degree of independence still available under present circumstances.

After an eighteen-year interlude, Einstein reestablished contact with Queen Elisabeth of the Belgians with whom he gladly shared his personal reflections on the state of the world. On 6 January 1951, he wrote: "While it proved eventually possible, at an exceedingly heavy cost, to defeat the Germans, the dear Americans have vigorously assumed their place. Who shall bring them back to their senses? The German calamity of years ago repeats itself: people acquiesce without resistance and align themselves with the forces for evil. And one stands by, powerless" (N & N 1960, 554).

The Queen and her friend regularly exchanged New Year's greetings. When Einstein did so for the last time, America's postwar foreign policy was still very much on his mind.

LETTER TO THE QUEEN MOTHER OF THE BELGIANS, 2 JANUARY 1955
EINSTEIN ARCHIVES 32-413; N & N 1960, 615–616

Your telegram illustrates the virtue that has been described as characteristic of your trade—punctuality. More than that, it is an expression of that warmth of human feeling which is so seriously neglected in our age of mechanization. I am ever aware of this quality in you when reading of your public activities. It must require a great deal of courage and independence, especially from one in your position, with its peculiar restraints on freedom of action.

When I look at mankind today, nothing astonishes me quite so much as the shortness of man's memory with regard to political developments. Yesterday the Nuremberg trials, today the all-out effort to rearm Germany. In seeking for some kind of explanation, I cannot rid myself of the thought that this, the last of my fatherlands, has invented for its own use a new kind of colonialism, one that is less conspicuous than the colonialism of old Europe. It achieves domination of other countries by investing American capital abroad, which makes those countries firmly dependent on the United States. Anyone who opposes this policy or its implications is treated as an enemy of the United States. It is within this general context that I try to understand the present-day policies of Europe, including England. I tend to believe that these policies are less the result of a planned course of action than the natural consequence of objective conditions.

Such thoughts are likely to come to one's mind when one delves into the writings of the dreamers and thinkers of the past. I am particularly taken with Lichtenberg. Now, with so many years behind me, the man makes an ever greater impression on me. I know no one else who so plainly hears the grass grow . . .

The Göttingen physicist Georg Lichtenberg (1742–1799) is best remembered for his aphorisms from his *Waste Books*, which deal with nearly all aspects of human affairs.

Hiroshima Redux

In 1952, seven years after the event, the people of Japan were finally shown photographs of the devastation caused by the atomic bombs dropped over Hiroshima and Nagasaki. Confronted with these vivid reminders of the war, the Japanese public was curious to learn about Einstein's role in the production of atomic weapons and the prospects for their use in the future. To gain some insight into these matters, Katusu Hara, the editor of *Kaizo,* sent Einstein a letter on 15 September 1952 in which he posed four questions: 1. What is your reaction to photographs showing the destructive effect of the bomb? 2. What do you think of the atomic bomb as an instrument of human destruction? 3. The next war, it is commonly predicted, will be an atomic war. Does this mean the destruction of mankind? 4. Why did you co-operate in the production of the atomic bomb although you were well aware of its tremendous destructive power? Einstein replied five days later.

REPLY TO THE EDITOR OF *KAIZO*, 20 SEPTEMBER 1952
EINSTEIN ARCHIVES 60-039; N & N 1960, 584

My participation in the production of the atomic bomb consisted of one single act: I signed a letter to President Roosevelt, in which I emphasized the necessity of conducting large-scale experimentation with regard to the feasibility of producing an atom bomb.

I was well aware of the dreadful danger for all of mankind were the experiments to prove successful. Yet I felt compelled to take the step because it seemed probable that the Germans might be working on the same problem with every prospect of success. I saw no alternative but to act as I did, *although I have always been a convinced pacifist.*

I believe that the killing of human beings in a war is no better than common murder; but so long as nations lack the determination to abolish war through common action and find means of solving their disputes and safeguarding their interests by peaceful arrangements according to existing laws, they will continue to consider it necessary to prepare for war. They will feel compelled to engage in the manufacture of even the most detestable weapons in their fear that they may lag behind in the general arms race. Such an approach can only lead to war, and warfare today would mean universal annihilation of human beings.

There is little point, therefore, in opposing the manufacture of *specific* weapons; the only solution is to abolish both war and the threat of war. That is the goal toward which we should strive. We must be determined to reject all activities which in any way contradict this goal. This is a harsh demand for any individual who is conscious of his dependence upon society; but it is not an impossible demand.

Gandhi, the greatest political genius of our time, indicated the path to be taken. He gave proof of what sacrifice man is capable once he has discovered the right path. His work in behalf of India's liberation is living testimony to the fact that man's will, sustained by an indomitable conviction, is more powerful than material forces that seem insurmountable.

Sidney Hook was moved by Einstein's statement to write him of his misgivings with regard to Gandhi's doctrine of nonviolent civil disobedience. Hook was similarly attracted to this brand of political protest, but cautioned that its success depended on an opponent like the British who upheld high human values. Gandhi would have failed miserably if faced "with the Japanese military, with the Gestapo and SS, and the Soviet MVD" (Hook 1987, 484). Regarding the Holocaust, Hook admitted that "when I think of how many millions of Jews permitted themselves to be slaughtered in what was in effect a passive resistance to evil, I find myself wishing that they had died like the Jews in the Warsaw Ghetto" (ibid.). For once, Einstein found himself in at least partial agreement with the combative political philosopher.

Letter to Sidney Hook, 12 November 1952
Einstein Archives 59-1032; Hook, 485

You are right that under the circumstances that have developed in the last twenty years (not *only* through the fault of the Russians) a sudden transition to the methods of Gandhi could not be ventured by any responsible statesman. I see the only possibility of improving the situation in abandoning the armament race and issues productive of conflict. Naturally the first step would be the neutralization and demilitarization of Germany. Despite the evil methods of the Russians, I regard as completely false the point of view that would represent them or treat them as common criminals. At the very least, from an education standpoint, any such attitude is impermissible.

Einstein's statement for *Kaizo* was translated by the Japanese pacifist, Seiei Shinohara, who afterward struck up a correspondence with the physicist. Shinohara gently criticized Einstein's decision to write President Roosevelt in 1939, an action he believed was inconsistent with rigorous pacifism. He also felt certain that Gandhi would have acted differently had the Indian statesman been confronted with this problem. Einstein's wavering commitment to absolute pacifism, he noted, undermined the position of those opposed to the resurgence of Japanese rearmament under the pressure of American interests. Shinohara received the following reply:

LETTER TO SEIEI SHINOHARA, 22 FEBRUARY 1953
EINSTEIN ARCHIVES 61-295; N & N 1960, 585, 588

Your reproach is well taken from the viewpoint of an absolute i.e., unconditional, pacifist. But in my letter to *Kaizo* I did not say that I was an *absolute* pacifist, but, rather, that I had always been a *convinced* pacifist. While I am a convinced pacifist, there are circumstances in which I believe the use of force is appropriate— namely, in the face of an enemy unconditionally bent on destroying me and my people. In all other cases I believe it is wrong and pernicious to use force in settling conflicts among nations.

This is why I believe that the use of force was indicated and justified in the case of Nazi Germany. With regard to Russia, it is quite a different matter. In the present conflict between the United States and the Soviet Union it is far from clear which country threatens the existence of the other—indeed, it is even doubtful whether any such threat actually exists. In such a situation I am convinced that an aggressive attitude on the part of either side is unjustified and that, therefore, no other nation has the right to assume a partisan role in the conflict. India's attitude seems to me an exemplary one; I believe that every true Japanese pacifist should seek to appreciate fully the Indian position and should adopt it as a model for Japan.

After a four-month interlude, Shinohara responded by criticizing Einstein's motives as well as his actions. Believing that Einstein had supported the use of the bomb against Germany, Shinohara implied that he shared responsibility for the American decision to employ it against innocent civilians in Hiroshima and Nagasaki. Einstein answered these charges with a sharp rebuttal.

LETTER TO SEIEI SHINOHARA, 23 JUNE 1953
EINSTEIN ARCHIVES 61-297; N & N 1960, 589

I am a *dedicated* [*entschiedener*] but not an *absolute* pacifist; this means that I am opposed to the use of force under any circumstances, except when confronted by an enemy who pursues the destruction of life as an *end in itself*. I have always condemned the use of the atomic bomb against Japan. However, I was completely powerless to prevent the fateful decision for which I am as little responsible as you are for the deeds of the Japanese in Korea and China.

I have never said I would have approved the use of the atomic bomb against the Germans. I did believe that we had to avoid the contingency of Germany under Hitler being in *sole* possession of this weapon. This was the real danger at the time.

I am not only opposed to war against Russia but to all war—with the above reservation.

P.S. You should endeavor to form an opinion of others and of their actions only on the basis of sufficient information!

In another letter from 30 June 1953, Shinohara wrote that he had not meant to imply that Einstein was responsible for the decision to bomb Hiroshima and Nagasaki. He also asked whether he would be willing to send a message for the commemoration of the anniversary of this tragic event. In doing so, Einstein took the opportunity once again to trumpet the cause of world government.

LETTER TO SEIEI SHINOHARA, 18 JULY 1953
EINSTEIN ARCHIVES 61-300; N & N 1960, 589–590

It is good that the memory of the disasters of Hiroshima and Nagasaki is kept alive in the hearts of all men of good will by means of regularly recurrent ceremonies. Yet such memorials will be of real value only if they succeed in strengthening the belief that it is necessary to establish a world government based on peaceful agreement among the nations of the world. This belief must be based on the realization that, in the absence of a supranational authority, war cannot be avoided in the long run and circumstances will ever compel the contending parties to employ the most effective, that is to say, most murderous, weapons of war.

It must be pointed out again and again that any effort to secure peace through military alliances will inevitably lead to war and universal destruction. The greatest danger to the future of mankind lies in man's faith in unworkable methods which are falsely put forward in the name of practical politics.

This pronouncement was published in *Yomiuri Shimbun*, Japan's third-largest daily newspaper. Einstein sounded the same note in his last private communication to Shinohara.

LETTER TO SEIEI SHINOHARA, 7 JULY 1954
EINSTEIN ARCHIVES 61-306; N & N 1960, 590

... The only comfort which may be derived from the development of atomic weapons is the hope that *this* weapon may act as a deterrent and give impetus to a movement to establish supranational safeguards. Unfortunately, at the present time, the insanity of nationalism seems more powerful than ever before. I shall not send a message to the Hiroshima commemoration ceremonies this year; everyone knows my thoughts on these matters.

The Fight against McCarthyism

After the Soviets successfully tested an atomic bomb in September 1949, the prevailing fear of communism within the United States quickly intensified. Suspected Communists and their sympathizers were called before the House Un-American Activities Committee (HUAC) and other congressional bodies. Many refused to testify by invoking the protection provided by the Fifth Amendment; others risked going to jail by charging that the hearings themselves constituted a violation of their First Amendment rights.

William Frauenglass, a Brooklyn teacher who was subpoenaed to testify before Congress, wrote Einstein about his plight, saying that a statement from him "at this juncture would be most helpful in rallying educators and the public to meet this new obscurantist attack." He was encouraged to do so after reading a remark by Einstein in which he described himself as "an incorrigible nonconformist whose nonconformism in a remote field of endeavor [physics] no senatorial committee has as yet felt impelled to tackle" (N & N 1960, 546).

The version presented here follows the text in the *New York Times*, which differs somewhat from that in N & N 1960.

Open Letter to William Frauenglass, 16 May 1953
New York Times, 12 June 1953; N & N 1960, 546–547

. . . The problem with which the intellectuals of this country are confronted is very serious. The reactionary politicians have managed to instill suspicion of all intellectual efforts into the public by dangling before their eyes a danger from without. Having succeeded so far, they are now proceeding to suppress the freedom of teaching and to deprive of their positions all those who do not prove submissive, i.e., to starve them.

What ought the minority of intellectuals to do against this evil? Frankly, I can see only the revolutionary way of non-cooperation in the sense of Gandhi's. Every intellectual who is called before one of the committees ought to refuse to testify, i.e., he must be prepared for jail and economic ruin, in short, for the sacrifice of his personal welfare in the interest of the cultural welfare of his country.

This refusal to testify must be based on the assertion that it is shameful for a blameless citizen to submit to such an inquisition and that this kind of inquisition violates the spirit of the Constitution.

If enough people are ready to take this grave step they will be successful. If not, then the intellectuals of this country deserve nothing better than the slavery which is intended for them.

P.S. This letter need not be considered "confidential."

The original draft of the letter included Einstein's observation that the refusal to testify should not be "based on the well-known subterfuge of invoking the Fifth Amendment against possible self-incrimination." It was the very position that Frauenglass had taken at his appearance before a congressional committee in April, and he requested that the passage be omitted in a second version. Einstein agreed, though the deleted passage is included in N & N 1960.

Einstein's response stirred considerable public controversy. In a letter to Carl Seelig, his Swiss biographer, Einstein noted that "all the important newspapers have commented in a more or less politely negative tone" (N & N 1960, 547) about this statement. He also received a flood of mail, most of it positive. A sharply critical editorial appeared in the *New York Times*, questioning the wisdom of Einstein's counsel and calling the use of "the unnatural and illegal forces of civil disobedience" merely an attempt "to attack one evil with another" (N & N 1960, 550).

Bertrand Russèll rose to Einstein's defense in a letter to the newspaper, published in its 26 June issue. The editors' position, he argued, seemed to suggest that one should always obey the law, no matter how unjust it may be. If so, he concluded, "I am compelled to suppose that you condemn George Washington and hold that your country ought to return to allegiance to Her Gracious Majesty, Queen Elizabeth II. As a loyal Briton, I of course applaud this view; but I fear it may not win much support in your country" (N & N 1960, 550). Einstein responded in turn.

LETTER TO BERTRAND RUSSELL, 28 JUNE 1953
EINSTEIN ARCHIVES 33-195; N & N 1960, 550–551

Your fine letter to *The New York Times* is a great contribution to a good cause. All the intellectuals in this country, down to the youngest student, have become completely intimidated. Virtually no one of "prominence" besides yourself has actually challenged these absurdities in which the politicians have become engaged. Because they have succeeded in convincing the masses that the Russians and the American Communists endanger the safety of the country, these politicians consider themselves so powerful. The cruder the tales they spread, the more assured they feel of their reelection by the misguided population. This also explains why Eisenhower did not dare to commute the death sentence of the two Rosenbergs, although he well knew how much their execution would injure the name of the United States abroad.

Julius and Ethel Rosenberg were linked with a spy ring that stole classified material from the Manhattan Project and passed it on to Soviet agents. They were convicted of conspiracy to commit espionage and sentenced to die by electrocution. Pope Pius XII appealed to President Eisenhower to spare their lives, but on 11 February 1953 he refused to grant a stay of execution. After losing all further appeals, they were put to death on 19 June, just nine days before Einstein wrote this letter. Although many on the left were convinced of their innocence, Einstein was not among them. He believed, however, that the sentence they received was far too harsh.

The following year, as recipient of the Chicago Decalogue Society of Lawyers' award for his contributions to human rights, Einstein reflected on the erosion of basic rights in American society.

HUMAN RIGHTS, 20 FEBRUARY 1954
EINSTEIN ARCHIVES 28-1012; EINSTEIN 1954, 34–36

You are assembled today to devote your attention to the problem of human rights. You have decided to offer me an award on this occasion. When I learned about it, I was somewhat depressed by your decision. For in how unfortunate a state must a community find itself if it cannot produce a more suitable candidate upon whom to confer such a distinction?

In a long life I have devoted all my faculties to reach somewhat deeper insight into the structure of physical reality. Never have I made any systematic effort to ameliorate the lot of men, to fight injustice and suppression, and to improve the traditional forms of human relations. The only thing I did was this: in long intervals I have expressed an opinion on public issues whenever they appeared to me so bad and unfortunate that silence would have made me feel guilty of complicity.

The existence and validity of human rights are not written in the stars. The ideals concerning the conduct of men toward each other and the desirable structure of the community have been conceived and taught by enlightened individuals in the course of history. Those ideals and convictions which resulted from historical experience, from the craving for beauty and harmony, have been readily accepted in theory by man—and at all times have been trampled upon by the same people under the pressure of their animal instincts. A large part of history is therefore replete with the struggle for those human rights, an eternal struggle in which a final victory can never be won. But to tire in that struggle would mean the ruin of society.

In talking about human rights today, we are referring primarily to the following demands: protection of the individual against arbitrary infringement by other individuals or by the government; the right to work and to adequate earnings from work; freedom of discussion and teaching; adequate participation of the individual in the formation of the government. *These* human rights are nowadays recognized theoretically, although, by abundant use of formalistic, legal maneuvers, they are being violated to a much greater extent than even a generation ago. There is, however, one other human right which is infrequently mentioned but which seems to be destined to become very important: this is the right, or the duty, of the individual to abstain from cooperating in activities which he considers wrong or pernicious. The first place in this respect must be given to the refusal of military service. I have known instances where individuals of unusual moral strength and integrity have, for that reason, come into conflict with the organs of the state. The Nuremberg Trial of the German war criminals was tacitly based on the recognition of the principle: criminal actions cannot be excused if committed on government orders; conscience supersedes the authority of the law of the state.

The struggle of our days is being waged primarily for the freedom of political conviction and discussion as well as for the freedom of research and teaching. The fear of Communism has led to practices which have become incomprehensible to the rest of civilized mankind and exposed our country to ridicule. How long shall we tolerate that politicians, hungry for power, try to gain political advantages in such a way? Sometimes it seems that people have lost their sense of humor to such a degree that the French saying, "Ridicule kills," has lost its validity.

In connection with his seventy-fifth birthday celebration, Einstein received a long letter from Norman Thomas, one of America's foremost socialists and a firm opponent of communism. Thomas began by expressing his deep admiration for Einstein's scientific achievements, but also for the way he "stood with prophetic earnestness for the dignity and freedom of man." Nevertheless, he felt compelled to decline the invitation he had received from the Emergency Civil Liberties Committee (ECLC) to attend a conference and luncheon at the Nassau Inn in Princeton to honor Einstein. In Thomas's opinion, this organization had over the years shown "anything but a consistent love of liberty, in or out of the academic field" due to its pro-Communist bias (Hook 1987, 486).

While he heartily agreed with Einstein's general views regarding academic freedom, he differed strongly with regard to the rights of alleged Communists. "I cannot agree with you," he wrote, "that it is an infringement on liberty for proper authorities in the state, in the university, or in the schools, to raise a question concerning the *allegiance* of men who seek posts in which it is of the utmost importance that their allegiance should be solely to their consciences in search of truth. It is allegiance to Communism as a dictatorial conspiracy, not as a heresy, which warrants proper inquiries under proper circumstances." Thomas went on to express regret that the ECLC had managed to appropriate Einstein's name to promote "a concept of freedom far different from yours." Conceding that Communists, fascists, and their fellow travelers are clearly entitled to certain rights, Thomas added that "civil liberties cannot effectively be defended by Americans who through the years have condoned its absolute denial in the Soviet Union, while rushing hurriedly to the aid of men into whose primary political allegiance some sort of inquiry may be warranted" (ibid.).

In responding, Einstein noted that he knew nothing about the people who headed the ECLC, nor was he aware of the committee's past activities. As for the gathering at the Nassau Inn, the ECLC had merely used his name and birthday to advertise their meeting, a ploy he found "not to my taste." He then turned to the substantive issues that Thomas had raised.

LETTER TO NORMAN THOMAS, 10 MARCH 1954
EINSTEIN ARCHIVES 61-549; SWANBERG, 369–370

I was very pleased to receive a letter from you, for I felt instinctively that you are one of the few whose every word carries true conviction, untarnished by hidden intentions. One feels, as well, your good will toward all. This encourages me to answer you as if I were speaking to an old friend . . .

I see with a great deal of disquiet the far-reaching analogy between Germany of 1932 and the U.S.A. of 1954.

On *one* point we are of the same opinion. Russia is, in a very clear sense, a "politically underdeveloped country," about like Europe at the time of the Renaissance and a bit later. Murder, with and without legalistic accouterments, has become a commonplace means of daily politics. The citizen enjoys no rights and no security against arbitrary interference from the power of the state. Science and art have become wards of those who govern.

All this is certainly abominable to the taste of modern civilized man. But I believe that it is the problem of the Russian people to make changes there. We cannot advance a progressive development there by threatening Russia from the outside. Similarly, our well-intentioned criticism cannot help because it will not come to the ears of the Russians.

It seems to me, therefore, more useful to confine ourselves to the following question: How about the danger which America faces from the side of its own Communists? Here is the principal difference of opinion between you and me. In short, I believe: America is incomparably less endangered by its own Communists than by the hysterical hunt for the few Communists there are here (including those fellow citizens whose red tinge is weaker, à la Jefferson). Why should America be so much more endangered than England by the English Communists? Or is one to believe that the English are politically more naïve than the Americans so that they do not realize the danger they are in? No one there works with inquisitions, suspicions, oaths, etc., and still "subversives" do not go unchecked. There, no teachers and no university professors have been thrown out of their jobs, and the Communists there appear to have even less influence than formerly.

In my eyes, the "Communist conspiracy" is principally a slogan used in order to put those who have no judgment and who are cowards into a condition which makes them entirely defenseless. Again, I must think back to Germany in 1932, whose democratic social body had already been weakened by similar means, so that shortly thereafter Hitler was able to deal it the death-blow with ease. I am similarly convinced that those here will go the same way unless men with vision and willingness to sacrifice come to the defense.

Now you will clearly see our difference of opinion. Who is right cannot be decided through a logical process of proof. The future will tell . . .

Promoting Peace

Einstein's last months were devoted to two peace initiatives, the first of which was launched by Bertrand Russell. In December of 1954 the BBC broadcast a speech by the British philosopher on "Man's Peril from the Hydrogen Bomb," the text of which Russell enclosed along with a letter to Einstein of 11 February. Hoping to build on the favorable response to this broadcast, he proposed that Einstein and a handful of other prominent physicists join him in signing a public statement challenging the nations of the world to repudiate the use of this horrible new weapon.

Einstein reacted enthusiastically to this proposal, counseling Russell that they should seek to win over Niels Bohr to the cause.

LETTER TO BERTRAND RUSSELL, 16 FEBRUARY 1955
EINSTEIN ARCHIVES 33-201; N & N 1960, 625–626

I agree with every word in your letter of February 11. Something must be done in this matter, something that will make an impression on the general public as well as on political leaders. This might best be achieved by a public declaration, signed by a small number of people—say, twelve persons whose scientific attainments (scientific in the widest sense) have gained them international stature and whose declarations will not lose any effectiveness on account of their political affiliations. One might even include men who, like Joliot, are politically labeled provided they were counterbalanced by men from the other camp.

The neutral countries ought to be well represented. For example, it is absolutely vital to include Niels Bohr, and surely there is little doubt that he would join. Indeed, he might even be willing to visit you beforehand and take part in formulating the text of the document to be signed. He might also be helpful in proposing and enlisting signatories.

I hope you will consent to my sending your letter to a few people here in America, men I think may prove useful to the project. The choice is particularly difficult. As you probably know, this country has been ravaged by a political plague that has by no means spared scientists.

I suggest that the text to be offered for signature should be composed by at most two or three people—indeed, preferably by you alone—but in such a way as to insure in advance that there will be full agreement on the part of at least a few of the signers. This will make it easier for the others to sign without offering time-consuming amendments. Of course, we should also obtain signatures from Russia, which should not prove too difficult. In this respect, my colleague L. Infeld, professor at the University of Warsaw, could possibly be of help.

Here in America, in my opinion, Whitehead and Urey should be considered. We should try to see to it, however, that half the signatories are citizens of neutral countries, because that will impress the "hotheads" (*Kriegerischen*) and emphasize the neutral character of the whole project.

Einstein's initial assessment proved overly optimistic; for although his erstwhile collaborator, the Pole Leopold Infeld, supported this effort, he was unable to gain any signatories from the Soviet Union. Nor did any of the others mentioned by Einstein ultimately sign the Russell-Einstein Manifesto (A. N. Whitehead was already deceased).

Russell replied on 25 February, welcoming the suggestion that they gain Bohr's support. Einstein then contacted the Danish physicist, hoping to disarm him by alluding to their old running argument regarding the foundations of quantum physics.

LETTER TO NIELS BOHR, 2 MARCH 1955
EINSTEIN ARCHIVES 8-112; N & N 1960, 629–630

Don't frown like that! This has nothing to do with our old controversy on physics, but rather concerns a matter on which we are in complete agreement. Bertrand Russell recently wrote me a letter, of which I enclose a copy. He seeks to bring together a small group of internationally renowned scholars who would join in a statement to all nations and governments warning of the perilous situation created by atomic weapons and the arms race. This declaration is to coincide with political action initiated by the neutral countries.

Bertrand Russell knows and desires that I write you. Of course, he is well aware that you could greatly aid the project because of your influence, your experience and your personal relationships with outstanding people; indeed, he realizes that your counsel and active participation are virtually indispensable to the success of the project.

The proposed action of the scholars is *not* to be limited to representatives of neutral countries, although the choice of participants should demonstrate clearly the absence of political partisanship. Unless I misinterpret Russell's purpose, he seeks to do more than merely emphasize the existing danger in the world; he proposes to *demand* that the governments publicly acknowledge the necessity for renouncing the use of military force as a means of solving international disputes.

Should you approve of the plan in principle, would you be kind enough to communicate with Bertrand Russell and advise him that you are disposed to participate? The two of you could then decide which individuals would be most desirable as participants. Among those over here, I have been thinking of Urey, Szilard, and James Franck, but there probably should not be too many physicists. I am ready to write to anyone whom the two of you consider suitable, but I am reluctant to undertake the initial (and irrevocable) step until I know your feelings in the matter.

In America, things are complicated by the likelihood that the most renowned scientists, who occupy official positions of influence, will hardly be inclined to commit themselves to such an "adventure." My own participation may exert some favorable influence abroad, but not here at home, where I am known as a black sheep (and not merely in scientific matters).

Much will be gained if you can reach agreement with Bertrand Russell on the main points. For the time being there is no need to write me at all.

Bohr also received a letter from Russell and responded on 23 March, expressing doubts that such a public declaration could lead to any real political progress. He was also concerned that it might have an adverse effect on the forthcoming U.N. Conference on the Peaceful Uses of Atomic Energy.

Russell mailed the text of the manifesto to Einstein on 5 April, and the latter signed it six days later. By the time Russell received it in the return mail, Einstein had died. After obtaining eight other signatures—Einstein's recommendations (Urey, Szilard, and Franck) were not among them—Russell unveiled the statement at a press conference in London on 9 July 1955 during which he noted that initially it had been conceived in cooperation with Einstein, who had signed it during the last week of his life.

During the last months of his life, Einstein was preoccupied with a second major peace initiative that concerned the state of Israel. Until the very end, he held fast to his dream that Jews and Arabs would one day find a way to live together in peace and harmony in the Middle East. At the same time, he urged that Israel adopt a neutral position with regard to the Cold War conflict. Privately, he expressed these opinions to Zvi Lurie, a prominent member of the Jewish Agency in Israel.

LETTER TO ZVI LURIE, 4 JANUARY 1955
EINSTEIN ARCHIVES 60-388; N & N 1960, 637

We must adopt a policy of neutrality concerning the international antagonism between East and West. By adopting a neutral position, we would not only make a modest contribution to the curtailment of the conflict in the world as a whole, but would, at the same time, also facilitate the development of sound, neighborly relations with the various governments in the Arab world.

The most important aspect of our policy must be our ever-present, manifest desire to institute complete equality for the Arab citizens living in our midst, and to appreciate the inherent difficulties of their present situation. If we pursue such a policy, we shall gain loyal citizens and even more, we shall, slowly but surely, improve our relations with the Arab world. In this respect, the Kibbutz movement is an excellent example. The attitude we adopt toward the Arab minority will provide the real test of our moral standards as a people.

In April 1955 Reuven Dafni, the Israeli consul in New York, asked Einstein whether he would be willing to speak about Israel's cultural and scientific achievements as part of the forthcoming celebration of the country's independence. Einstein preferred instead to address the overall political situation and, in particular, the Arab-Israeli conflict. Dafni met with him in Princeton on 11 April and again two days later, as Einstein began to draft the speech he would not live to deliver. It began with the following words.

A FINAL UNDELIVERED MESSAGE TO THE WORLD, APRIL 1955
EINSTEIN ARCHIVES 28-1098; N & N 1960, 639–640

I speak to you today not as an American citizen and not as a Jew, but as a human being who seeks with the greatest seriousness to look at things objectively. What I seek to accomplish is simply to serve with my feeble capacity truth and justice at the risk of pleasing no one.

At issue is the conflict between Israel and Egypt. You may consider this a small and insignificant problem and may feel that there are more serious things to worry about. But this is not true. In matters concerning truth and justice there can be no distinction between big problems and small; for the general principles which determine the conduct of men are indivisible. Whoever is careless with the truth in small matters cannot be trusted in important affairs.

This indivisibility applies not only to moral but also to political problems; for little problems cannot be properly appreciated unless they are understood in their interdependence with big problems. And the big problem in our time is the division of mankind into two hostile camps: the Communist World and the so-called Free World. Since the significance of the terms *Free* and *Communist* is in this context hardly clear to me, I prefer to speak of a power conflict between East and West, although, the world being round, it is not even clear what precisely is meant by the terms *East* and *West*.

In essence, the conflict that exists today is no more than an old-style struggle for power, once again presented to mankind in semireligious trappings. The difference is that, this time, the development of atomic power has imbued the struggle with a ghostly character; for both parties know and admit that, should the quarrel deteriorate into actual war, mankind is doomed. Despite this knowledge, statesmen in responsible positions on both sides continue to employ the well-known technique of seeking to intimidate and demoralize the opponent by marshaling superior military strength. They do so even though such a policy entails the risk of war and doom. Not one statesman in a position of responsibility has dared to pursue the only course that holds out any promise of peace, the course of supranational security, since for a statesman to follow such a course would be tantamount to political suicide. Political passions, once they have been fanned into flames, exact their victims.

As the editors of *Einstein on Peace* fittingly wrote: "Here the hand that changed the world, and yet, in so many ways, could not change it, faltered and wrote no more" (N & N 1960, 640).

BIBLIOGRAPHY

Anshen, Ruth Nanda, ed. 1940. *Freedom, Its Meaning*. New York: Harcourt, Brace.

Bennett, Edward W. 1979. *German Rearmament and the West, 1932–1933*. Princeton: Princeton University Press.

Beyerchen, Alan. 1977. *Scientists under Hitler: Politics and the Physics Community in the Third Reich*. New Haven: Yale University Press.

Blumenfeld, Kurt. 1962. *Erlebte Judenfrage. Ein Vierteljahrhundert deutscher Zionismus*. Stuttgart: Deutsche Verlags-Anstalt.

Böhme, Klaus, ed. 1975. *Aufrufe und Reden deutscher Professoren im Ersten Weltkrieg*. Stuttgart: Reclam Universal-Bibliothek.

Born/Einstein 2005. *The Born-Einstein Letters: Friendship, Politics and Physics in Uncertain Times*. Gustav Born, ed. New York: Macmillan.

Braunbuch 1933. *Braunbuch über Reichstagsbrand und Hitler-Terror*. Basel: Universum-Bücherei.

Brenner, Arthur. 2001. *Emil J. Gumbel: Weimar German Pacifist and Professor*. Boston: Humanities Press.

Brenner, Michael. 1996. *The Renaissance of Jewish Culture in Weimar Germany*. New Haven: Yale University Press.

Calaprice, Alice, ed. 2005. *The New Quotable Einstein*. Princeton: Princeton University Press.

Churchill, Winston. 1948. *The Second World War: The Gathering Storm*. Boston: Houghton Mifflin.

Clark, Ronald W. 1971. *Einstein: The Life and Times*. New York: World Publishing.

CPAE 1. 1987. *Collected Papers of Albert Einstein*, vol. 1: *The Early Years: 1879–1902*. John Stachel et al., eds. Princeton: Princeton University Press.

CPAE 6. 1996. *Collected Papers of Albert Einstein*, vol. 6: *The Berlin Years: Writings, 1914–1917*. A. J. Kox et al., eds. Princeton: Princeton University Press.

CPAE 7. 2002. *Collected Papers of Albert Einstein*, vol. 7: *The Berlin Years: Writings, 1918–1921*. Michel Janssen et al., eds. Princeton: Princeton University Press.

CPAE 8A. 1998a. *Collected Papers of Albert Einstein*, vol. 8A: *The Berlin Years: Correspondence, 1914–1917*. Robert Schulmann et al., eds. Princeton: Princeton University Press.

CPAE 8B. 1998b. *Collected Papers of Albert Einstein*, vol. 8B: *The Berlin Years: Correspondence, 1918*. Robert Schulmann et al., eds. Princeton: Princeton University Press.

CPAE 9. 2004. *Collected Papers of Albert Einstein*, vol. 9: *The Berlin Years: Correspondence, January 1919–April 1920*. Diana Buchwald et al., eds. Princeton: Princeton University Press.

CPAE 10. 2006. *Collected Papers of Albert Einstein*, vol. 10: *The Berlin Years: Correspondence, May 1920–December 1920*. Diana Buchwald et al., eds. Princeton: Princeton University Press.

Craig, Gordon. 1980. *Germany, 1866–1945*. Oxford: Oxford University Press.

Déclarations. 1915. *Déclarations de l'institut et des universités de France à propos du manifeste des intellectuels d'Allemagne*. Paris: [Institut de France].

Dirks & Simon. 2005. Dirks, Christian, and Simon, Hermann, eds. *Relativ jüdisch. Albert Einstein: Jude, Zionist, Nonkonformist*. Berlin: Text.Verlag.

Doty, Paul. 1982. "Einstein and International Security." In Holton and Elkana, 347–367.

Einstein, Albert. 1916. "My Opinion on the War." In Berliner Goethebund, ed., *Das Land Goethes 1914/1916. Ein vaterländisches Gedenkbuch*. Stuttgart: Deutsche Verlags-Anstalt.

—— 1920. "On the Contribution of Intellectuals to International Reconciliation." In *Thoughts on Reconciliation*. New York: Deutscher Gesellig-Wissenschaftlicher Verein.

—— 1929. *Gelegentliches: Zum fünfzigsten Geburtstag 14. März 1929*. Berlin: Soncino-Gesellschaft der Freunde des Jüdischen Buches.

—— 1931a. *About Zionism: Speeches and Letters*. Leon Simon, ed. and trans. New York: Macmillan.

—— 1931b. *Cosmic Religion with Other Opinions and Aphorisms*. New York: Covici Friede.

—— 1933. *The Fight against War*. Alfred Lief, ed. New York: John Day.

—— 1934. *The World as I See It*. Alan Harris, ed. New York: Covici Friede.

—— 1946. *Testimony before the Anglo-American Committee of Inquiry on Jewish Problems in Palestine and Europe. Washington, D.C. State Department Building, 11 January 1946*. Einstein testimony, vol. 5, 118–135. Washington, D.C.: Ward and Paul (Electreporter, Inc.).

—— 1949. "Autobiographical Notes." In Schilpp, ed., 1949, 1–95.

—— 1954. *Ideas and Opinions.* New York: Bonanza Books.

—— 1956. *Out of My Later Years.* Secaucus, N.J.: Citadel Press.

—— 2001. *Mein Weltbild.* Carl Seelig, ed. Munich: Ullstein.

—— 2003. *Verehrte An- und Abwesende! Originaltonaufnahmen 1921–1951,* Klaus Sander, ed. [2 CDs of Einstein's speeches]. Cologne: Supposé.

Einstein/Besso. 1972. *Correspondance 1903–1955.* Pierre Speziali, ed. and trans. Paris: Hermann.

Einstein/Freud. 1933. *Why War?* Paris: League of Nations.

Einstein/Murphy/Sullivan. 1930. "Science and God: A German Dialogue." In *The Forum* 83 (June 1930): 373–379.

Einstein/Sommerfeld. 1968. *Briefwechsel. 60 Briefe aus dem goldenen Zeitalter der modernen Physik.* Armin Hermann, ed. Basel: Schwabe.

Erikson, Erik. 1982. "Psychoanalytic Reflections on Einstein's Centenary." In Holton and Elkana, 151–174.

Fadiman, Clifton, ed. 1939. *I Believe: The Personal Philosophies of Certain Eminent Men and Women of Our Time.* New York: Simon and Schuster.

Fölsing, Albrecht. 1993. *Albert Einstein. Eine Biographie.* Frankfurt am Main: Suhrkamp.

Frank, Philipp. 1947. *Einstein: His Life and Times.* New York: Alfred Knopf.

—— 1949. *Einstein, sein Leben und seine Zeit.* Munich: Paul List.

Freud, Sigmund. 1927. *The Future of an Illusion.* In *Standard Edition,* 21: 3–56.

—— 1930. *Civilization and Its Discontents.* In *Standard Edition,* 21: 59–145.

—— 1953–1974. *The Standard Edition of the Complete Psychological Works of Sigmund Freud.* James Strachey, ed., 24 vols. London: Hogarth Press.

Fromm, Erich. 1990. *Ethik und Politik: Antworten auf aktuelle politische Fragen.* Rainer Funk, ed. and trans. Weinheim: Beltz.

Galilei, Galileo. 1953. *Dialogue Concerning the Two Chief World Systems, Ptolemaic and Copernican.* Stillman Drake, trans. Berkeley: University of California Press.

Gilbert, Martin, ed. 1997. *Winston Churchill and Emery Reves, Correspondence, 1937–1964.* Austin: University of Texas Press.

Goenner, Hubert. 2003. "Albert Einstein and Friedrich Dessauer: Political Views and Political Practice." In *Physics in Perspective* 5 (2003): 21–66.

—— 2005. *Einstein in Berlin.* Munich: Beck Verlag.

Grundmann, Siegfried. 1998. *Einsteins Akte.* Berlin: Springer-Verlag.

Gülzow, Erwin. 1969. *Der Bund "Neues Vaterland." Probleme der bürgerlich-pazifistischen Demokratie im ersten Weltkrieg (1914–1918).* Dissertation, Humboldt University, Berlin.

Gumbel, Emil Julius. 1922. *Vier Jahre politischer Mord*. Berlin: Verlag der neuen Gesellschaft.

—— 1927. *Vom Russland der Gegenwart*. Berlin: Laubsche Verlagsbuchhandlung.

—— 1931. *Lasst Köpfe Rollen. Faschistische Morde, 1924–1931*. Berlin: Deutsche Liga für Menschenrechte.

—— 1991. *Auf der Suche nach Wahrheit*. Annette Vogt, ed. Berlin: Dietz.

Hentschel, Klaus. 1990. *Interpretationen und Fehlinterpretationen der speziellen und der allgemeinen Relativitätstheorie durch Zeitgenossen Albert Einsteins*. Basel: Birkhäuser.

Hobohm, Martin. 1931. "Einstein und Gumbel." In *Die Menschenrechte* 6 (15 July 1931): 122–124.

Holton, Gerald. 2002. "Einstein's Third Paradise." In *Daedalus,* Fall 2002, 26–34.

Holton, Gerald, and Elkana, Yehuda, eds. 1982. *Albert Einstein: Historical and Cultural Perspectives*. Princeton: Princeton University Press.

Hook, Sidney. 1987. *Out of Step: An Unquiet Life in the 20th Century*. New York: Harper and Row.

Jammer, Max. 1999. *Einstein and Religion*. Princeton: Princeton University Press.

Jeansonne, Glen. 1996. *Women of the Far Right: The Mothers' Movement and World War II*. Chicago: University of Chicago Press.

Jerome, Fred. 2002. *The Einstein File: J. Edgar Hoover's Secret War against the World's Most Famous Scientist*. New York: St. Martin's Press.

Johnson, Paul. 1991. *Modern Times: The World from the Twenties to the Nineties*. New York: HarperCollins.

Kahler, Erich. 1967. *The Jews among the Nations*. New York: Frederick Unger.

Kisch, Frederick Hermann. 1938. *Palestine Diary*. London: V. Gollancz.

Kleinert, Andreas. 1993. "Paul Weyland, der Berliner Einstein-Töter." In *Naturwissenschaft und Technik in der Geschichte. 25 Jahre Lehrstuhl für Geschichte der Naturwissenschaften und Technik am Historischen Institut der Universität Stuttgart*, 198–232. Helmuth Albrecht, ed. Stuttgart: Verlag für Geschichte der Naturwissenschaften und der Technik.

Lanouette, William. 1992. *Genius in the Shadows: A Biography of Leo Szilard*. New York: Charles Scribner's Sons.

Lenard, Philipp. 1933. "Ein großer Tag für die Naturforschung. Johannes Stark zum Präsidenten der Physikalisch-technischen Reichsanstalt in Berlin berufen." In *Völkischer Beobachter* 46 (13 May 1933).

—— 2003. *Wissenschaftliche Abhandlungen*. Vol. 4. Charlotte Schönbeck, ed. Berlin: Diepholz.

Levenson, Thomas. 2003. *Einstein in Berlin*. New York: Bantam Books.

Mach, Ernst. 1904. *Die Mechanik in ihrer Entwickelung. Historisch-kritisch dargestellt*. 5th rev. ed. Leipzig: Brockhaus.

McMeekin, Sean. 2003. *The Red Millionaire: A Political Biography of Willi Münzenberg, Moscow's Secret Propaganda Tsar in the West*. New Haven: Yale University Press.

N & N 1960. Nathan, Otto, and Norden, Heinz. *Einstein on Peace*. New York: Avenel Books.

N & N 1975. —— *Albert Einstein über den Frieden: Weltordnung oder Weltuntergang?* Bern: Herbert Lang & Cie.

Nicolai, Georg Friedrich. 1982. *Die Biologie des Krieges*. Originally published 1917, Zurich: Orell Füssli. Republished, Darmstadt: Verlag Darmstädter Blätter.

Noel-Baker, Philip. 1979. *The First World Disarmament Conference, 1932–33, and Why It Failed*. Oxford: Pergamon Press.

Ossietzky, Carl von. "Professor Gumbel." In *Weltbühne* 27 (27 January 1931): 150–151.

Pais, Abraham. 1982. '*Subtle is the Lord ...*': *The Science and the Life of Albert Einstein*. Oxford: Oxford University Press.

Poppel, Stephen M. 1977. *Zionism in Germany, 1897–1933: The Shaping of a Jewish Identity*. Philadelphia: Jewish Publication Society.

Pressebericht. 1937. *Pressebericht über die Strafsache des Sowjetfeindlichen Trotzkistischen Zentrums*. Moscow: Volkskommissariat für Justizwesen der UdSSR.

Radek, Karl. 1918. "Die Entwicklung des Sozialismus von der Utopie zur Wissenschaft." Republished 1971 in *Marxismus und Politik. Dokumente zur theoretischen Begründung revolutionärer Politik*, 156–165. Karl-Heinz Neumann, ed. Frankfurt am Main: Makol Verlag.

Reiser, Anton (pseud. Kayser, Rudolf). 1930. *Albert Einstein: A Biographical Portrait*. New York: Albert and Charles Boni.

Remmert, Volker. 2005. "Galileo, God, and Mathematics." In *Mathematics and the Divine: A Historical Study*, 347–360. Luc Bergmans and Teun Koetsier, eds. Amsterdam: Elsevier.

Reves, Emery. 1942. *A Democratic Manifesto*. New York: Random House.

—— 1945. *The Anatomy of Peace*. New York: Harper.

Riesenberger, Dieter. 1985. *Geschichte der Friedensbewegung in Deutschland, von den Anfängen bis 1933*. Göttingen: Vandenhoeck & Ruprecht.

Rhodes, Richard. 1988. *The Making of the Atomic Bomb*. New York: Touchstone.

—— 1995. *Dark Sun: The Making of the Hydrogen Bomb*. New York: Touchstone.

Rolland, Romain. 1952. *Journal des années de guerre, 1914–1919*. Paris: Michel.

Rozelle, Robert. 1985. "Emery Reves, the Early Years." In *The Wendy and Emery Reves Collection*, 15–30. Dallas: Dallas Museum of Art.

Russell, Bertrand. 1922. *Politische Ideale.* E. J. Gumbel, ed. and trans. Berlin: Deutsche Verlagsgesellschaft für Politik und Geschichte.

Samuel, Herbert Viscount. 1945. *Memoirs.* London: Cresset Press.

Sayen, Jamie. 1985. *Einstein in America: The Scientist's Conscience in the Age of Hitler and Hiroshima.* New York: Crown Publishers.

Scheideler, Britta. 2002. "The Scientists as Moral Authority: Albert Einstein between Elitism and Democracy, 1914–1933." In *Historical Studies in the Physical and Biological Sciences* 32, Part 2 (2002): 319–346.

Schilpp, Paul Arthur, ed. 1949. *Albert Einstein: Philosopher-Scientist.* Vol. 1. New York: Harper Torchbooks.

Schönbeck, Charlotte. 2000. "Albert Einstein und Philipp Lenard." In *Schriften der Mathematisch-naturwissenschaftlichen Klasse der Heidelberger Akademie der Wissenschaften,* Nr. 8.

—— 2003. "Philipp Lenard und die frühe Geschichte der Relativitätstheorien." In Lenard 2003, 323–375.

Segev, Tom. 1999. *One Palestine, Complete: Jews and Arabs under the British Mandate.* New York: Henry Holt.

Smith, Alice Kimball. 1965. *A Peril and a Hope: The Scientists' Movement in America, 1945–47.* Chicago: University of Chicago Press.

Stachel, John. 2002. *Einstein from "B" to "Z."* Boston: Birkhäuser.

Sulloway, Frank J. 1983. *Freud, Biologist of the Mind.* New York: Basic Books.

Swanberg, W. A. 1976. *Norman Thomas: The Last Idealist.* New York: Charles Scribner's Sons.

Vucinich, Alexander. 2001. *Einstein and Soviet Ideology.* Stanford, Calif.: Stanford University Press.

Wallace, Henry A. 1948. *Toward World Peace.* New York: Reynal and Hitchcock.

Wazeck, Milena. 2004. "Einstein in the Daily Press: A Glimpse into the Gehrcke Papers." In Max Planck Institute for the History of Science Preprint 271, 67–85.

Weart, Spencer R. 1988. *Nuclear Fear: A History of Images.* Cambridge, MA: Harvard University Press.

Welles, Sumner. 1946. "The Atomic Bomb and World Government." In *Atlantic Monthly,* January 1946, 39–42.

Wirth, Joseph. 1925. *Reden während der Kanzlerschaft.* Berlin: Germania.

Zuelzer, Wolf. 1982. *The Nicolai Case.* Detroit: Wayne State University Press.

INDEX

PLATE CREDITS

Plate 1.   Courtesy of The Hebrew University of Jerusalem, Israel.

Plate 2.   Courtesy of ullsteinbild/The Granger Collection, New York.

Plate 3.   Courtesy of 2007 Artists Rights Society (ARS), New York/VG Bild-Kunst, Bonn, and Bildarchiv Preussischer Kulturbesitz/Art Resource, New York.

Plate 4.   Courtesy of UPI and The Hebrew University of Jerusalem, Israel.

Plate 5.   Courtesy of The Hebrew University of Jerusalem, Israel.

Plate 6.   Courtesy of Library of Congress.

Plate 7.   Courtesy of akg-images.

Plate 8.   Courtesy of Central Zionist Archives, Jerusalem.

Plate 9.   Courtesy of Central Zionist Archives, Jerusalem.

Plate 10.   Courtesy of Max Planck Institute of the History of Science, Berlin.

Plate 11.   Courtesy of ullsteinbild/The Granger Collection, New York.

Plate 12.   Courtesy of Underwood & Underwood/CORBIS.

Plate 13.   Courtesy of The Hebrew University of Jerusalem, Israel.

Plate 14.   Courtesy of ullsteinbild/The Granger Collection, New York.

Plate 15.   Courtesy of *Deutsche Tageszeitung*, 1 April 1933.

Plate 16.   Courtesy of Library of Congress.

Plate 17.   Courtesy of Bettmann Archive/CORBIS.

Plate 18.   Courtesy of The Hebrew University of Jerusalem, Israel.

Plate 19.   Courtesy of Bettmann Archive/CORBIS.

Plate 20.   Courtesy of Time & Life Pictures/Getty Images.

Plate 21.   Courtesy of Franklin Delano Roosevelt Library, Hyde Park, NY.

Plate 22.   Time Inc.

Plate 23.   Courtesy of Library of Congress.

Plate 24.   Courtesy of Bettmann Archive/CORBIS.

Plate 25.   Courtesy of Bettmann Archive/ CORBIS.